REVIEWS OF UNITED KINGDOM STATISTICAL SOURCES

Volume XXIV

# LOCAL GOVERNMENT

REVIEWS OF UNITED KINGDOM STATISTICAL SOURCES

*Editor:* W. F. Maunder
*Assistant Editor:* M. C. Fleming

| | |
|---|---|
| Volume I | *Personal Social Services*, B. P. Davies<br>*Voluntary Organizations in the Personal Social Services Field*, G. J. Murray |
| Volume II | *General Government Routine Health Statistics*, Michael Alderson<br>*Social Security Statistics*, Frank Whitehead |
| Volume III | *Housing in Great Britain*, Stuart Farthing<br>*Housing in Northern Ireland*, Michael Fleming |
| Volume IV | *Leisure*, F. M. M. Lewes and S. R. Parker<br>*Tourism*, L. J. Lickorish |
| Volume V | *General Sources of Statistics*, G. F. Lock |
| Volume VI | *Wealth*, A. B. Atkinson and A. J. Harrison<br>*Personal Incomes*, T. Stark |
| Volume VII | *Road Passenger Transport*, D. L. Munby<br>*Road Goods Transport*, A. H. Watson |
| Volume VIII | *Land Use*, J. T. Coppock<br>*Town and Country Planning*, L. F. Gebbett |
| Volume IX | *Health Surveys and Related Studies*, M. Alderson and R. Dowie |
| Volume X | *Ports and Inland Waterways*, R. E. Baxter<br>*Civil Aviation*, Celia M. Phillips |
| Volume XI | *Coal*, D. J. Harris<br>*Gas*, H. Nabb<br>*Electricity Industries*, D. Nuttall |
| Volume XII | *Construction and the Related Professions*, M. C. Fleming |
| Volume XIII | *Wages and Earnings*, Andrew Dean |
| Volume XIV | *Rail Transport*, D. H. Aldcroft<br>*Sea Transport*, Derrick Mort |
| Volume XV | *Crime*, Monica A. Walker |
| Volume XVI | *Iron and Steel*, David W. Heal<br>*Shipbuilding*, Anthony Slaven |
| Volume XVII | *Weather*, B. W. Atkinson<br>*Water*, E. C. Penning-Rowsell and D. J. Parker |
| Volume XVIII | *Posts and Telecommunications*, S. Wall and P. Nicholson |
| Volume XIX | *Intellectual Property Rights*, D. L. Bosworth |
| Volume XX | *Religion*, L. Barley, C. Field, B. Kosmin and J. Nielsen |
| Volume XXI | *Finance*, K. Phylaktis and G. Kaye |
| Volume XXII | *Printing and Publishing*, W. D. McClelland |
| Volume XXIII | *Agriculture*, G. H. Peters |
| Volume XXIV | *Local Government*, J. M. Gillespie |
| Volume XXV | *Family Planning*, P. F. Selman |

Volumes I–XXII are available from Pergamon Press, Headington Hill Hall, Oxford OX3 0BW.
Volumes XXIII–XXV are available from Chapman and Hall Ltd, 11 New Fetter Lane, London EC4 4EE.

REVIEWS OF UNITED KINGDOM STATISTICAL SOURCES

Edited by W. F. MAUNDER
*Professor Emeritus of Economic and Social Statistics,*
*University of Exeter*

Assisted by M. C. FLEMING
*Professor of Economics,*
*Loughborough University*

Volume XXIV

# LOCAL GOVERNMENT

J. M. GILLESPIE
*Formerly Head of Central Information Service,*
*Devon County Council*

Published for The Royal Statistical Society and
The Economic and Social Research Council

London New York
CHAPMAN AND HALL

First published in 1988 by Chapman and Hall Ltd
11 New Fetter Lane, London EC4P 4EE
Published in the USA by Chapman and Hall
29 West 35th Street, New York NY 10001



Printed in Great Britain at the University Press, Cambridge

ISBN 0 412 31680 3



*British Library Cataloguing in Publication Data*

Reviews of United Kingdom statistical sources.
Vol. 24. Local government
1. Great Britain. Social conditions.
Statistics. Information sources
I. Maunder, W.F. (Wynne Frederick), *1920-*
II. Fleming, M.C. (Michael Carl), *1933-*
III. Gillespie, J.M. *1919-* IV. Royal Statistical Society
V. Economic and Social Research Council
314.1′07

ISBN 0-412-31680-3

*Library of Congress Cataloging in Publication Data*

Gillespie, J.M., 1919-
Local government/J.M. Gillespie.
p. cm. – (Reviews of United Kingdom statistical sources: V. 24)
"Published for the Royal Statistical Society and the Economic and Social Research Council."
Bibliography: p.
Includes index.
ISBN 0-412-31680-3
1. Great Britain – Statistical services. 2. Local government – Great Britain – Statistical services. I. Title. II. Series.
HA37.G7G55 1988 87-34695
001.4′22′0941 — dc19 CIP

# CONTENTS OF VOLUME XXIV


*Foreword* vii

*Membership of Joint Steering Committee* ix

*Introduction* xi

Review No. 42: LOCAL GOVERNMENT 1

Subject Index 335

# FOREWORD

*The Sources and Nature of the Statistics of the United Kingdom*, produced under the auspices of the Royal Statistical Society and edited by Maurice Kendall, filled a notable gap on the library shelves when it made its appearance in the early post-war years. Through a series of critical reviews by many of the foremost national experts, it constituted a valuable contemporary guide to statisticians working in many fields as well as a bench-mark to which historians of the development of Statistics in this country are likely to return again and again. The Social Science Research Council* and the Society were both delighted when Professor Maunder came forward with the proposal that a revised version should be produced, indicating as well his willingness to take on the onerous task of editor. The two bodies were more than happy to act as co-sponsors of the project and to help in its planning through a joint steering committee. The result, we are confident, will be judged a worthy successor to the previous volumes by the very much larger 'statistics public' that has come into being in the intervening years.

Mrs SUZANNE REEVE
*Secretary*
*Economic and Social Research Council*

Mrs E.J. SNELL
*Honorary Secretary*
*Royal Statistical Society*

*SSRC is now the Economic and Social Research Council (ESRC).

# MEMBERSHIP OF JOINT STEERING COMMITTEE

(November 1986)

*Chairman:* Miss S. V. Cunliffe

*Representing the Royal Statistical Society:*

Mr M. C. Fessey

Dr S. Rosenbaum

Mrs E. J. Snell

*Representing the Economic and Social Research Council:*

Mr I. Maclean

Miss J. Morris

*Secretary:* Mr D. E. Allen

# INTRODUCTION TO VOLUME XXIV

The first review in this series was devoted to one very important part of local government statistics – namely, personal social services – and it has long been the intention to publish further reviews dealing with other aspects. This volume fulfils a major part of that objective although it has been a close run race to achieve it within the final phase of the ESRC's sponsorship. The chief remaining area of omission from local government sources is education statistics and it is regrettable that this should be so despite the considerable amount of work which was undertaken. Many similar examples could be quoted and that such key topics remain to be dealt with highlights the extremity of need for a successor to assume the ESRC's mantle, thus enabling the continuation of the series.

The primary aim of this series is to act as a work of reference to the sources of statistical material of all kinds, both official and unofficial. It seeks to enable the user to discover what data are available on the subject in which he is interested, from where they may be obtained, and what the limitations are to their use. Data are regarded as available not only if published in the normal printed format but also if they are likely to be released to a *bona fide* enquirer in any other form, such as duplicated documents, computer print-out or even magnetic tape. On the other hand, no reference is made to material which, even if it is known to exist, is not accessible to the general run of potential users. The distinction, of course, is not clear-cut and mention of a source is not to be regarded as a guarantee that data will be released; in particular cases it may very well be a matter for negotiation. The latter caution applies with particular force to the question of obtaining computer print-outs of custom specified tabulations. Where original records are held on magnetic media it might appear that there should be no insuperable problem, apart from confidentiality, in obtaining any feasible analysis at a cost; in practice, it may well turn out that there are capacity restraints which override any simple cost calculation. Thus, what is requested might make demands on computer and programming resources to such an extent that the routine work of the agency concerned would be intolerably affected.

The intention is that the source for each topic should be reviewed in detail, and the brief supplied to authors has called for comprehensive coverage at the level of 'national interest'. This term does not denote any necessary restriction to statistics collected on a national basis (still less, of course, to national aggregates) but it means that sources of a purely local character, without wider interest in either content or methodology, are excluded. Indeed, the mere task of identifying all material of this latter kind is an impossibility. The interpretation of the brief has obviously involved discretion and it is up to the users of these reviews to say what unreasonable gaps become apparent to them. They are cordially invited to do so by communicating with us.

To facilitate the use of the series as a work of reference, certain features have been incorporated which warrant a word or two of explanation. First, the text of each review is designed, in so far as varying subject matter permits, to follow a standard form of arrangement so that users may expect a similar pattern to be followed throughout the series. The starting point is a brief summary of the activity concerned and its organisation, in order to give a clear background understanding of how data are collected, what is being measured, the stage at which measurements are made, what the reporting units are, the channels through which returns are routed and where they are processed. As a further part of this introductory material, there is a discussion of the specific problems of definition and measurement to which the topic gives rise. The core sections on available sources which follow are arranged at the author's discretion – by origin, by subject subdivision, or by type of data; there is too much heterogeneity between topics to permit any imposition of complete uniformity on all authors. The final section is devoted to a discussion of general shortcomings and possibly desirable improvements. In case a contrary expectation should be aroused, it should be said that authors have not been asked to produce a comprehensive plan for the reform of statistical reporting in the whole of their field. However, a review of existing sources is a natural opportunity to make some suggestions for future policy on the collection and publication of statistics within the scope concerned and authors have been encouraged to take full advantage of it.

Secondly, detailed factual information about statistical series and other data is given in a Quick Reference List (QRL). The exact nature of the entries is best seen by glancing at the list and accordingly they are not described here. Again, the ordering is not prescribed except that entries are not classified by publication source since it is presumed that it is this which is unknown to the reader. In general, the routine type of information which is given in the QRL is not repeated verbally in the text; the former, however, serves as a search route to the latter in that a reference (by section number) is shown against a QRL entry when there is a related discussion in the text.

Third, a subject index to each review acts as a more or less conventional line of enquiry on textual references; it is a computerised system and, for an individual review, the only advantage is the possibility of easily permuting entries. The object at this level is merely to facilitate search by giving as many variants as possible. In addition, however, it also makes possible selective searches by keyword over any combination of reviews and a printout of the entries found may then be prepared.

Fourth, each review contains two listings of publications. The QRL Key gives full details of the publications shown as sources and text references to them are made in the form [QRL serial number]; this list is confined essentially to data publications. The other listing is a general bibliography of works discussing wider aspects; text references in this case are made in the form [B serial number].

Finally, an attempt is made to reproduce the more important returns or forms used in data collection so that it may be seen what tabulations it is possible to make as well as helping to clarify the basis of those actually available. Unfortunately, there are severe practical limitations on the number of such forms that it is possible to append to a review and authors perforce have to limit their choice. The generous selection of forms presented with this volume is thought to be particularly notable.

As editors, we must express our very grateful thanks to all the members of the Joint Steering Committee of the Royal Statistical Society and the Economic and Social Research Council. It would be unfair to saddle them with any responsibility for shortcomings in execution but they have directed the overall strategy with as admirable a mixture of guidance and forbearance as any editors of such a series could desire. Especial thanks are due to the Secretary of the Committee who has been involved with the project almost as long and almost as closely as anybody. The author joins us in thanking all those who gave up their time to attend the seminar held to discuss the first draft of the review and which contributed materially to improving the final version.

We are most grateful to Messrs. Chapman and Hall for their support and in particular to the Production Department who put all the pieces together. The subject index entries have been compiled by Mrs. Marian Guest who has also acted as editorial assistant throughout. Special thanks are due to Mr.Ray Burnley of the Social Studies Data Processing Unit who again has masterminded our use of the Lasercomp System at Oxford University Computer Service and to the latter for the use of this facility. Finally, we also wish to record our appreciation of the permission granted us to reproduce certain copyright material by the Controller of Her Majesty's Stationery Office, CIPFA and the Local Authorities Conditions of Service Advisory Board.

M.C.Fleming
University of
Loughborough

W.F.Maunder
University of Exeter

September 1987

# 42: LOCAL GOVERNMENT

JOHN GILLESPIE
*formerly of Devon County Council*
with the assistance of
John Hughes
*Chartered Institute of Public Finance and Accountancy*

# REFERENCE DATE OF SOURCES REVIEWED

This review is believed to represent the position as it obtained at the end of 1986. Later revisions have been inserted up to the proof reading stage (mid-1987) taking account, as far as possible, of any major changes in the situation.

# LIST OF ABBREVIATIONS

| | |
|---|---|
| CB | County Borough |
| CIPFA | Chartered Institute of Public Finance and Accountancy |
| CIPFA/SIS | Chartered Institute of Public Finance and Accountancy, Statistical Information Service |
| CIPFA/FIS | Chartered Institute of Public Finance and Accountancy, Financial Information Service |
| CO | Capital Outturn |
| CSO | Central Statistical Office |
| COSLA | Convention of Scottish Local Authorities |
| DES | Department of Education and Science |
| DLO | Direct Labour Organisation |
| DOE | Department of the Environment |
| DVLC | Driver and Vehicle Licensing Centre |
| GLC | Greater London Council |
| GREA | Grant Related Expenditure Assessment |
| GRO | General Register Office |
| IDLG | Information Development and Liaison Group |
| IEHO | Institute of Environmental Health Officers |
| ILEA | Inner London Education Authority |
| JMW | Joint Manpower Watch |
| LACSAB | Local Authority Conditions of Service Advisory Board |
| LARIA | Local Authority Research and Intelligence Association |
| LFR | Local Financial Return |
| MAFF | Ministry of Agriculture, Fisheries and Food |
| MLH | Minimum List Heading |
| NES | New Earnings Survey |
| NGR | National Grid Reference |
| NJC | National Joint Council |
| NLGSLC | National and Local Government Statistical Liaison Committee |
| OPCS | Office of Population Censuses and Surveys |
| RER | Return of Expenditure and Rates |
| RO | Revenue Outturn |
| RSG | Rate Support Grant |
| RV | Rateable Value |
| SCT | Society of County Treasurers |
| SED | Scottish Education Department |
| SF | Special Funds |

| | |
|---|---|
| SHHD | Scottish Home and Health Department |
| SIC | Standard Industrial Classification |
| SOLACE | Society of Local Authority Chief Executives |
| SSLC | Scottish Statistical Liaison Committee |
| TSG | Transport Supplementary Grant |
| TSR | Trading Services Return |

# ACKNOWLEDGEMENTS

The preparation of this review, which spans a number of years, would not have been remotely possible without the help of many people prepared to share with me their time, specialist knowledge and experience. They, to whom I am deeply indebted, include staff and members of CIPFA, DOE, the Welsh Office, the Department of Employment, OPCS, the Home Office, the Department of Transport, CSO, LACSAB, The NJC (Scottish Councils), the SCT, the Scottish Office (statistics, finance and library sections), the National Library of Scotland, COSLA, Municipal Publications Limited, the Commission for Local Administration in Scotland and a number of local authorities throughout Great Britain.

My thanks are also due to all those who not only read the first draft of the review but attended the ESRC seminar at which the draft was discussed. For their helpful comments, criticisms and suggestions, both oral and written, I am very grateful. Although the major impact of the seminar is to be found in the expansion of those chapters devoted to the statistical sources of particular services, let me emphasise that I alone am responsible for the final content of the review.

To a number of individuals I am bound to express a special word of thanks. At the Scottish Office, David Wann (statistician) and Carol John (librarian) were exceptional. For their diligence, forbearance and kindness in dealing with a barrage of requests, I am everlastingly grateful. To C.J Morris of the Department of Finance and Personnel at Stormont and to W.G Laverty of the Department of Environment of Northern Ireland I must acknowledge a substantial debt for their tolerance, industry and helpful guidance over a long period during part of which the very definition of the boundary of "local government" in Northern Ireland continued to elude me.

A major part of the research was carried out by John Hughes of CIPFA. I am indebted to him for his scrupulous attention to detail, for his industry, for his providing many helpful suggestions and not least for his having guided me through some of the more esoteric particulars of local government finance.

Throughout the long period of perparation of this review Pat Musgrave was responsible for all of the typing work. Here I wish to record my appreciation of her endless patience, her imperturbability, her sheer hard work and above all for her having conveyed throughout that nothing was too much trouble for her.

I must also thank the editor, Professor W.F Maunder for his willing help, wise guidance and encouragement especially in the early stage of preparing the review.

Finally let me express my unbound gratitude to my wife Margaret without whose support, enduring good temper, unrewarded chores and endless self-sacrifice the project could never have been completed.

# CONTENTS OF REVIEW 42


| | | |
|---|---|---|
| **1.** | **Introduction** | 11 |
| 1.1 | Coverage and Arrangement of Subjects | 11 |
| 1.2 | Previous Coverage | 11 |
| 1.3 | Recent Changes | 12 |
| 1.4 | Some Specific Points | 12 |
| **2.** | **Organisation and Functions of Local Authorities** | 14 |
| 2.1 | Introduction | 14 |
| 2.2 | England and Wales | 15 |
| 2.3 | Scotland | 16 |
| 2.4 | Isles of Scilly | 16 |
| 2.5 | Northern Ireland | 16 |
| **3.** | **Financial Statistics: General Considerations** | 18 |
| 3.1 | Important Characteristics | 18 |
| 3.2 | The Publication of Local Authority Financial Statistics | 19 |
| 3.3 | The Structure of Local Authority Accounts | 21 |
| 3.4 | Northern Ireland | 22 |
| **4** | **Financial Statistics: Expenditure** | 24 |
| 4.1 | Introduction | 24 |
| 4.2 | Revenue (or current) Expenditure | 25 |
| 4.3 | Capital Expenditure | 28 |
| 4.4 | Local Authority Expenditure in Context of Public Expenditure | 30 |
| 4.5 | Northern Ireland | 35 |
| **5** | **Financial Statistics: Rates** | 37 |
| 5.1 | Introduction | 37 |
| 5.2 | Rate Income | 38 |
| 5.3 | Rate Poundage and Average Rate Payment | 39 |
| 5.4 | Rate Collection | 42 |
| 5.5 | Rate Rebates | 42 |
| 5.6 | Northern Ireland | 43 |
| **6** | **Financial Statistics: Government Grants** | 46 |
| 6.1 | Introduction | 46 |
| 6.2 | Specific and Supplementary Grants | 46 |

6.3 Rate Rebate Grants 48
6.4 Rate Support Grant 48
6.5 Northern Ireland 52

**7 Financial Statistics: Borrowing and Debt** 55
7.1 Introduction 55
7.2 Borrowing 56
7.3 Outstanding Debt 58
7.4 Net Financial Transactions 59
7.5 Northern Ireland 60

**8 Financial Statistics: Miscellaneous Income** 61
8.1 Introduction 61
8.2 Rate Fund Services 62
8.3 Housing Revenue Account 63
8.4 Trading Services 63
8.5 Special Funds 64
8.6 Superannuation Funds 65
8.7 Northern Ireland 66

**9 Rateable Values and Penny Rate Products** 68
9.1 Rateable Values 68
9.2 Penny Rate Product 70
9.3 Northern Ireland 72

**10 Manpower Earnings and Hours** 74
10.1 Manpower 74
10.2 Earnings and Hours 78
10.3 Northern Ireland 79

**11 Elections** 80
11.1 Incidence of Elections 80
11.2 Electoral Statistics 81
11.3 Election Results 81
11.4 Northern Ireland 82

**12 Public Protection** 83
12.1 Police 83
12.2 Fire 86
12.3 Probation 89
12.4 Administration of Justice and Public Protection Items 91
12.5 Consumer Protection (including Trading Standards) 95
12.6 Northern Ireland 96

**13 Transport** 97
13.1 Highways and Transportation (England and Wales) 97
13.2 Highways and Transportation (Scotland) 99

13.3 Ports 101
13.4 Airports 101
13.5 Northern Ireland 102

**14 Environmental Services** 103
14.1 Definition and Scope 103
14.2 Environmental Health (England and Wales) 104
14.3 Refuse Collection (England and Wales) 106
14.4 Refuse Disposal (England and Wales) 107
14.5 Cemeteries and Crematoria (England and Wales) 109
14.6 Scotland 110
14.7 Northern Ireland 114

**15 Miscellaneous** 115
15.1 Museums, Art Galleries and Library Services 115
15.2 Direct Labour Organisations 117
15.3 Smallholdings 119
15.4 Surveys 120
15.5 Northern Ireland 120

**16 Complaints** 121
16.1 Great Britain 121
16.2 Northern Ireland 122

**17 Conclusion** 123
17.1 Recent Developments 123
17.2 Shortcomings 126
17.3 Suggestions 127
**Quick Reference List Contents** 129
**Quick Reference List** 132
**Quick Reference List Key to Publications** 190

**Bibliography** 200

**List of Appendices** 203
**Appendices** 205

**Subject Index** 335

CHAPTER 1

# INTRODUCTION

## 1.1 Coverage and Arrangement of Subjects

### 1.1.1

In this publication we do not purport to review the whole body of local government statistics in the United Kingdom. The fact is that subject reviews of several local government services have either appeared in companion volumes of this series or are now in prospect. Details are set out in Appendix 1. While some overlap is unavoidable, it is hoped that this has been kept to a minimum.

### 1.1.2

The subjects included in this review fall into two broad categories. First, there are four general subjects (finance, manpower and earnings, elections, complaints) which encompass the entire local government scene. Second, there are statistics of specific services or subjects which have either not been reviewed in previous volumes or are not currently under review. These are: public protection; transport; environmental services; miscellaneous subjects and services (i.e. libraries, direct labour, small holdings, surveys). With the publication of this review, those in prospect and those already published, coverage of local government statistics in the UK should be complete.

## 1.2 Previous Coverage

### 1.2.1

It is fitting at this stage to refer to the most recent previous review of UK local government statistics undertaken in this country. That is a research study entitled *Local Government Statistics* [B.1] by W. Barker, who at the time was an Assistant City Treasurer in Manchester. The study, published by the Institute of Municipal Treasurers and Accountants–the forerunner of CIPFA–was subtitled "a guide to statistics on local government finance and services in the United Kingdom at August 1964". An introduction to the study containing a catalogue of previous guides to local government statistics is reproduced in Appendix 2. While it is true that Barker's research study was all embracing and is perhaps the most comprehensive ever undertaken on the subject of local government statistics, it differs from the present review (and indeed companion volumes in the present series) in that the latter identifies the primary sources of *data*, supplies specimen copies of source documents, wherever possible describes the mechanics of data collection and manipulation, provides much more by way of explanation of terms and elaboration of concepts and above all, furnishes quick references to the statistics themselves. Thus the present volumes are not strictly comparable with the earlier review which was mainly limited to the identification of *publication* sources.

## 1.3 Recent Changes

### 1.3.1

Not the least of the difficulties encountered in the preparation of this review arose from the fact that the local government scene is perpetually changing. Within the historical range of the review, the structure, functions and geographical boundaries had been completely revised in Northern Ireland (1972), England and Wales (1974) and Scotland (1975). It was therefore, thought useful to include a chapter reciting the up to date details of the organisation and functions of the revised authorities. However, even while the review was being written a further major change in organisation was effected: the Greater London Council and the metropolitan county councils were replaced (see section 2.1).

### 1.3.2

One other major change to be accommodated was the implication of the new method of disbursing the major part of central government funds to local authorities in England and Wales (the Block Grant–see Section 6.4). This change spawned a whole set of new publications containing relevant statistical and explanatory material the content and form which are still changing. The replacement of Rate Rebate Grants by Housing Benefits (see Section 5.5) may also affect the presentation of statistics although in a much less dramatic way.

## 1.4 Some Specific Points

### 1.4.1

As a general rule in this review we have sought to concentrate on *original data and publication sources only*. However one exception merits special mention. In the case of some of the CIPFA/SIS or SCT service statistics dealt with in Chapters 12 to 15, the statistical tabulations and accompanying comments may appear first of all in the local government press e.g. the County Council Gazette. This circumstance has been ignored throughout. Since the CIPFA/SIS or SCT booklets usually contain definitions and other statistical material omitted from press coverage, they have been treated as the authentic source publications.

### 1.4.2

Given the content and structure of the statistics contained in the review it proved impossible to avoid a certain amount of *inter-chapter overlap*. As explained in paragraph 1.1.2 the review is mixed in the sense that part of it relates (a) to the statistics of general subjects such as manpower and finance which have a bearing on all services and (b) to the statistics of a series of specific services which inevitably include the manpower and financial aspects of these services. However, much of the overlap is more apparent than real in as much as the manpower and financial statistics of services relate to service levels of activity only.

**1.4.3**

As the structure of local government and the publication of statistics vary so fundamentally between Great Britain on the one hand and Northern Ireland on the other, it has been thought useful throughout to separate consideration of the Northern Ireland scene from that of England, Wales and Scotland. Thus, in the review chapters proper i.e. chapters 2 to 16, Northern Ireland is invariably considered in a separate and final section.

**1.4.4**

The organisation of the Scottish Office and one of its practices in the matter of statistics publication may present confusion especially to those outside Scotland. To avoid such confusion, an explanatory note is contained in Appendix 3.

**1.4.5**

The emphasis in this review is on straight description of primary publication and data sources. Wherever practicable basic data source documents and accompanying completion instructions and data definitions have been included as appendices. This approach is thought to have the merit of satisfying both the needs of the general user of local government published statistics and those of the specialist statistician working in the local authority area.

CHAPTER 2

# ORGANISATION AND FUNCTIONS OF LOCAL AUTHORITIES

## 2.1 Introduction

Major changes in the organisation and functions of local authorities in Greater London and the metropolitan areas of England came into effect on 1st April, 1986. For precise details of these changes, the Local Government Act, 1985 should be consulted. Broadly, they reduce to:

(a) the abolition of the GLC and the metropolitan county councils;

(b) the transfer to the London Boroughs and metropolitan districts of some of the functions of the abolished councils (e.g. planning, highways and roads, waste regulation and disposal, courts and probation service, licensing powers, trading standards;

(c) the creation of new joint authorities (e.g. for police, fire, education in Inner London, passenger transport in metropolitan counties);

(d) the transfer of functions to existing bodies (e.g. land drainage and flood prevention in London, certain arts sponsorship in London).

The description of organisation and functions contained in this Chapter relates to the situation up to 31st March, 1986 which effectively underlies the statistical sources considered in this review.

### 2.1.1

With the exception of part of Scotland and the Isles of Scilly, the present local authority structure throughout Great Britain is two tier. It was established in England and Wales, outside Greater London, in April 1974 and in Scotland in May 1975. In Greater London the current structure dates from 1965. The relevant enactments are:

(1) the Local Government Act 1972;

(2) the Local Government (Scotland) Act 1973;

(3) the London Government Act 1963.

It should be noted that the systems in the Isles of Scilly and Northern Ireland are covered in sections 2.4 and 2.5 respectively.

### 2.1.2

An index to individual local authorities in Great Britain, to the structure prior to 1965, 1974 and 1975 respectively and a correlation index (i.e. a means of translating the areas of the new authorities to those which pre-dated re-organisation) are contained in the *Municipal Year Book* [QRL.44]. In *Local Government Trends* [QRL.43] there are quick reference tables to the functions of each type of authority while the Central Office of Information's Reference Pamphlet *Local Government in Britain* [B.17] contains a concise guide to the organisation and working of present local authorities. Many aspects of local government in Scotland are fully described in *Local Government in Scotland* [B.28]. Unfortunately none of these sources is committed to automatic updating to accommodate minor amendments to function and title, some of which have taken place since the re-organisations were effected.

## 2.2 England and Wales

### 2.2.1

The *metropolitan areas* comprise the six conurbations of Greater Manchester, Merseyside, South Yorkshire, West Yorkshire, West Midlands, Tyne and Wear. Within each of these six areas the upper tier authority is known as the *metropolitan county council.* The principal services for which these are responsible are: overall planning, transport, police, consumer protection, fire services and waste disposal. Second tier authorities in the metropolitan areas are known as *metropolitan district councils* (of which there are 36) and are mainly responsible for: education, personal social services, housing, local planning, environmental health, leisure services and waste collection.

### 2.2.2

In the remainder of England and Wales (outside Greater London), the upper tier authorities are *non metropolitan* or *shire county councils.* Of these there are 47; the principal services they manage are: strategic planning, transport, police, consumer protection, fire services, education, personal social services and waste disposal. The second tier authorities in the non metropolitan areas are the *district councils* of which there are 333 in England and Wales. In the main they are responsible for housing, local planning, environmental health, leisure and waste collection. In Wales, the districts deal with refuse disposal as well as collection and four district councils have been temporarily designated as library authorities. Throughout the non metropolitan areas of England there are over 8000 local (parish or town) councils responsible for local amenities such as allotments, bus shelters, footway lighting, footpaths and so on. In Wales there are about 500 community councils with powers broadly similar to those of the English parish councils.

### 2.2.3

In Greater London, the upper tier authority is the *Greater London Council* which principally has responsibilities for: overall planning, transport, fire services, waste disposal and some housing. There are 32 second tier or London *borough councils* plus the *City of London* providing the following principal services: housing, local planning, environmental health, personal social services, waste collection and education (in 20 boroughs). Education in the remaining 12 boroughs plus the City of London is provided by the *Inner London Education Authority* (ILEA).

## 2.3 Scotland

### 2.3.1

As in England and Wales, mainland Scotland has a two tier local authority structure. The upper tier authorities, of which there are 9, are the *regional councils* the main responsibilities of which are: overall planning, transport, police, fire services, education, personal social services, sewerage, water and drainage. There are 53 second tier or *district councils*, which have responsibility for housing, local planning, environmental health, leisure services, tourism, waste collection and disposal. In the island areas of Orkney, Shetland and the Western Isles there are 3 all purpose authorities called *island councils* responsible principally for those services which on the mainland, are provided by regional and district councils. In addition there is in Scotland, statutory provision for the establishment of over 1200 community councils, where there is a demand for them. These councils are not to be regarded as a third tier of local government in Scotland but rather as bodies which express the views of the community they represent. They may receive funds from the local authorities to carry out particular projects.

## 2.4 Isles of Scilly

### 2.4.1

Local government in the Isles of Scilly is carried out by the Council of the Isles of Scilly which is an all purpose authority. This arrangement derives from the Local Government Act, 1972 and the Isles of Scilly Orders, 1979 all legislation being specifically applied to the islands by Special Orders. The council has the powers and responsibilities of a county and district in the rest of England, but a few special arrangements should be noted. Unlike the position in the rest of England, the council is responsible for water supply: police services are administered by the Devon and Cornwall Police Authority and libraries are run jointly with Cornwall County Council.

## 2.5 Northern Ireland

### 2.5.1

Under the Local Government Act (Northern Ireland) 1972, the reorganisation of the functions and structure of authorities which took effect in October 1973 was much more fundamental than that in Great Britain around the same time. Certain major functions (education, health and welfare services, water and sewerage, highways and bridges, electricity supply, passenger transport) were transferred to central government and other public bodies. (Housing had been transferred in 1972 to the Northern Ireland Housing Executive.) The functions and responsibilities which remained were transferred to 26 *district councils* which now administer the following services: museums and art galleries; environmental health services such as refuse collection and disposal; swimming pools; parks and recreation grounds; regulatory services such as consumer protection, licensing places of public entertainment, building regulations, dog control, local by-laws; administration of housing grants for small dwellings acquisition; trading services such as gas undertakings, cemeteries and crematoria, markets, harbours and abattoirs. A very detailed description of the working of the district councils is provided in the *Ulster Year Book* [B.32]. There is a concise description of the councils' duties in *Local Government in Britain* [B.17] and a listing of the Councils alongside their predecessors (prior to October 1973) is printed in the *Municipal Year Book* [QRL.44].

CHAPTER 3

# FINANCIAL STATISTICS: GENERAL CONSIDERATIONS

## 3.1 Important Characteristics

### 3.1.1

All local authorities in the UK are obliged by law to make available, in the form of published accounts, information about their financial activities. This review of local government financial statistics is not about such publications. It deals only with those "accounts" and tabulations which have a national or regional interest or application and it is this width of application which, along with other factors, gives them their character of statistics. In describing the statistics of the National Accounts, Harold Copeman [B.2] makes this point: "In all cases it must be remembered that we are not describing the operations of a firm under a single or coherent management, that produces audited accounts, but with very many independent firms and individuals, plus public authorities, and that although accounts are used when possible, we are dealing with *statistics* and with a considerable degree of estimation". This, *mutatis mutandis*, is equally true of local authority financial statistics. Again and again, in producing them we have recourse to a host of statistical devices such as: the rounding of figures; allowing for non-response; producing provisional outturn figures to effect timely publication; interpolating part year figures from complete year data. Unlike National Accounts statistics however, local authority financial statistics come in two distinct forms: (a) summary statistics (e.g. at national, regional, class of authority level and (b) individual authority statistics provided in national or regional publications. In interpreting *national* summary statistics it has to be remembered that since 1978/79 these have related to the local authority *sector* of the economy ("local authority" sector as distinct from personal, central government, public corporation, companies and financial institution sectors). Thus, inter authority transactions or "within sector flows of funds" have, as far as possible, been eliminated.

### 3.1.2

Some further important features of local authority financial statistics are worth mentioning. First, the data itself may relate to *estimates* (e.g. advance estimates of expenditure; income from government grants in a later financial year) or may be *actual/outturn* data (i.e. statistics collected for a period now completed). Indeed, some of the latter are themselves "estimates" in the sense that they are produced before outturn or actual figures are finally available; they are thus referred to as provisional outturn figures. Second, the quality of outturn/actual data: having been produced for accounting purposes and therefore subject to audit and other accounting disciplines, it is of a very high order.

### 3.1.3

There is one important *accounting convention* which may affect the interpretation of some local government financial statistics and which is therefore worthy of special mention. Statistics relating to revenue or current accounts are recorded on an accruals or income and expenditure basis. Thus, at the end of a financial year, some figures will relate to debts incurred but not settled and to credits outstanding. By contrast, capital accounts usually (but not universally), are recorded purely on a cash basis. That is to say, they show only actual payments and receipts of cash.

### 3.1.4

Most local authority financial statistics in the UK have a clear regional dimension. This is true of coverage, source, publication and form of presentation with England and Wales following a rather similar pattern, but Scotland often being quite distinctive; (Northern Ireland which is unique is dealt with later). As a result it will be found that most subjects discussed in the following text require separate regional consideration.

### 3.1.5

It is worth noting that CIPFA has now published a *Guide to Local Authority Finance* [B.23] which provides in detail and in layman's language, a comprehensive description of the operation of local government spending, income raising, borrowing, budgeting, accounting, financial control and auditing. While the publication applies primarily to local government finance in England and Wales, its principles can be taken as applying to Scotland also. It is invaluable as background to the comprehension of the financial statistics discussed in this review.

## 3.2 The Publication of Local Authority Financial Statistics

### 3.2.1

It may be useful to consider the actual publication of local authority financial statistics in Great Britain under two headings: (a) central government produced and (b) "others". Under the various statutes referred to paragraph 2.1.1 above, local authorities in Great Britain make returns of their actual income and expenditure to central government and the latter is obliged to produce summaries; an obligation which has existed since the second half of last century. The summaries are published in: *Local Government Financial Statistics in England and Wales* [QRL.41]; *Welsh Local Government Financial Statistics* [QRL.92] and *Scottish Local Government Financial Statistics* [QRL.77]; in the case of [QRL.92] and [QRL.77] a good deal of individual authority coverage is also provided. Responsibility for publication lies respectively with the DOE, the Welsh Office and the Scottish Office. Also, regular financial statistics about the local authority sector of the economy are published by the Central Statistical Office in *United Kingdom National Accounts* [QRL.88], *Financial Statistics* [QRL.22] and *Economic Trends* [QRL.16]. Whereas all the publications named so far are purely statistical in content, it should be remembered that central government also provides a mass of local authority financial data in House of Commons Papers not produced primarily for statistical purposes. These include: *The Government's Expenditure Plans* [QRL.27]; *Rate Support Grant Report–England* [QRL.60]; *Rate Support Grant Report–Wales* [QRL.61]; *Rate Support Grant (Scotland) Order Report* [QRL.72], *Public Expenditure to (??)–A Commentary on the Scotland Programme* [QRL.55].

**3.2.2**

Notwithstanding the range of financial statistics produced by central government, the most extensive coverage in England and Wales is provided by the Statistical Information Service (SIS) of the Chartered Institute of Public Finance and Accountancy. The establishment, functions, management and methods of working of SIS are described in an article by G.R. Prentice in *The Statistical Information Service* [B.34]. A feature of the SIS method of preparing statistics publications is that processing, vetting and publications are often the responsibility of financial officers of individual authorities, under the guidance of working parties of practitioners. In other cases the work is carried out by CIPFA whole time staff also guided by working parties. Details of the present arrangement pertaining to publications mentioned in this Review are set out in Appendix 5. SIS publications are far too numerous to list here, covering as they do general finance; education; housing; leisure; planning; police and fire; social services; transport and environmental health. However a detailed and up to date list of publications is contained in the brochure *Statistical Information Service* [B.34]. In Scotland where, with a few minor exceptions, SIS does not operate, the main provider outside central government is the Scottish Branch of CIPFA which is responsible for the annual publications *Rating Review Part 1* [QRL.64] and *Rating Review–Part 2* [QRL.65]. Outside CIPFA, the only other body producing local government financial statistics to any great extent, is the Society of County Treasurers, whose publications cover: finance and general; trading standards; small holdings; block grant indicators. The last of these (see [QRL.6]) is produced in association with the Society of Metropolitan Treasurers and the Association of District Council Treasurers. (It is understood that SCT statistics publications are in the process of being incorporated within SIS). Other bodies publishing particular sets of statistics (e.g. statistical material relevant to Rate Support Grant–see chapter 6) are identified at the appropriate points in the review.

**3.2.3**

Among the principal suppliers of local authority financial statistics (DOE/Welsh Office/Scottish Office on the one hand and CIPFA/SIS on the other) a very crude division of labour may be discerned. Whereas the central government suppliers have hitherto tended to concentrate on providing summary statistics of actual/outturn expenditure and income, (overall and in respect of the various accounts discussed in section 3.3 below), CIPFA/SIS has published advance expenditure estimates for individual authorities at the same time rather concentrating on Rate Fund Services. It has also produced statistics (estimates and actuals) for individual services and authorities and has covered other aspects of local government finance untouched by central government (e.g. rate collection, direct labour organisations and individual service indicators). The point needs to be made that the division just described is far from precise. In both Wales and Scotland for instance, central government provide outturn expenditure data not only in summary form but at county area and region/islands level. Furthermore the Welsh Office is now publishing forward estimates of expenditure. While, in the interest of comprehension the division of responsibility has to be recognised, it should be emphasised that central government and CIPFA/SIS do not operate in isolation from one another. On the contrary, as this review will repeatedly show, there is close cooperation between the parties extending often to the use of common sources of data.

**3.2.4**

A recent innovation in the publication of local government (mainly financial) statistics calls for special mention. For each authority in England and Wales, the Audit Commission now produces a statistical *Profile* [B.29] and a *Cost Tree* [B.16]. In the former, which also includes social and manpower data, the individual authority is compared to a "family" of broadly similar authorities. The latter [B.16] which contains only financial data, displays a set of "logic trees" for each of the authority's main services and compares costs with the average for its "family" or class of authority. These publications are intended primarily for internal use.

## 3.3 The Structure of Local Authority Accounts

**3.3.1**

The presentation of a good deal of local authority financial statistics is governed by the general form in which authorities maintain their various accounts. The following *separate accounts* are maintained:

(a) Rate Fund Services Account;
(b) Housing Revenue Account;
(c) Trading Services.

Accounts. Each of these is subdivided into a revenue and capital account. Local authorities also have power to set up Special Funds covering a wide range of activities; only one of these–the Common Good Fund in Scotland–has a capital account in addition to a revenue account.

**3.3.2**

*The Rate Fund Services Account* relates to all those services which are financed by income from local rates and central government grants and only to a very limited extent by fees and charges imposed on the actual users of the services. The Rate

Fund Services are those which are regarded as typical council services (i.e. education, social services, police, fire, libraries, roads, refuse collection and disposal, consumer protection, town and country planning, parks, swimming baths).

**3.3.3**

*The Housing Revenue Account* despite its misleading title is subdivided into a current income and expenditure account and a capital account. Although local authority housing is partly supported by Rate Fund revenue it is managed as a separate service and accounted for as such via the Housing Revenue Account. As the subject of local authority housing statistics, including finance, is dealt with in a companion review (see volume III) it need not detain us further here.

**3.3.4**

*The Trading Services Accounts* arise from the fact that local authorities may engage in a range of "business" activities such as passenger transport, ports and piers, civic restaurants, aerodromes, industrial estates, markets, slaughterhouses; all of which are financed mainly by income from the customers and users of the services themselves. Although any of these services may be supported by a contribution from the Rate Fund, a separate account for each service is maintained. It is worth noting that not all authorities empowered to provide these services, do so.

**3.3.5**

In addition to the *Superannuation Funds* which the larger authorities maintain on behalf of their own employees and the employees of smaller authorities (but not teachers, firemen and police) all maintain a number of *Special Funds* the most important of which are (a) the capital and reserve funds and (b) the repairs and renewals funds. The recorded transactions of these accounts consist mainly of transfers to and from the other principal accounts mentioned above and are in the form of current income and expenditure only, the one exception to the rule, being the Common Good Fund in Scotland. This fund which is peculiar to certain Scottish Districts is discussed at length in G.B. Esslemont's article *The Common Good in Scotland* [B.3].

## 3.4 Northern Ireland

**3.4.1**

There are two *general characteristics* of Northern Ireland's local authority financial statistics which are worth pointing out. The first is that, like the rest of the United Kingdom the province follows the accounting convention that revenue or current accounts are on an income and expenditure basis whereas capital accounts relate to cash receipts and payments only (see 3.1.3 above). The second characteristic, which contrasts with what happens in Great Britain, is the practice of not eliminating inter authority transactions in producing summary statistics for the province. This omission is almost certainly of no great consequence in an area of single tier authorities, transfer payments being much more likely to take place in the double tier situation in which two authorities share responsibilities for a given area.

**3.4.2**

*Provision* of local authority financial statistics in Northern Ireland is almost entirely statutory. Since 1973–under the terms of the Local Government Act (Northern Ireland) 1972–the body responsible for publication is the Northern Ireland Department of the Environment; the only other body providing primary statistics being CIPFA/SIS which includes the Northern Ireland districts in the publication *Capital Expenditure and Debt Financing* [QRL.8]. (Note: even prior to 1973 Northern Ireland library statistics were, and still are, published by CIPFA. We return to the subject in Chapter 15.) Before 1973, the government of Northern Ireland's Ministry of Development undertook the publication of financial statistics pertaining to the local authorities then in existence (i.e. counties, boroughs, urban districts, rural districts, the city and county borough of Belfast). The Northern Ireland component of United Kingdom public sector and National Accounts statistics is of course included in the statistics produced by the Central Statistical Office and referred to in paragraph 3.2.1 above.

**3.4.3**

*The Structure of Accounts* in Northern Ireland as set out in [QRL.15] differs from that in Great Britain. The following revenue and capital accounts are maintained and published in summary form: (a) Rate Fund Service Account; (b) Special Funds; (c) Gas Accounts. The first relates to virtually the whole range of services for which the 26 District Councils are responsible. (These are listed in paragraph 2.5.1 above.) There are two Principal Special Funds: (a) capital fund and (b) renewals and repairs fund; in addition unspecified funds are established from time to time. The Gas Accounts relate to town gas undertakings operated by nine of the 26 District Councils.

**3.4.4**

Prior to 1973, the structure of Northern Ireland Accounts, reflected in statistics published in *Local Authority Financial Returns* [QRL.36] was as follows. Each authority held a General Revenue Account (roughly equivalent to rate fund services in Great Britain). However, authorities other than county councils were obliged to contribute to certain county wide services and this necessitated the keeping of separate accounts which were identified in the statistics as "Accounts held by the County Councils". In addition to general revenue accounts, there were in each authority, accounts for: housing, renewal and repairs separately; trading undertakings such as electricity, gas, transport; education, health and welfare committees which provided their respective services at County and Belfast County Borough level. In each authority there was also a "Separate Charge Account" which related to the provision of services to part of an authority's area. Capital Accounts for all services were also published in [QRL.36].

CHAPTER 4

# FINANCIAL STATISTICS: EXPENDITURE

## 4.1 Introduction

### 4.1.1

While public attention is usually focussed on one category of expenditure *viz* Rate Fund expenditure–both revenue and capital (see section 3.3)–it is important for the user of local government financial statistics to realise that the term "expenditure" has a wide variety of meanings and is used in at least 3 differing broad contexts. The *first* of these is the straight forward accounting context into which most financial statistics fall. Clear definitions of all the accounting categories, couched in layman's language are to be found in the Glossary and Notes in *Welsh Local Government Financial Statistics* [QRL.92]. See also glossary of terms in *Issues in Public Sector Accounting No.3* [B.25]. In many of the statistics publications mentioned below, a definition of the terms as used in a specific tabulation is often provided; but the practice is by no means universal. See in addition CIPFA *Guide to Local Authority Finance* [B.23].

### 4.1.2

*A second* broad context to which a unique classification of "expenditure" is applied is that of Central Government *grant distribution.* While this classification is defined in various Acts of Parliament, Statutory Instruments and so on, it is useful to note that the elements have been conveniently set out in great detail in a recent CIPFA publication *A General Guide to the Block Grant System in England 1983/84* [B.4]. The publication [QRL.92] also contains definition of the appropriate terms (total expenditure; relevant expenditure; grant related expenditure) albeit in somewhat less detail. In chapter 6 below, we return to a fuller consideration of the subject of grant distribution.

### 4.1.3

A *third* context in which still further concepts of expenditure are introduced is in the area of *public expenditure* of which local government expenditure is a component. There, a 3 way classification is adopted corresponding to 3 common sense types of questions which may be asked of any body of public expenditure:
(a) who spends it? (Classification by spending authority);
(b) why is it spent? (Classification by functional programme);
(c) what is it spent on? (Classification by economic category).
This type of classification is met, most notably, in the annual White Paper *The*

*Government's Expenditure Plans* [QRL.27] and is fully explained in a set of explanatory and technical notes in Volume 2 of [QRL.27]. Perhaps it should be pointed out that many local authority published statistics, from whichever source, are tabulated in such a way as to make it possible to answer these 3 questions either at summary or individual authority level.

## 4.2 Revenue (or current) Expenditure

### 4.2.1

The following truncated *definition* of local government revenue expenditure, taken from the Glossary and Notes of the 1982 edition of [QRL.92] is worth repeating at this point: "Revenue expenditure (is) expenditure required to be financed from revenue in the current year, covering both the continuing cost of maintaining local services including salaries and wages of employees and purchases of goods and services...and the cost of financing capital expenditure including the debt repayment and interest payable on loans raised to finance past capital expenditure in the current year". It is usual to regard current expenditure as that part of revenue expenditure exclusive of the cost of financing capital expenditure in any form, although public expenditure White Papers have their own definition of current expenditure–see [QRL.92] and section 4.5 below.

### 4.2.2

*Actual or outturn* revenue expenditure summaries (inclusive of all funds) are published respectively in *Local Government Financial Statistics–England and Wales* [QRL.41]; *Welsh Local Government Financial Statistics* [QRL.92] and *Scottish Local Government Financial Statistics* [QRL.77]. In the case of [QRL.41], the figures relate to England only and appear within a table which refers to "gross expenditure". It should be noted that a long run series of these statistics for England and Wales combined, appeared in the 1974/75 and 1982/83 issues of [QRL.41] along with a medium term series for England and Wales separately. In Scotland, the Scottish Branch of CIPFA now publish outturn revenue expenditure statistics for individual authorities in *Rating Review–Part 2* [QRL.65].

### 4.2.3

The source of the published summaries in [QRL.41] and [QRL.92] are the statutory Revenue Outturn forms completed by individual authorities in England and Wales and for [QRL.77] the Local Financial Returns (LFR's) in Scotland. (see lists in Appendix 4). As the forms themselves are too numerous to reproduce in this review, selected specimens plus completion instructions are shown in Appendix 6. The summary net expenditure figures for [QRL.65] are collected by the Scottish Office by means of a Final Outturn form which is consistent with but contains much less detail than the LFR's being completed at the same time; but the detailed figures on individual services in [QRL.65] are obtained from LFR's. As [QRL.65] usually contains partial information –not all authorities have made returns in time–the

CIPFA publication date precedes the Scottish Office publication of [QRL.77] by a wide margin.

**4.2.4**
As explained in Section 3.3 the Rate Fund Services which constitute by far, the major part of local authority activities are those financed in the main, from local rates and by central government grants. *Rate Fund Expenditure* statistics draw a distinction between that expenditure which is net of income from local sources (e.g. sales, fees, charges, interest receipts) and total revenue expenditure. The former is usually referred to as "net expenditure to be financed from grants, rates and balances". *Actual or outturn* statistics of both are published in [QRL.41]; [QRL.92] and [QRL.77]. Publication [QRL.41] contains a summary for England and for "class of authority" only; the others provide data down to "county areas", "region/islands" levels. The sources of the statistics are the DOE/Welsh Office Revenue Outturn forms–for [QRL.41] and [QRL.92]–and the Scottish Office Local Financial Returns (see Appendix 4). In respect of English and Welsh *Counties* only, the Society of County Treasurers publishes rate and grant borne expenditure statistics in it's *Financial and General Statistics of County Councils* [QRL.21]. The source of the SCT data is a specially designed questionnaire in which individual authorities provide the outturn figures produced for DOE/Welsh Office. An additional publication containing England/Wales outturn summaries is *Local Government Trends* [QRL.43] produced by CIPFA/SIS. While the original sources of the figures quoted are [QRL.41] and [QRL.92], the publication is worthy of special mention for its simple, clear cut presentation and valuable accompanying comments. In Scotland, the Scottish Branch of CIPFA publishes outturn data in *Rating Review–Part 2* [QRL.65].

**4.2.5**
*Estimated* Rate Fund revenue expenditure data (for one year ahead) collected from *English and Welsh* authorities jointly by DOE/Welsh Office and CIPFA/SIS is published by the latter in [QRL.20]. The source of the figures is the Return of Expenditure and Rates (RER) shown in Appendix 7. These estimates are published early in the financial year to which they relate. Using the same source, the Welsh Office now publish summary estimates in [QRL.92] for the financial year just completed and the preceding year. Clearly, these statistics are to be regarded by users, as best estimates of likely outturn and are designed to bridge the time gap between the end of a financial year and publication of final outturn statistics.

**4.2.6**
*Estimates* of Rate Fund expenditure one year ahead for all authorities in Scotland, are published by the Scottish Branch of CIPFA in [QRL.64]. While the Scottish Office collects data from authorities on lines rather similar to the RER in England and Wales (see Appendix 7), they do so for financial planning purposes only. CIPFA in Scotland assemble estimates data independently of the Scottish Office, (see Appendix 7).

**4.2.7**

Clearly, estimates of expenditure for a period which will not be completed for approximately 15 months from the time at which they are made, must allow for *expected price changes*. The practice in local government is, in the first instance, to estimate on a November price base, the November in question being that prior to the beginning of the financial year to which the estimates relate. Thereafter each authority estimates a repricing factor or expected rate of increase above the November price based expenditure. In completing RER form, authorities in *England and Wales* are expected to conform to a centrally prescribed definition of the November price base. (In Appendix 7, a definition and worked example for the 1982/83 returns are shown in RER Annex 1 and Appendix B). Where authorities do not make returns on the November price base, their figures–for statistics publication purposes –are repriced centrally using methods agreed between DOE/Welsh Office, CIPFA and the Local Authority Associations. The results however approximate, are seen as preferable to publishing expenditure statistics on varying price bases.

**4.2.8**

It may be useful to note that later in the financial year, *revised estimates* of Rate Fund expenditure in England and Wales by authorities are sent to the DOE/Welsh Office and CIPFA/SIS on form RER (see Appendix 7). Although not published in England these statistics (by individual authority) can now be obtained from the CIPFA computerised data base (see paragraph 17.1.5).

**4.2.9**

In Scotland, authorities, having estimated expenditure at November price base, revalue their estimates to take account of price increases agreed by the Costs Sub-Group of the Local Government Finance Working Party which comprises officers of the Scottish Office Finance Division and local government officers representing COSLA. Details of increased costs and prices taken into account in revaluing estimates are set out in the COSLA *Report on Rate Support Grant Order* [QRL.71].

**4.2.10**

Statistics of the *actual or outturn* revenue expenditure of the other local authority *accounts* and *funds* (Housing Revenue Account; Trading Services Account; Special Funds and Superannuation Funds) are also published in [QRL.41] and [QRL.92] for England and Wales, the sources being the DOE/Welsh Office Revenue Outturn forms listed in Appendix 4–Specimen Appendix 6. Corresponding Scottish statistics are published in [QRL.77] based on the Scottish Office Local Financial Returns (see Appendix 4). Unlike Rate Fund practice, statistics of estimated expenditure for these other accounts and funds are not published. (Note: Superannuation funds do not constitute part of the local government sector–see paragraph 3.1.1 and section 4.4. Their inclusion in this review is justified on the ground that details of the funds in the summary are published regularly in [QRL.41], [QRL.92] and [QRL.77] as part of the body of local government statistics).

### 4.3 Capital Expenditure

**4.3.1**
While capital expenditure is clearly expenditure on the acquisition or creation of assets with a long term value (e.g. the purchase of land and buildings; the construction of new buildings; the purchase of plant, furniture, vehicles and equipment), in a local authority context, it has –since 1981–included spending on certain types of long term leasing of property, vehicles, plant and machinery. This extension of the boundary is fully developed in the 1981 version of [QRL.27]–*The Governments' Expenditure Plans* Command Paper 8175. It is worth noting that capital spending includes the cost of the services of certain employees engaged in planning and supervising capital projects.

**4.3.2**
It is essential to a proper understanding of the statistics of local authorities capital expenditure to appreciate that capital spending limits are imposed by central government and that the method of control has changed recently. Prior to the passing of the *Local Government Planning and Land Act* (1980), all capital spending in England and Wales, was subdivided into (a) key sector and (b) locally determined sector categories. Key sector services were those for which borrowing approval, for individual projects, had to be obtained from central government and included those services over which central control was exercised (e.g. education, housing and principal roads). The locally determined sector consisted of those services for which borrowing approval was given as a block, each authority being free to determine which projects to undertake. Included in the latter were services such libraries, car parking and swimming baths. (This categorisation of capital expenditure did not apply in Scotland). Since the 1980 Act, each council has received from central government an annual capital allocation which is net of the proceeds of any sales of its own assets (land, buildings and vehicles). Statistics of those allocations in England and Wales are now published by CIPFA/SIS in [QRL.20], the source being the DOE/Welsh Office.

**4.3.3**
Statistics of *actual/outturn* capital expenditure for Rate Fund Services, Housing and Trading Services in England and Wales are published in [QRL.41] and [QRL.92], the level of geographical detail varying between the two publications. However, because of the long delay in publishing, CIPFA/SIS and the Welsh Office respectively produce statistics of *provisional outturn* capital expenditure and capital *payments* for years for which final outturn data is not available. The CIPFA/SIS figures are published in *Capital Expenditure and Debt Financing Statistics* [QRL.8]; the Welsh Office publishes in [QRL.92]. The former relates not only to England but to each authority in the United Kingdom whereas the latter contains Wales summary figures only. In interpreting provisional outturn and payment statistics and in particular, in comparing them with final outturn figures, it should be noted that definitions vary–although not widely. The variations are brought out in detail in "notes for guidance

in completing table" (Form CPA 4 in Appendix 9). At this point attention is drawn to the fact that long run series of outturn capital expenditure statistics–summaries for England and Wales combined–are contained in the 1974/75 and 1982/83 editions of [QRL.41].

### 4.3.4

The sources of final outturn capital expenditure statistics in [QRL.41] and [QRL.92] are the Capital Outturn returns made to DOE/Welsh Office by authorities in England and Wales and listed in Appendix 4; notes for completion and a specimen CO form are also shown in Appendix 6. The source of the CIPFA statistics in [QRL.8] is the CP form completed by all authorities–see Appendix 9.

### 4.3.5

In *Scotland, outturn* capital payments statistics for Rate Fund Services, Housing and Trading Services are published by the Scottish Office in [QRL.77]–see also paragraph 4.3.6. As in England and Wales, CIPFA also publishes in [QRL.8] provisional outturn capital payments statistics for all authorities in Scotland, albeit in rather less detail. Recently the Scottish Branch of CIPFA has begun to publish outturn statistics of authorities' "capital payments requiring consent" [QRL.65]. However it employs a different definition of "payments" and a different source from that used in producing its sister publication [QRL.8]. Capital payments requiring consent may differ significantly from total capital payments in as much as the latter may be partially offset by capital receipts, whereas the former represent capital payments over and above receipts from sales of capital assets. One other respect in which Scottish practice differs from that in the rest of the UK is that the Scottish Branch of CIPFA publishes *estimates* of capital expenditure for each authority; these appear in [QRL.64].

### 4.3.6

The *sources* of the final outturn figures in [QRL.77] are the Scottish Local Financial Returns listed in Appendix 4; a specimen form containing completion instructions is shown in Appendix 6. CIPFA's [QRL.8] statistics in relation to Scotland are collected, as in the rest of the UK via CP form (Appendix 9), (Note: in Scotland, Capital Payments Returns are in use similar to those made to the DOE/Welsh Office). The source of the much earlier outturn figures in [QRL.65] are set out in detail in the publication itself. Sources of capital expenditure *estimates* are locally devised CIPFA forms which are completed by individual authorities. The statistics relate to expenditure on each service and are doubtless influenced by capital expenditure on cash allocations which have been notified to authorities by the Scottish Office.

**4.3.7**
As in the case of England and Wales (see paragraph 4.3.2) statistics of capital *consents* (basic, supplementary and adjusted) are now published in [QRL.65], the source being set out in the publication itself.

**4.3.8**
With one exception the summary expenditure statistics of Special Funds and Superannuation Funds which are published in [QRL.41], [QRL.92] and [QRL.77] do not contain capital accounts, although one of the largest of these Special Funds is itself a capital account, payments from which are mere transfers to the Rate Fund and the other principal accounts already identified. The exception referred to is the Common Good Fund in Scotland (see section 3.3). This fund does have clearly defined capital assets of its own (land and property) to which it adds from time to time and income from which is used independently of other Fund activities. The source of outturn capital expenditure data is the Scottish Office form LFR 25 listed in Appendix 4, which is completed by the District and Islands authorities concerned.

## 4.4 Local Authority expenditure in the context of public expenditure and the national accounts

**4.4.1**
Tabulations in the annual public expenditure White Paper *The Government's Expenditure Plans* [QRL.27], bring out the relationship between local authority expenditure and other public expenditure, while the contribution of the local authority sector to the national economy is shown in *UK National Accounts* [QRL.88], *Financial Statistics* [QRL.22] and *Economic Trends* [QRL.16]. Both public expenditure and the national accounts figures are based on essentially the same definitions, but there are marginal differences which are listed in the *White Paper* [QRL.27] and the *Financial Statistics Explanatory Handbook* [B.22]. Of particular interest here is that [QRL.27] includes in the local authority sector services in Northern Ireland such as education, police and fire, which are administered by local authorities in the rest of the United Kingdom.

**4.4.2**
The statistics of *outturn* and *provisional outturn* expenditure (current and capital) as part of public expenditure, are published in the annual White Paper [QRL.27]. This contains an annual series covering the estimated outturn for the current and the previous financial year and outturn figures for the four prior years. A definition of the public sector is set out in detail in the explanatory and technical notes in Volume II of the White Paper. In various tabulations in both volumes, the statistics are ordered by the 3 categories described in paragraph 4.1.3. Condensed versions of the tabulations are published in [QRL.41] which also contains (a) the ratio of Local Authority to Public Expenditure and (b) notes defining the difference between the current and capital expenditure figures in the White Paper and those contained

elsewhere in [QRL.41]. A secondary publication with brief comments, is CIPFA's *Local Government Trends* [QRL.43] which also relates local authority aggregate expenditure to public spending as a whole.

**4.4.3**
The data *sources* for the *outturn* expenditure statistics in [QRL.27] are the DOE/Welsh Office and Scottish Office revenue and capital outturn forms (see Appendix 4). (As to the Northern Ireland part of the UK, the statistics in [QRL.27] do not relate to district council expenditure only. There is therefore no point in identifying sources.) With respect to (a) the *provisional or estimated outturn* expenditure statistics for the immediate past year and (b) the estimated outturn or budget data for the current year in [QRL.27]: these are derived from the Returns of Expenditure and Rates (see Appendix 7) "revised estimates" and "budgets" columns respectively–and the Capital Payments Return (see Appendix 9), 4th quarter (cumulative) and projections based on historical trends taken from the 1st and 2nd quarter CPR's respectively. There are corresponding capital payments returns (see Appendix 4) and current expenditure returns in use in Scotland (see Appendix 4). The statistics in [QRL.41] from which the ratios are computed are simply the White Paper series. By contrast the expenditure data published by CIPFA and attributed to CSO, are taken from tabulations in [QRL.22] the data sources of which are described in paragraph 4.4.7.

**4.4.4**
With respect to Scotland only, the publication *Public Expenditure to (Year)* [QRL.55] contains matching outturn/provisional outturn expenditure statistics, both revenue and capital. The data sources for the outturn figures are those referred to in paragraph 4.4.3; provisional/estimated outturn data is obtained from PO/BE form (see Appendix 7).

**4.4.5**
In addition to the statistics of UK local authority actual/outturn current and capital expenditure, central government publishes *planned* expenditure estimates of these entities in the annual White Paper [QRL.27]. These planned figures relate to the financial year which commences in the April following publications of the White Paper a few months previously. By their nature as plans they are not derived directly from any statistical returns. It is worth noting that in [QRL.27] not only are planned expenditure figures for the United Kingdom published but separate estimates by programme are shown for England, Wales and Scotland. The publication [QRL.55] also contains, with accompanying commentary, summary statistics of planned expenditure in Scotland. Northern Ireland district council expenditure plans do not receive similar treatment. It should also be noted that prior to the White Paper of 1982, the planned expenditure figures were expressed at constant or survey prices to enable comparison to be made over time of the quantity of resources used up in providing local authority services; the statistics were said to be made in volume

terms. Since 1982, the figures have been expressed in cash terms, that is to say, they have related to the amount of money which central government expects to be spent in the plan years. The transition to cash planning and the methodology of computing survey prices are set out in the explanatory and technical notes which are part of [QRL.27]. A useful background paper on the transition to cash planning is the Treasury's *Public Expenditure: Planning in Cash–Economy Progress Report No. 139* [B.30].

**4.4.6**

The *national expenditure* context is that in which broadly, all expenditure of the nation's resources is categorised according to whether the resources are used up ('final consumption" category) or add to the stock of the nation's wealth ('gross domestic fixed capital formation"/'additions to stocks and work in progress" category). It must be remembered that local authorities do not calculate additions to stock nor value work in progress. Thus their only addition to the nation's wealth is *via* gross domestic fixed capital formation. In the national income and expenditure framework, a basic distinction is made on the one hand between payments which are made for the supply of goods and services (e.g. fuel, telephones, paper), factors of production (labour, land, buildings) which enter into national income and on the other hand, transfers or unrequited payments (e.g. interest payments, capital grants, student grants) which represent a redistribution of national income without adding to its total. It is this distinction which, in the main, marks the difference between local authority expenditure in its public and national expenditure contexts respectively; spending in the national framework excluding such transfer payments. The precise correlations between the local authority component of national expenditure and the other expenditure categories discussed in this chapter, are described in notes in [QRL.41] and the *National Accounts: A Short Guide* [B.2]. In the national accounts context expenditure can in principle be measured in two different ways, at market prices or at factor cost. The former includes all taxes on expenditure with subsidies treated as negative taxes; the latter can be derived by deducting taxes on expenditure and adding back subsidies to arrive at the income of the factors of production. The concepts are discussed in *National Accounts: A Short Guide* [B.2], *Financial Statistics Explanatory Handbook* [B.22], and *United Kingdom National Accounts–Sources and Methods* [B.36]. Local authorities expenditure is shown at market prices and includes (for comparability with other sectors' expenditure) an estimate of net value added tax (which is in practice refunded to the authorities). One further feature of national accounts expenditure statistics is that Superannuation Funds are excluded from the local authority sector as the Funds are the property of the members and thus part of the personal sector of the national accounts.

**4.4.7**

The statistics of *estimated annual* United Kingdom local authority expenditure (i.e. final consumption and gross domestic fixed capital formation) set in a national expenditure context are published in the CSO's Blue Book: *United Kingdom National Accounts* [QRL.88]. It should be noted that the estimates in this case are of actual expenditure and are not forward estimates. But since the figures are computed on a calendar as distinct from financial year basis they must remain estimates, that is to say, they will never be "proved" as financial year data may be in time. The Blue Book figures are reproduced in handy summary form in [QRL.41] and in [QRL.43]. Both publications also provide ratios of local authority expenditure to national expenditure, but the fact that there is a wide discrepancy between the ratios derived calls for comment. We return to this subject in paragraph 4.4.13 below.

**4.4.8**
*Quarterly estimates* of United Kingdom local authority expenditure in the national context are published by the CSO in [QRL.22] and [QRL.16]. The latter also publishes annual calendar year series while the former provides both calendar and financial year figures. Definitions and useful notes on the components of the expenditure aggregates are set out in the annual publication *Financial Statistics–Explanatory Handbook* [B.22]. These notes are invaluable as an aid to the understanding of the sources of the figures described below.

**4.4.9**
In addition to the basic national expenditure categories–"final consumption" and "gross domestic fixed capital formation"–in the CSO publications [QRL.88], [QRL.22] and [QRL.16] the transfer or unrequited payment categories of expenditure are also identified. As all three use common expenditure classifications and data sources it might be useful to focus on each expenditure item across the publications, when considering original sources.

**4.4.10**
With regard to *current expenditure* of national resources by the local authority sector, the main category is of course "final consumption", which excludes debt charges and is measured net of related receipts, other than interest or government grants. In the Blue Book [QRL.88] this item is analysed by service and in England and Wales the source document is the RS 1B (Appendix 8) and the RO forms where necessary (Appendix 4); in Scotland the LFR form (Appendix 4) and in Northern Ireland the Abstract of Accounts are used (see paragraph 4.5.2). However, since national expenditure statistics relate to calendar years whereas the returns mentioned cover financial years, some adjustment or interpolation is called for. This is effected for Great Britain authorities via the quarterly return QRR (see Appendix 9) which calls for "all services" wages and salaries data, accounting for some three quarters of all current expenditure. The services allocations are made by interpolation. The Northern Ireland District Council portion of United Kingdom expenditure is so small that interpolation from and extrapolation of *annual* data is employed. The quarterly statistics of final consumption in [QRL.22] and [QRL.16] are based on the QRR (see Appendix 9) for England, Wales and Scotland with the Northern Ireland data being interpolated from the District Councils General Estimates of Rates (see paragraph 4.5.3). Before leaving the subject of final consumption, there is one special

point that has to be mentioned. The Blue Book does, within the local authority current account, identify the item "non trading capital consumption". This item is estimated by using a CSO capital stock model employed elsewhere in the national income and expenditure estimates, there being no direct estimate of this entity by the local authorities themselves.

**4.4.11**

As to the *transfer payments* items in the current account: the first of these, subsidies, is analysed under three headings–housing, passenger transport and other trading. The source of the housing figures is an interpolation from form RO 4 (Appendix 4) in England and Wales, LRF 20 (Appendix 4) in Scotland; and Abstract of Accounts or General Estimates of Rates (see paragraph 4.5.2) in Northern Ireland. The second item–passenger transport subsidy–is obtained from returns made directly to CSO by British Rail and the Passenger Transport Executive while the final part–other trading–is obtained in England and Wales by interpolating from RS 1B data (Appendix 8) and LFR 22 in Scotland (see list Appendix 4). There does not appear to be a Northern Ireland component. *Current grants to the personal sector* are of two kinds: (a) scholarship and grants to universities and colleges, the figures for which are based on the pattern of educational grant payments known to central government; (b) rent rebates and allowances are interpolated from RO4 and Housing Revenue Account outturn form in England and Wales, LFR 20 form in Scotland (see Appendix 4 list for all forms). The Northern Ireland component is also obtained by interpolating from the Abstract of Accounts (see paragraph 4.5.2). With regard to the final part of the transfer payments statistics, *debt interest*: quarterly returns of external interest payments are obtained from individual authorities via the QRR form shown in Appendix 9 (Note: this return is made by all authorities in Great Britain and on behalf of Northern Ireland authorities). The breakdown of this figure into "central government" and "other" parts is obtained not from the individual authority returns but via CSO other sources (i.e. banks, financial institutions, central government)

**4.4.12**

The local authority sector's *capital account* in the national expenditure context divides into *gross domestic fixed capital formation* and the two transfer payments items: *capital grants to the personal sector* and *capital grants to public corporations* in [QRL.22]. These grants are referred to as capital transfers. The capital formation figures are gathered from the quarterly capital payments returns (see Appendix 9) made by Great Britain authorities or on behalf of Northern Ireland District Councils. The capital grants (or transfers) data for England is obtained quarterly from the PI Housing Activity returns adjusted to the financial year outturns based on a reconciliation of the Housing Investment Programme return and the Capital Outturn forms (see Appendix 4). For Scotland and Wales the figures are interpolated from the respective LFR's and capital outturn figures (see Appendix 4). In respect of Northern Ireland the figures are simple interpolations from the capital outturn data in the Abstract of Accounts (see paragraph 4.5.2).

**4.4.13**
In paragraph 4.4.7 we undertook to return to the subject of the wide discrepancy between the CSO's calculation of the local authority sector's share of United Kingdom national expenditure, reproduced annually in [QRL.41], and in CIPFA estimate published in [QRL.43]. The first point to note is that the latter estimate relates to the financial year where as the former refers to the calendar year; some discrepancy therefore is inevitable. However, the major difference between the estimates arises from contrasting definitions. The CSO defines the ratio of local authority sector expenditure to national expenditure as: local authority final consumption plus its domestic fixed capital formation all divided by total domestic expenditure, figures computed at market prices for the calendar year in question. The CIPFA calculation is defined as: local authority sector expenditure (less debt interest) plus capital expenditure all divided by total domestic expenditure–the denominator computed at factor cost and deflated to 1975 price level. Clearly the main difference stems from two sources: (a) the treatment of transfer payments and (b) the contrasting use of market price and factor cost data in the calculations. The controversial subject of defining public expenditure/GDP ratios is discussed extensively by David Heald in *Public Expenditure* [B.7]).

## 4.5 Northern Ireland

**4.5.1**
In local authority statistics publications in Northern Ireland, there is nothing like the variety of *definition* of the term "expenditure" which prevails in the rest of the United Kingdom. This state of affairs would appear to arise either because the distribution of central government's General Grant or General Exchequer Contributions (pre-1973) does not or did not demand special categorisation of expenditure or alternatively that the figures in which grant distribution to individual authorities are based, are not published. In fact, the only classification of "expenditure" used in present district council statistics are "gross" and "net"–the "revenue" and "capital" boundary of each being clear from the content of the tabulations. Prior to 1973, the term "expenditure" was not used at all in local government statistics publication. Expenditure figures were recorded as "payments", despite the fact that accounts were kept on an accruals basis.

**4.5.2**
The statistics of *actual* or *outturn* revenue and capital expenditure of district councils (summary only) are published by the Northern Ireland Department of the Environment in *District Councils Summary of Statement of Accounts* [QRL.15]. The source of the figures is the Abstract of Accounts completed by district councils (specimen not provided). Unlike summary statistics in Great Britain, these figures do not comprise sector accounts, i.e. they simply aggregate the transactions of the 26 district councils and do not eliminate inter-district or transfer transactions. The Abstract of Accounts returns by district councils are made under the authority of Section 77(2) of the Local Government Act (Northern Ireland) 1972. *Provisional*

*outturn* capital expenditure (payments) statistics for each district council are published annually by CIPFA in *Capital Expenditure and Debt Financing Statistics* [QRL.8]. The source of the figures is the CIPFA/SIS return (see Appendix 10) but note that a special print of the form is used for Northern Ireland.

**4.5.3**
Published *estimates* of district council expenditure in Northern Ireland relate only to Rate Fund current or revenue expenditure. The estimates appear in the Northern Ireland Department of the Environment's *District Council Rate Statistics* [QRL.14] immediately prior to the commencement of the financial year to which they relate. The source is the General Estimate of Rates returned to the Departments by district councils Clerks (specimen not provided). In making their estimates, districts are expected to have regard to the assumptions on price increases included in the Government White Paper [QRL.27].

**4.5.4**
Figures of central government's *plans* for local authority expenditure –current and capital combined–appear alongside a matching outturn series in [QRL.27]. It is worth repeating that these statistics do not relate only to the expenditure of district councils–see paragraph 4.4.1.

**4.5.5**
*Prior to 1973*, statistics of *actual/outturn* expenditure by the then local authorities were published in *Local Authority Financial Returns* [QRL.36] by the Northern Ireland Ministry of Development. Both revenue and capital expenditure were encompassed in [QRL.36]. *Estimates* of current expenditure (referred to as "payments") on General Revenue Account only (see paragraph 3.4.4) were published by the Ministry of Development in *Local Authority Rate Statistics* [QRL.38]. As the principal object of this publication was to set out for each authority, its estimated notional charge on ratepayers, 3 categories of estimated "payments" were published (i.e. total payments, net payments, notional charge on ratepayers). Net payments were net of specific government grant and miscellaneous receipts. Notional charge on ratepayers (i.e. estimated expenditure to be borne by ratepayers) was deduced by taking into account likely income from General Exchequer Contribution.

**4.5.6**
It should be noted that in the examination of local authorities expenditure in the context of public expenditure and the national accounts, Northern Ireland has had to be treated as a component part of the United Kingdom (see paragraph 4.4).

CHAPTER 5

# FINANCIAL STATISTICS: RATES

## 5.1 Introduction

### 5.1.1

In this review it has been thought preferable to separate consideration of rates (i.e. rates income; rate poundages; rate collection) from the subject of rateable values and penny rate products which are dealt with in Chapter 9. The reason for the separation is that whereas the rate is a component of local government income, which is the subject of this and the three succeeding chapters, RV and penny rate product measure (a) *capacity* to raise income or (b) the magnitude of an authority's tax base.

### 5.1.2

The rate is virtually the only form of taxation available to local authorities in this country and is paid by the occupier of non-agricultural land and buildings. In the case of Crown properties located within a local authority area, payments are made to the authority in lieu of rates–a point worth noting as in some parts of the country (e.g. garrison towns) this type of payment forms a large part of an authority's income. In England and Wales rates are collected by district councils, London Boroughs, the City of London and the Council of the Isles of Scilly. Other authorities–county councils, parish councils, county and metropolitan police authorities–raise their income from rates by precepting or requiring the collecting authorities in their areas to gather rates on their behalf (see *Guide to Local Authority Finance* [B.23]). In Scotland the regional/islands councils and district councils determine the annual rate necessary to pay for the services they provide; rates are levied by the regional councils (for both regional and district rates) and by the islands councils.

### 5.1.3

In addition to the rate levied by the collecting authority in England and Wales–referred to as the "general rate"–and the separate rates of the various precepting authorities, there is provision (see [QRL.20]) for a "special rate" which is used only in a small number of cases. The special rate is a levy imposed on an area for special works deemed to be non-statutory (e.g. footpath lighting in one small part of a council's administrative area). In Scotland, because the regional councils have responsibility for water supply a separate *domestic water rate* is levied and separately identified in the statistics. In England and Wales the power to levy a supplementary rate has now been rescinded; Scottish authorities never enjoyed that power.

**5.1.4**
By statute, rating authorities are required to levy lower rates on *domestic* than non domestic properties. The resulting loss of income to a local authority is made up by central government in England and Wales, by means of the domestic rate relief grant, formerly the domestic element of RSG. In Scotland the grant is still designated "domestic element of RSG". This subject is dealt with in greater detail in section 6.4.

## 5.2 Rate Income

**5.2.1**
Early in the financial year, two forward *estimates* of income from rates are published. First, for each authority in England and Wales, CIPFA/SIS provide in [QRL.20] figures designated "authority's own rate" for each rating authority or as "precept" for all others. Second, the Rating and Valuation Association in *General Rate Poundages and Products* [QRL.26] publishes statistics for each rating authority in England and Wales and for each district and islands council in Scotland showing the expected rate income, as "total product", that is the total income expected to be raised by each rate collecting authority for all purposes including county precepts in England and Wales but excluding parish council or other precepts; in Scotland, the figure includes the rates to be collected for regional purposes. (It is worth noting that the [QRL.20] data may include a precept from a parish council or other body, both of which are also separately identified.) At the end of each financial year, *revised estimates* of income from rates for that year are provided in [QRL.26] and until the year 1975/76 were published by the DOE/Welsh Office in *Rating and Rateable Values in England and Wales* [QRL.63]. Up to and including the financial year 1978/79 the DOE continued to make available in note form, the rates income data hitherto provided in [QRL.63].

**5.2.2**
Data for [QRL.20] is collected from councils by means of the Return of Expenditure and Rates (RER) "budget column" shown in Appendix 7. The source document for [QRL.26] is a Rating and Valuation Association questionnaire not reproduced here. For [QRL.63] the data source was the RER "revised estimate" column–see Appendix 7.

**5.2.3**
In England, statistics of *actual/outturn* income from rates (national summary only) appear in [QRL.41]; for Wales, the relevant figures not only in summary but at individual authority level, are published in [QRL.92]. Additionally, for counties only in England and Wales, the SCT publishes, somewhat earlier than [QRL.41] and [QRL.92], rate income (actual) statistics in *Financial and General Statistics of County Councils* [QRL.21]. In Scotland, actual/outturn rate income figures are published in [QRL.77] in summary and by rating authority. In the latter case the data contains

not only income raised from ratepayers but also grants from central government by way of (a) domestic element of RSG and (b) Rate Rebate Grant–see paragraph 6.4.9 and section 6.3.

**5.2.4**
For calendar years as distinct from financial years, to which all the above statistics relate, the CSO publishes in summary for the United Kingdom, statistics of rates income in its Blue Book *United Kingdom National Accounts* [QRL.88].

**5.2.5**
The data source of the actual/outturn rates income statistics in [QRL.41] and [QRL.92] is the return RS 1B (see Appendix 8). From counties in England and Wales, the SCT collects data for [QRL.21] by means of a questionnaire in which respondents are asked to provide their RS 1B figures. For [QRL.77] in Scotland, the data source is the Scottish Office form LFR 11 (see Appendix 11). With respect to the calendar year rates income statistics (summary) for the United Kingdom in [QRL.88]: the source documents for authorities in England, Wales and Scotland are the quarterly QRR forms (see Appendix 9). The Northern Ireland component of the statistic being very small in relation to the UK total, is derived by interpolation/extrapolation, of financial year data in the Northern Ireland Abstract of Accounts (see paragraph 4.5.2).

## 5.3 Rate Poundage and Average Rate Payment

**5.3.1**
The rate poundage is the number of pence in the pound which each occupier has to pay on the rateable value of his property. Average rate payment is expressed, in published statistics, as the average rate bill (usually per domestic rate payer) or alternatively as average rate poundage per property. The significance of both statistics resides in the fact that they represent rates of local property tax or alternatively measures of the local rate burden imposed for the purpose of paying for local authority services.

**5.3.2**
The levels of *rate poundage* are set by local authorities in the early part of a calendar year, i.e. at the point at which they are budgeting for their financial year beginning in April. In *England and Wales*, the earliest national publication of rate poundage statistics is undertaken by CIPFA/SIS in [QRL.20] and the Rating and Valuation Association in [QRL.26]. In both cases poundages must be regarded strictly as estimates since changes can take place later in the year. In respect of completed financial years, both bodies provide final rate poundage figures in [QRL.26] for the Rating and Valuation Association and for CIPFA/SIS *Rate Collection Statistics–Actuals* [QRL.58]. Central government (DOE/Welsh Office) also publish actual rate

poundage statistics in [QRL.41] for England and [QRL.92] for Wales. Until the year 1975/76, the DOE/Welsh Office also published rate poundages in [QRL.63]; up to and including 1978/79 the DOE continued to make available on request all [QRL.63] figures in note form.

**5.3.3**
For clarity's sake, features of some of the statistics referred to in paragraph 5.3.2 need to be mentioned. In [QRL.20] rate poundages actually levied (as distinct from national poundages which are referred to as "equivalent rates in the pound" and explained in an accompanying text) are shown for each authority as a range–i.e. maximum and minimum values. This is to accommodate the circumstance that in England and Wales–in addition to the general rate,–special rates may be levied for particular areas within a district (see paragraph 5.1.3). In [QRL.41] the published poundages are derived or secondary statistics for each class of authority and arrived at by weighting the rate poundage of each authority by its rateable value and averaging thereafter. By contrast, [QRL.92] and [QRL.58] provide primary data for individual authorities, although in the former instance, average rate poundages are also published. In [QRL.58] the poundage figures exclude parish and other precepts (e.g. for land drainage, joint crematoria).

**5.3.4**
In *Scotland*, the Scottish branch of CIPFA publishes rate poundage statistics in *Rating Review Part 1* [QRL.64] as does the Rating and Valuation Association in [QRL.26]. Both appear early in the financial year to which the figures relate and, since changes can be made later, must be regarded as provisional. However, the publication [QRL.26] does also contain final rate poundage statistics for the previous financial year. Later in *Scottish Local Government Financial Statistics* [QRL.77] final rate poundages are published for regions and islands, i.e. the rating authorities in Scotland. The Scottish branch of CIPFA now publishes final figures for each authority in *Rating Review–Part 2* [QRL.65].

**5.3.5**
Differences in scope and definition between the poundages published by CIPFA in [QRL.64] and [QRL.65] and the Scottish Office in [QRL.77] should be noted. CIPFA publishes separate general rate poundages for: (a) regions/islands; (b) districts; (c) domestic water (regional and islands councils are water authorities). But the figures published for domestic ratepayers do not represent the actual levels of taxation to be borne. Domestic rate relief has to be deducted and the level of this relief is referred to in a footnote to the relevant tables as "the differential rate applicable to domestic property". With respect to [QRL.77]: the statistics provided for regions are secondary in that published poundages are average district and regional rates, there being of course, more than one district in each region. Note that in the case of domestic rate poundages, allowance has been made for domestic rate reduction.

**5.3.6**
The *sources* of the rate poundage data published by CIPFA in [QRL.20] and [QRL.58] are respectively, the RER form shown in Appendix 7 and a CIPFA questionnaire which is completed by authorities in England and Wales at the end of each financial year (specimen not reproduced in this Review). The Rating and Valuation Association's questionnaire, which is, the source of the data for [QRL.26] is not reproduced here. For the DOE/Welsh Office publications [QRL.41] and [QRL.92] the definitive source documents are the final Rate Demand Notes issued to ratepayers by individual rating authorities, copies of which are received by DOE/Welsh Office.

**5.3.7**
In Scotland, the source of the statistics in [QRL.64] and [QRL.65] are questionnaires sent out to all authorities by the Scottish Branch of CIPFA. As in England and Wales, the Rating and Valuation Association collects its Scottish data from district and islands authorities by means of the questionnaire referred to in the previous paragraph; this for [QRL.26]. For [QRL.77] the Scottish Office assembles data from the individual authorities' Rate Demand Notes.

**5.3.8**
As a measure of the rates burden on domestic ratepayers, statistics are computed of average domestic rate bills or *average payment* per domestic hereditament. With respect to English and Welsh authorities, *estimates* of average payment per domestic hereditament are published by CIPFA/SIS in [QRL.20] using average rates supplied by authorities on RER forms (see Appendix 7) and discounting to allow for domestic rate relief (see 6.4.1). *Outturn* figures of average domestic rate payments are subsequently published for authorities in Wales, in the Welsh Office's [QRL.92]. In this case the sources are the Rate Demand Notes issued to ratepayers by rating authorities and the rateable value data provided by the Inland Revenue (see Appendix 14). In Scotland, *estimated* average domestic rate bills are published by the Scottish Branch of CIPFA in [QRL.64] using the information on rate poundages already supplied by authorities together with rateable values data provided by Land Valuation Assessors (see paragraph 9.1.3). Thereafter, average domestic rate bills are calculated by taking into account the rate reduction due to the domestic element of Rate Support Grant. In later years [QRL.64] and [QRL.65] provide revisions of these statistics using later issues of form RRV/ASS (Appendix 14). In [QRL.77] the Scottish Office publish average domestic rate payments and average domestic water rate payment statistics using the sources as above.

**5.3.9**
Before leaving the subject of rate poundage and related statistics, reference must be made to a term which will be included frequently in certain England/Wales statistics publications, i.e. "grant related poundage". This is defined and discussed at some length in an Appendix dealing with income from government grants (see Appendix 12 paragraph 6).

## 5.4 Rate Collection

### 5.4.1

In England and Wales, statistics are published annually by CIPFA/SIS relating to the *actual collection* of income from rates in respect of each rating authority. The aspects of collection covered are:

(a) amount of rates collectable;
(b) amount of rate arrears;
(c) constraints on collection or percentage of rates uncollectable because of factors such as empty properties, rate rebates, bankruptcies, discounts;
(d) method of collection (e.g. cash, standing orders, P.O. Giro);
(e) proportion of rate income collected per quarter;
(f) number of reminders, summonses etc;
(g) cost of collection of rates;
(h) staff employed on collection of rates.

### 5.4.2

The publication containing the data referred to above is *Rate Collection Statistics* [QRL.58] and the source of the data is a CIPFA questionnaire (specimen not reproduced). Publication normally occurs about 12 months after the end of the financial year to which the statistics relate.

### 5.4.3

In [QRL.65] the Scottish branch of CIPFA now publish for each rating authority in Scotland, a statistic of the actual cost of rate collection and the cost per head of population. This publication sets out its data sources in an accompanying text.

## 5.5 Rate Rebates

### 5.5.1

It may seem odd to treat the subject of rate rebates in a chapter devoted to income from rates, since by definition a rebate is a species of non income. The reason for inclusion is that, in the language of [QRL.58], a rate rebate is best treated conceptually as a "constraint on income" from rates or income forgone. In fact in local authority statistics terms, rate rebates have 3 separate aspects: (a) the part of the rebate to be borne by a rating authority itself which is treated as net expenditure; (b) the major part of the rebate being compensated by central government is of course grants income and is dealt with in chapter 6 below; (c) the total rebate regarded as "constraint on income or income forgone" has to be distributed over a number of recipients dispersed throughout the country, and it is with the statistics of this *distribution* that this section of the chapter deals.

**5.5.2**
Up to and including the financial year 1982/83 the power to reduce the rates payable by certain domestic ratepayers on the basis of family circumstances–some 90% of the income foregone being reimbursed by central government–was conferred on authorities in England and Wales by the Local Government Act 1972 and in Scotland by the Local Government (Scotland) Act 1973. Also the Rating Disabled Persons Act 1978 extended the power to houses and institutions in which a room was required to meet the special needs of the disabled. Now, the granting of rate rebates to domestic ratepayers rests on the authority of the Social Security and Housing Benefits Act 1982. The impact of this legislative change on the content of local government rate rebates statistics is likely to be minimal. So far however it has led to the cancelling of one statistics publication (see below).

**5.5.3**
Rate rebate statistics (i.e. amounts granted, number of recipients etc.) are published annually by CIPFA/SIS in [QRL.58]; the figures relate to English and Welsh rating authorities. In [QRL.92] the Welsh Office published somewhat similar details for Welsh rating authorities only. Up to and including the financial year 1982/83 the DOE/Welsh Office published *Rate Rebates* [QRL.59] which provided data for individual rating authorities in England and Wales. This publication has been discontinued since responsibility for the distribution of rebates passed to DHSS under the 1982 Act. In Scotland, for each district within the area of a region and for each islands council, the Scottish Office publishes relevant statistics in [QRL.77]. It may seem puzzling that rate rebate statistics are published for districts in Scotland which are not the rating authorities. The fact is that districts act as agents of the regions in the matter of collecting rates. One other peculiarity of rate rebate statistics in Scotland is that the [QRL.77] tabulation contains regional data of the amount of rate rebate granted under the Rating (Disabled Persons) Act 1978.

**5.5.4**
The data source for the publication [QRL.58] is the CIPFA/SIS questionnaire previously referred to. Both [QRL.92] and [QRL.59] when published employ or employed the form DOE 12504 which is not reproduced in this review. The sources of the Scottish data are the form RRS (see Appendix 14) and LFR 11 (see Appendix 11). The sources of the data on rebates granted in respect of the disabled, are grant claim forms received by authorities. As these are purely administrative documents, no specimen is reproduced.

## 5.6 Northern Ireland

**5.6.1**
Since 1973, two distinct types of local "rates" have been levied and collected in Northern Ireland. There is a *district rate* which is determined by the district council but collected centrally by the Rating Division of the Department of the Environment

(N1). The proceeds of this rate go towards financing district council services. In addition, there is a *regional rate* which is determined by the Department of Finance and Personnel and collected by the Rating Division of the Department of the Environment (N1), the income from which goes towards the financing of local services which are provided by central government either directly or *via* grants to area boards of various sorts. This review is concerned with the district rate only.

**5.6.2**

Prior to 1973, rates were levied as in Great Britain for services provided by councils, i.e. exclusive of Trading Undertakings and Housing. With the exception of the County Borough of Belfast, which levied a consolidated or all purpose rate, the ratepayers in Northern Ireland, although receiving only one demand note, paid two types of rate: (a) a borough, town or district rate and (b) a county-at-large rate for those services not provided at district, town or borough level.

**5.6.3**

Statistics of *actual* or *outturn income* from district rates in Northern Ireland are available only in national summary in [QRL.15]. *Estimates* of *income* to be raised from district council rates are published by the Local Government Branch of the Northern Ireland DOE in *District Council Rate Statistics* [QRL.14]. The source of the estimates is the General Estimate of Rates completed by district councils in the February prior to the beginning of the financial year to which the estimates relate (see paragraph 4.5.3). Prior to 1973, statistics of *actual* income from rates were published annually by the Northern Ireland Department of Development in *Local Authority Financial Returns* [QRL.36]. *Estimates* of rates income (showing a notional charge on ratepayers) were produced by the Department in its *Local Authority Rate Statistics* [QRL.38].

**5.6.4**

Since 1973, district council *rate poundage* statistics have been published by the DOE for Northern Ireland in *District Council Rate Statistics* [QRL.14]. Figures are shown for each of the districts alongside figures for the previous year. The source of the data is the General Estimate of Rates (see paragraph 4.5.3). (Note: those statistics are also published in [QRL.44].) Before 1973, local authority rate poundage statistics were published by the Ministry of Development in [QRL.38]. Except in the case of Belfast CB separate rate poundages are shown for borough, town or district rate and county-at-large rates (see paragraph 5.6.2). In addition notional rate poundages were computed for individual services within each local authority area. In an explanatory note in [QRL.38], it is pointed out that the poundages published should be reduced by a specified amount for domestic and mixed properties. An average rate for Northern Ireland was also published in [QRL.38] with the method of calculation explained in an accompanying text.

**5.6.5**
While since 1973 in Northern Ireland there have not been national statistics published which show the figures of *rate collection*, this was not so prior to 1973. Although there was not a publication (or indeed a separate tabulation) given over to the subject, as there is in England and Wales today, [QRL.36] did include in its Abstract of the General Revenue Account for each authority, data relating to the cost of collecting rates along with a calculation "irrecoverable rates, discounts etc.".

CHAPTER 6

# FINANCIAL STATISTICS: GOVERNMENT GRANTS

## 6.1 Introduction

### 6.1.1

A major source of the annual income of local authorities in Great Britain is the set of grants and subsidies distributed to them by central government. Perhaps the easiest way to grasp the nature and nomenclature of these grants is to begin with a consideration of the way they are set down, at Great Britain aggregate level, in volume II of the Annual White Paper *The Governments' Expenditure Plans* [QRL.27] and at England, Wales and Scotland levels in respectively, *Rate Support Grant Report (England)* [QRL.60], *Rate Support Grant (Wales)* [QRL.61] and *Report on Rate Support Grant (Scotland) Order* [QRL.72]. See also *Guide to Local Authority Finance* [B.23]. In these publications, *Aggregate Exchequer Grant* (Aggregate Exchequer Assistance in Scotland) is identified. These are (almost) the total amounts which central government intends to disburse to authorities in Great Britain in the course of the following financial year; the grants not included are dealt with in paragraph 6.1.2. These aggregates, which are subject to modification in Supplementary Reports are subdivided in [QRL.60], [QRL.61] and [QRL.72] into their various "regional" components. In the case of England and Wales these are: (a) Specific and Supplementary Grants; (b) Rate Support Grant (further subdivided into block grant and domestic rate relief grant). As to Scotland, the components are: (a) Specific Grants and (b) Rate Support Grant (further subdivided into domestic, resources and needs elements).

### 6.1.2

The grants and subsidies omitted from the Aggregates are: Rate Rebate Grant and one or two other items the principal of which are housing subsidies and mandatory awards to students. As the subjects of Housing and Education are dealt with in companion volumes (see Appendix 1), they will not be further considered in this review. The subject of Rate Rebate Grant is examined below.

## 6.2 Specific and Supplementary Grants

### 6.2.1

*Specific Grants* are payable for services within all categories of local authority accounts (i.e. Rate Fund Services; Housing Revenue Account; Trading Services Accounts and Special Funds). They are payable in respect of capital and revenue accounts and are provided, in the main, to support expenditure on law and order services (e.g. police, probation and after-care, and courts).

**6.2.2**

*Supplementary Grants* (England and Wales only) are paid to help meet the cost of providing particular services. These grants are of two kinds: (i) Transport Supplementary Grant (TSG) and (ii) National Park Supplementary Grant. In England, TSG was until recently received for revenue and capital purposes. However it is now, as it always was in Wales, payable for capital purposes only.

**6.2.3**

As mentioned in paragraph 6.1.1, the total *estimated* amounts of each *specific grant* at the England, Wales and Scotland level are published in [QRL.60], [QRL.61] and [QRL.72] respectively. Statistics of the estimated total specific grants likely to be received by individual authorities are shown in CIPFA's *Finance and General Statistics* [QRL.20] for authorities in England and Wales, and *Rating Review–Part 1* [QRL.64] for Scottish Authorities. The respective sources of these figures within the individual authorities, are the Budget columns of the Return of Expenditure and Rates (see Appendix 7) for England and Wales: in Scotland, the source is the individual authority's budget estimate made to the Scottish Branch of CIPFA–see paragraph 4.2.6.

**6.2.4**

*Planned Supplementary Grants* (not estimates since they are cash limited by central government) are published in detail in [QRL.60] and [QRL.61]. Statistics for one of the grants (Transport Supplementary Grant) are given by county authority to which payable. The tabulation also shows in some detail, the permitted expenditure of individual authorities, on which the grants rest. With respect to the other supplementary grant (National Parks Supplementary Grant) the apportionment is shown, not to the authority to which it will be paid, but to the national park for which it will be made.

**6.2.5**

Statistics of *actual* income for all specific and supplementary grants are published for England in [QRL.41], Wales in [QRL.92] and for Scotland in [QRL.77] and [QRL.65]. The respective sources of the data are (England and Wales) Revenue and Capital Outturn Form (see Appendix List 4) which are sent by each authority to DOE/Welsh Office. The corresponding sources in Scotland for [QRL.77] are the Scottish Office Local Financial Returns listed in Appendix 4. The sources of [QRL.65] are set out in the publication itself.

## 6.3 Rate Rebate Grants

### 6.3.1

As explained in paragraph 5.5.1 above, rate reductions or housing benefits are allowed to certain domestic ratepayers on the basis of income and family circumstances. Rate relief is also allowed for certain domestic premises occupied by disabled persons and for certain institutions for the disabled. These arrangements apply throughout Great Britain, central government reimbursing almost all of the consequent loss of income from rates. The amount of reimbursement appears in local government statistics as income, under the title Rate Rebate Grants.

### 6.3.2

The only published statistics of local authority income from Rate Rebate Grants relate to *outturn* amounts. Nothing is published by way of estimates. In England and Wales statistics are published in [QRL.41] and [QRL.92]. The source of the statistics is the return made by authorities to DOE/Welsh Office–RG 1B, Rate Fund Services Summary Form–see Appendix 8. In Scotland, the source of the statistics in [QRL.77] is the annual return made by the regional and island councils to the Scottish Office on Form LFR 11 (see Appendix 11).

## 6.4 Rate Support Grant

### 6.4.1

Unlike Specific and Supplementary Grants, Rate Support Grants are payable in aid of local government revenue generally and in accounting terms, are credited to the Rate Fund Revenue Account. Following the passage of the Local Government Planning and Land Act, 1980 a new system of distributing Rate Support Grants in England and Wales was introduced. From the financial year 1981/82, the RSG has consisted of two elements: (a) domestic rate relief grant and (b) the block grant. The former is payable to rating authorities only and replaces the old domestic element of RSG (see paragraph 6.4.14 below). It is designed to meet the cost to rating authorities of levying a lower rate on domestic than on non-domestic properties. The distribution of block grant is conceptually and statistically, a much more complex matter and we return to it later.

### 6.4.2

As explained in paragraph 6.1.1, *estimates* of the England and Wales aggregates of the RSG and its two components are first published, respectively, in the *Rate Support Grant Reports* [QRL.60] and [QRL.61]. With respect to *domestic rate relief grant*: these publications also contain rates of relief expressed as pence-in-the-pound of domestic rateable value. Statistics of *actual* or *outturn* domestic rate relief grant (England and Wales *summaries*) are published respectively in [QRL.41] and [QRL.92]; the sources of the figures are the RSIA and RSIB (Rate Fund Services Summary Form); see Appendix 8. Estimated income from this grant, for each rating authority in England and Wales is published by CIPFA in [QRL.20] using the Return of Expenditure and Rates (see Appendix 7) as the source document. Statistics of *actual* income from domestic rate relief grant for each county area are shown for Welsh authorities in [QRL.92].

**6.4.3**
It is beyond the scope of this review to explain in minute detail, the aims and mechanics of distributing the block grant component of Rate Support Grant to individual authorities. The subject has been adequately covered in the following publications: *A General Guide to the Block Grant System in England 1983/84* [B.4]; *Rate Support Grant (England) 1983/84* [B.32]; *Rate Support Grant/Wales* [B.62]; *Rate Support Grant Settlement 1983/84: a Guide* [B.33]; *Block Grant Indicators* [QRL.6]; a "professional" coverage is contained in the CIPFA *FIS Volume 8* [B.21], while the authoritative up to date description of the distribution mechanism appears annually in [QRL.60] and [QRL.61]; but see brief description in Appendix 12.

**6.4.4**
At this point it is worth noting that, superimposed on the block grant distributive method which applies to all authorities in England and Wales, there are special arrangements applying to London and the Metropolitan Police. These are explained in layman's language in [B.4]. Further reference for those who wish to pursue the matter in some depth are *Rate Support Grant; the System and its application to London* [B.9]; *Rate Equalisation in London* [B.8].

**6.4.5**
The estimated amount of block grant at England and Wales level is derived as a residual of the Aggregate Exchequer Grants Settlement–allowance having been made for the other grants just described. It is computed as a proportion of local authority estimated Rate Fund relevant expenditure in the year to which the grants relate. Relevant expenditure for grant purposes is defined in all of the publications mentioned in paragraph 6.4.3 and also in those containing expenditure figures. At England and Wales levels, *relevant expenditure estimates* are contained in [QRL.60] and [QRL.61], i.e. the main RSG *Reports* prior to the commencement of the financial year; amended assessments appear in Supplementary RSG *Reports.* However in respect of individual authorities, estimates are not published. At *outturn* or *actuals* stage, relevant expenditure data in England and Wales is published and compared with the RSG Settlement figures in [QRL.41], the source of outturn data being the series of Revenue Outturn forms listed in Appendix 4 (see also specimen in Appendix 6) and the RSIB Rate Fund Services Summary form shown in Appendix 8.

**6.4.6**
As mentioned in 6.4.2 above, the total amounts of the block grant to be distributed in England and Wales are published in the main RSG *Report* and amended in

Supplementary Reports [QRL.60] and [QRL.61]. With respect to individual authorities' own estimates of block grant amounts: these are published early in the financial year in CIPFA/SIS [QRL.20] under the heading "Anticipated Block Grant Receipt", the source of the figures being each authority's own calculation shown in the DOE/Welsh Office/CIPFA Return of Expenditure and Rates (see Appendix 7). In the same publication there appears also for each authority a "Calculated grant Claim–before and after abatement" (see Appendix 12). These are computed by CIPFA and may differ from an authority's anticipated receipt already referred to. The broad explanation of this kind of difference, which is marginal, is shown in the notes accompanying the tabulations. Later in the financial year, CIPFA/SIS publish for all authorities in England and Wales, estimates of block grant before holdback (or penalty) and block grant after holdback (for explanation of "holdback", "abatement" and "penalty" see Appendix 12). The CIPFA/SIS publications called *Block Grant Statistics* [QRL.7] and the basic data on which calculations rest is provided by the Return of Expenditure and Rates (Appendix 7).

**6.4.7**

The *outturn* statistics of block grant are published at England level in [QRL.41] and by individual Welsh authority both in [QRL.92] and *Rate Support Grant (Wales)* [QRL.62]. The SCT publish block grant outturn figures for all counties in Wales and England in [QRL.20]. (There does not appear to be any national publication of outturn block grant statistics for English Districts.) The source of the [QRL.41], [QRL.92] and [QRL.62] data is the RSIB Form (Appendix 8); the SCT obtains its data for [QRL.21] by means of a questionnaire in which a county is asked to repeat the Rate Fund Summary figures already supplied to DOE/Welsh Office. Before completing this section on outturn block grant statistics, it is worth pointing out that the total amount actually distributed to authorities in England and Wales in respect of relevant expenditure in a given financial year may "spill over" into a later financial year. The amount of spill over is shown in the outturn statistics in [QRL.41] and [QRL.92].

**6.4.8**

The mechanics of distributing block grant to individual authorities have led to the publication of a set of, so to speak, second order statistics and other numerical variables which fall to be described in this review. However, because of their extent and the fact that they are peripheral to the subject matter of this chapter, they have been set apart and are reviewed separately in Appendix 12.

**6.4.9**

In *Scotland*, Rate Support Grant is distributed on the authority of the Local Government (Scotland) Act, 1966 and is conducted on a different basis from that which now obtains in England and Wales. The grant consists of 3 main elements: (a) domestic element; (b) resources element; (c) needs element. The domestic element is similar to the domestic rate relief grant in England and Wales in that it is paid to rating authorities in compensation for the loss of income suffered by levying a lower rate on domestic than non-domestic properties. The *resources* element is dispensed to those authorities which, lacking highly valued industrial and commercial properties in their areas, would for this reason require to levy relatively high rates. The needs element portion is distributed on the basis of a needs formula to which we return below.

**6.4.10**

The *estimated* aggregate amounts available for distribution in Scotland as a whole for each of the elements of RSG are published in the *Report on Rate Support Grant (Scotland) Order* [QRL.72] which also sets out the basis of distribution of the various elements. The amounts listed in the Report are usually amended in later Orders and indeed the final distribution (a) is not determined until some time after the end of the financial year to which it relates and (b) is published in RSG Reports for later years. It should be noted that the Report also contains the estimated relevant expenditure (at Scotland level) on which the size of Aggregate Exchequer Assistance is based.

**6.4.11**

In respect of *individual authorities, estimates* of income from the resources and needs elements are published by the Scottish Branch of CIPFA in its *Rating Review Part 1* [QRL.64]. [QRL.72] contains estimated needs element per head figures for each authority. The source of the statistics in [QRL.64] is the return made by authorities to the Scottish Branch of CIPFA (see paragraph 4.2.6). Estimates of income from the domestic element of RSG are not identified in [QRL.64] but the publication does include statistics of average domestic rate bills for each authority, allowance having been made for receipt of the domestic element; the source of the data is also the CIPFA return referred to in paragraph 4.2.6.

**6.4.12**

The statistics of *actual/outturn* income from RSG (Scotland summary) are published in [QRL.77], the source of the figures for resources and needs elements being the final distribution of RSG published in [QRL.72]. In respect of the domestic element of RSG, the source is the Scottish Office form LFR 11 (see Appendix 11). Actual amounts of resources and needs elements received by individual authorities are now published in [QRL.65] data source forms are shown in the publication itself. Because, as explained previously, the distribution of RSG entitlement in any given financial year may spread over into a later year, the actual/outturn income for any financial year will usually consist of an initial entitlement for the year in question plus an adjustment to the entitlement for a previous financial year.

**6.4.13**
For the reasons stated in paragraph 6.4.8 the statistics lying behind the distribution of *resources* and *needs* elements of Rate Support Grant are separately considered in Appendix 12 together with a note on penalty provisions.

**6.4.14**
In England and Wales, *prior* to the financial year 1981/82 the method of distributing RSG was rather similar to that still practised in Scotland. The grant consisted of three elements: domestic, resources and needs elements. A single RSG Order and Report were published for England and Wales. This showed the amounts to be distributed and the determinants of the distribution. Supplementary Orders and Reports were usually published. *Estimates* of income from all three elements of RSG, by individual authorities, were provided by CIPFA/SIS in [QRL.20]. *Actual/outturn* statistics were published, at national level, in [QRL.41] and [QRL.92] and for individual *counties* in England and Wales in [QRL.21]. The source of the outturn data in [QRL.41] and [QRL.92] was the Rate Fund Services–Summary of Outturn Revenue Expenditure and Income to DOE/Welsh Office (see Appendix 8). Counties simply passed on the figures via a questionnaire to SCT for compilation of [QRL.21].

**6.4.15**
The basis of allocating the needs element of RSG to counties and metropolitan districts in the years immediately prior to the introduction of block grant was an estimate of need from a multiple regression equation derived from past expenditure. Distribution to non metropolitan districts (from 1979/80) was on a population and share-of-county-area expenditure basis. The variables and the coefficients were published in the RSG Order and Report. Statistics of the values used in the equations in respect of each authority (e.g. population, primary school pupils, housing starts, authority areas) were published in respect of 1979/80 and 1980/81 by the SCT in *Rate Support Grant–Needs Elements Statistics* [QRL.45]. The source of the figures was the DOE. The publication also contained statistics of estimated final income from needs element for each authority, for the year of publication and the following year plus outturn needs element statistics for each year back to 1974/75–all supplied by DOE.

## 6.5 Northern Ireland

**6.5.1**
At present there are two *categories of grant* paid by central government to help finance district council services in Northern Ireland. These are: (a) specific service grants (which may be paid either for revenue or capital purposes) and (b) a general grant paid for current or revenue purposes. The former includes assistance by the Department of Education (NI) towards the running cost of community centres operated or assisted by councils and community employment grants paid by the Department of Economic Development to encourage district councils to undertake

local improvement schemes to alleviate unemployment; specific grants are also paid to those authorities with gas undertakings. Specific capital grants are paid towards the cost of capital projects such as the provision of community centres, recreation facilities, acquisition of land for open space, tourist development, improvement of harbours, environmental improvement and so on. The general grant which is paid under the Local Government (Northern Ireland) Order 1972 (S.I. 1972/1999) (Northern Ireland 22) is divided into 2 parts: (i) derating (ii) resources. Derating is paid to all 26 district councils to compensate them for loss of rate income due to statutory derating of certain types of hereditament (e.g. industrial, sport and recreation, freight transport and commercial premises in Belfast and Derry City Enterprise Zones). The resources element of the general grant is paid only to councils whose rateable resources per head of population are below a certain standard determined by the DOE (Northern Ireland). Details of the system of grants distribution are fully set out in the *Ulster Year Book* [B.35].

**6.5.2**

*Prior to 1973* central government grants were of 3 kinds: (a) specific service grants (both for revenue and capital purposes); (b) a General Exchequer contribution; (c) a domestic rate aid grant. The nature of specific service grants is self explanatory; on the capital side the principal grant aided services were roads, housing, education and parks. The General Exchequer contribution consisted of a resource and a needs element, although these terms were not used in the relevant statistics tables. The two components approximately matched the derating and resources elements of the present general grant described above. The domestic rate aid grant was paid to a rating authority to compensate for loss of rate income arising from the statutory requirement to reduce the rate poundage on dwelling houses and mixed hereditaments in its area, by an amount-in-the-pound prescribed each year by the Ministry of Development.

**6.5.3**

Today, summary statistics of *actual* government grants paid to district councils are published in *District Councils Summary of Statements of Accounts* [QRL.15], the source of the figures being the Abstract of Accounts (see paragraph 4.5.2) completed by district councils for the Northern Ireland Department of the Environment. The statistics in [QRL.15] relate to revenue and capital transactions. *Estimates* of income from general and specific grants (for revenue purposes only) are published in *District Council Rate Statistics* [QRL.14] by DOE (Northern Ireland). The data relates to each of the district councils and is assembled from districts via the General Estimate of Rates (see paragraph 4.5.3).

**6.5.4**

*Prior to 1973*, statistics of local authority *actual* income from central government were published in *Local Authority Financial Returns* [QRL.36]. The publication contained various summary level statistics as well as individual council figures.

*Estimates* of likely or notional income from central government grants for current or revenue purposes only, were provided in *Local Authority Rate Statistics* [QRL.38]. Not only were capital grants omitted, but certain specific revenue grants were also left out. These were identified in a foreword to the tables. The expected income from domestic rate aid grant was not specifically identified but was included in the accounts as a receipt from the series "Irrecoverable Rates, Discounts etc.".

CHAPTER 7

# FINANCIAL STATISTICS: BORROWING AND DEBT

## 7.1 Introduction

### 7.1.1

While not strictly a component of local authority "income", borrowing (and thus incurring debt) has to be resorted to in anticipation of income, and this for two main reasons: (a) to smooth authorities' cash flow and (b) to finance capital expenditure. Borrowing for the first of these purposes is limited by law to the amount of outstanding revenue to be received. For capital spending purposes, borrowing is allowed only for purposes authorised by statute and within the limits of loan sanction issued by Central Government (see also Section 4.3 above and *Guide to Local Authority Finance* [B.23]). Local authorities are allowed to borrow by means of temporary loans, bank overdrafts, foreign currency, capital and revenue bills, stock issues, bonds and through the Public Works Loan Board which is financed by the Exchequer. Conceptually the statistics of borrowing fall into two broad categories:

(a) the statistics of borrowing over a period of, say, a year or a quarter ("flow" statistics) and
(b) the statistics of debt outstanding at a point in time, say, the end of a year or quarter ("stock" statistics).

This distinction has largely dictated the form of this chapter.

### 7.1.2

Relevant to this part of the Review is consideration of the statistics of the local authority sector's net financial transactions with all other sectors of the economy. A full explanation of these sector and inter sector financial statistics is contained in the CSO's *Financial Statistics: Explanatory Handbook* [B.22]. Briefly, they arise from the circumstance that the local authority sector–like all other sectors–not only makes payments for the supply of goods and services required to carry out its functions but must undertake accompanying transactions in financial assets and liabilities; in a word, it requires on the one hand to borrow from or issue securities to other sectors and on the other hand makes loans, buys securities, deposits with banks and so on. The sector accounts or financial transaction statistics record the *"net financial flows" between* the local authority sector and each of the other sectors which are:

(a) central government;
(b) public corporations such as the nationalised industries, water authorities and so on;
(c) the monetary or banking sector;

(d) other financial institutions such as insurance companies, authorities own superannuation funds, building societies, finance houses;
(e) industrial and commercial companies;
(f) the personal sector comprising individuals, trade unions, charities etc.;
(g) the overseas sector

The resultant of all these transactions in a quarter or a year is either a local authority sector financial surplus or more usually, a deficit leading to a sector borrowing requirement. We return to this subject in 7.4 below.

## 7.2 Borrowing

### 7.2.1

The statistics of *new loans* raised for capital spending purposes in the course of a financial year are contained in summary form for English, Welsh and Scottish Authorities respectively in central government's *Local Government Financial Statistics England and Wales* [QRL.41]; *Welsh Local Government Financial Statistics* [QRL.92] and *Scottish Local Government Financial Statistics* [QRL.77]. In each publication, the figures which include internal loans are assigned to (1) the Rate Fund Capital Account; (2) the Housing Capital Account; (3) the Trading Services Capital Account and in the case of Scotland only, the Common Good Capital Account (see paragraph 3.3.5). The sources of these statistics are, in the case of the English and Welsh authorities, the DOE/Welsh Office Capital Outturn or Capital Account forms the titles of which are shown in Appendix 4 (see specimen in Appendix 6). The source of the Scottish statistics is the set of Scottish Office Local Financial Returns also listed in Appendix 4 with a specimen of the form shown in Appendix 6.

### 7.2.2

New loan statistics are to be distinguished from the statistics of *total net borrowing* i.e. allowance having been made for any repayment of debt or the acquisition by authorities of financial assets (see section 7.4 below). (Total net borrowing relates to borrowing for all purposes.) The publication responsible for the primary provision of total net borrowing figures–by both financial and calendar years and by quarter–in respect of the *UK local authority sector* is the CSO's *Financial Statistics* [QRL.22]; the financial year adjusted figures for the UK are reproduced by DOE in [QRL.41]. The latter also provides statistics for England only, figures for Wales being published in [QRL.92]. It is worth noting that the publication [QRL.22] contains not only relevant statistics but copious explanatory notes.

### 7.2.3

At this stage the difference between the local authority sector's net borrowing and two closely related statistics should be pointed out. These are: (a) *the local authority sector borrowing requirement* and (b) *the local authority contribution to the public sector borrowing requirement*. The first equates to the local authority sector's total net borrowing less transactions in "other public sector debt"–both statistics to be found in [QRL.22] and partly reproduced in [QRL.41] and *United Kingdom National Accounts* [QRL.88]. The second–local authority contribution to the public sector borrowing requirement–comprises the local authorities' borrowing requirement less borrowings direct from central government. Figures for this latter statistic are published in [QRL.22] for quarters, financial years and calendar years; *Economic Trends* [QRL.16] reproduces calendar year ending figures for the former.

**7.2.4**
The source document for the statistics of borrowing referred to in paragraphs 7.2.2 and 7.2.3 is the DOE Quarterly Borrowing and Lending Inquiry (specimen not provided) which is completed by all authorities in the United Kingdom. Revision of the end year statistics are made following return of the various Capital Outturn forms (see appendix 4) to DOE, Welsh Office and Scottish Office. With respect to DOE (Northern Ireland), the source is the Abstract of Accounts (see paragraph 4.5.2).

**7.2.5**
An analysis is published of the local authority sector's net borrowing by *sector providing finance* in [QRL.22]. The figures are quarterly and annual (both financial year and calendar year). According to *Financial Statistics: Explanatory handbook* [B.22] "the quarterly data are based mainly on counterpart reporting by the sectors because local authorities cannot analyse their marketable debt by sector of holder". However, it should be noted that the total, as distinct from the sector analysis, is provided *via* a monthly (sample) borrowing survey conducted by CSO.

**7.2.6**
Net borrowing by the United Kingdom local authority sector is further analysed in [QRL.22] *by source of borrowing* and *term of borrowing* (i.e. long term/short term) with separate figures for England, Wales and Scotland. In [QRL.41] the "term of borrowing" statistics for the United Kingdom are reproduced along with a "source of borrowing" breakdown for England only. Welsh authority figures are published in [QRL.92] in respect of both source and term of borrowing. The common source document for all the statistics is the DOE Quarterly Borrowing and Lending Inquiry already referred to in 7.2.4.

**7.2.7**
Statistics of *borrowing rates* i.e. the various rate of interest at which local authorities borrow from different sources (including PWLB) and for different terms, are published in [QRL.22] and in CIPFA/SIS *Capital Expenditure and Debt Financing* [QRL.8]. The sources of the former, which are end-of-latest-month figures for a selected series of borrowing instruments, are the Treasury–for PWLB rates–and the Bank of England for all others. The CIPFA/SIS publication provides an *average interest rate* for each UK local authority on the basis of returns (see Appendix 10)

which contain a formula to be used in the calculation of the average. This is in fact a weighted average of interest rates which vary according to the date of repayment of the sums borrowed.

## 7.3 Outstanding Debt

### 7.3.1

The *outturn* statistics of *loans outstanding* at the end of each financial year are published in summary form for England and in greater detail for Wales and Scotland in respectively, [QRL.41], [QRL.92] and [QRL.77]. CIPFA/SIS publishes *provisional outturn* statistics for each local authority in the UK in its annual publication [QRL.8]. The term used is "outstanding debt" and it is broken down by *source of finance* and *maturity of the debt*. *Estimates* of "net debt" (England and Wales) and "capital debt" (Scotland) for all authorities are published annually in, respectively [QRL.20] and [QRL.64]. It is important to appreciate that all of these statistics relate to loans raised to finance capital spending and that they refer to each authority's total borrowing for this purpose, whether financed by internal capital funds or sinking funds or inter authority borrowing. Thus the statistics are to be distinguished from local authority *sector* figures which of course, exclude "within sector" debt (see paragraph 7.3.3).

### 7.3.2

*Sources* of the loans outstanding (outturn) statistics published by central government in [QRL.41], [QRL.92] and [QRL.77] are, for England and Wales, the DOE/Welsh Office Capital Outturn form (see Appendix 4) and for Scotland the Local Financial Returns (see Appendix 4). CIPFA/SIS collects its provisional outturn data for [QRL.8] *via* the form shown in Appendix 10. It is worth pointing out that prior to processing at CIPFA headquarters, all Appendix 10 forms for county councils, Scottish authorities and Northern Ireland Districts are vetted by senior financial officers "in the field", to whom the completed forms are sent in the first instance. Vetting of all others is carried out at CIPFA headquarters. With respect to the statistics of estimates in [QRL.20] and [QRL.64]: the source of the former is the Return of Expenditure and Rates (Appendix 7) which defines debt as total debt for which an authority is responsible and which relates to its own functions, whether the debt is administered by the authority or not, advances from an authority's own capital funds being included as debt. The Scottish data [QRL.64] is collected from authorities by means of a questionnaire designed by the Scottish Branch of CIPFA and not reproduced here.

### 7.3.3

The amount of *outstanding* UK local authority *sector borrowing* (i.e. debt) at the end of each financial year, is published in the CSO's *Financial Statistics* [QRL.22]. As explained above this borrowing is net of internal borrowing transactions and therefore can be seen as representing the *true* level of indebtedness of local government as a whole at a point in time. The statistic is analysed by *source of borrowing* and *term of borrowing*. Separate analyses are also published for England, Wales and Scotland in [QRL.22], the English and Welsh figures being reproduced respectively in [QRL.41] and [QRL.92].

**7.3.4**
These statistics are produced by CSO from the Quarterly Borrowing and Lending Inquiry previously referred to, which in effect updates the results of a DOE survey carried out in December 1981. The survey was directed to establishing the then, outstanding net borrowing categorised by source of finance and term of borrowing.

**7.3.5**
Clearly, debt management on the part of local authorities must give rise to substantial administration costs. The statistics of *debt management expenses*–with data for each authority in the UK–are provided by CIPFA/SIS in [QRL.8]. Expenses are calculated as a proportion of total debt outstanding. The source document is the CIPFA/SIS return shown in Appendix 10.

**7.3.6**
Statistics of the obverse of borrowing–*local authority investments*–are published by the CSO in [QRL.22]. The figures, which are local authority *sector* figures, relate to the position at the end of a period–quarter, financial year, calendar year; they are therefore "stock" statistics. The analyses are by *term* (long, short) and by type of asset or investment. The DOE's Quarterly Borrowing and Lending Inquiry (see paragraph 7.2.4) is the source of the data.

## 7.4 Net Financial Transactions

**7.4.1**
Worthy of separate identification in a borrowing and lending context is the tabulation by CSO of the UK local authority sector's *net financial transactions* (in the form of a "financial account and balancing item"). The nature of these transactions in financial assets and liabilities has been described in paragraph 7.1.2. The statistics enable us to see quickly the net creditor/debtor position of the local authority sector (to the other sectors of the economy). The statistics are a mixture of stock and flow figures, since they relate to end quarter, end financial year and end calendar year and to net borrowing and lending over a period. They are published in [QRL.22]. The CSO's *United Kingdom National Accounts* [QRL.88] also includes end of calendar year, "all sectors" net financial transaction figures leading to the identification of a local authority sector financial surplus or deficit in the year's transactions in financial assets and liabilities. Most of the quarterly inter sector statistics are based on reporting by counterpart sectors but the total borrowing is derived from the DOE's Quarterly Borrowing and Lending Inquiry (see paragraph 7.2.4).

## 7.5 Northern Ireland

### 7.5.1

Under the Local Government Act (Northern Ireland) 1972 the district councils make to the DOE (Northern Ireland), financial returns in the form of the Abstract of Accounts (see paragraph 4.5.2). The DOE (Northern Ireland) summary of these figures is published in the *District Councils Summary of Statements of Accounts* [QRL.15], which, in terms of the statistics of *borrowing* ("flow statistics") shows the amount of *new loans* raised during the financial year in question. Prior to 1973, the figures for *amount borrowed* during the financial year by each authority in Northern Ireland were published by the Ministry of Development in its annual *Local Authority Financial Returns* [QRL.36]. It should be noted that at present, and unlike the position in Great Britain, the statistics of net borrowing and borrowing requirement for the Northern Ireland local authority sector are not separately identified or published by the CSO.

### 7.5.2

The statistics of *outstanding debt* in Northern Ireland are produced in two separate publications: *District Councils Summary of Statements of Accounts* [QRL.15] and the CIPFA/SIS *Capital Expenditure and Debt Financing Statistics* [QRL.8]. The former contains summary figures of *loans outstanding* whereas the latter displays *outstanding debt* for each authority plus analyses by *source of finance* and *term*. Statistics of *average rate of interest* paid during the year and *debt management expenses* are also included. The sources of the figures in [QRL.15] and [QRL.8] are respectively the Abstract of Accounts (see paragraph 4.5.2) and the CIPFA/SIS form in Appendix 10. Before the changes of 1973, the statistics of *loan debt* at the end of a financial year for each authority in the province, were published in *Local Authority Financial Returns* [QRL.36]. Also provided for each authority, was a statement of *net indebtedness* derived from a comparison of loans outstanding with the value of sinking funds and other relevant balances.

CHAPTER 8

# FINANCIAL STATISTICS: MISCELLANEOUS INCOME

## 8.1 Introduction

### 8.1.1

In addition to income from ratepayers, taxpayers, (*via* government grants) and lenders (*via* borrowing), local authorities obtain substantial income from: (a) those who use their services; (b) outside bodies to which they sell capital assets such as land and buildings; (c) interest on temporarily idle cash balances. They may also run down cash balances which although not "income" in the generally accepted meaning of the term, has the same effect on their finances as the receipt of income. (Note that since end year cash balances can be increased as well as run down, the result in these circumstances would be to transform the cash balance change into an item of "expenditure"). For fuller details on the subject of this chapter, see *Guide to Local Authority Finance* [B.23].

### 8.1.2

Because the coverage of items of "miscellaneous" income is very uneven with respect to level of detail, precise category of income and geographical reference, the statistical presentation in local authority publications can be difficult to follow. However a useful point of entry is *via* the England, Wales and Scotland summaries of gross (outturn) income–separate figures for revenue and capital–published in [QRL.41], [QRL.92] and [QRL.77]. In these summaries the term "miscellaneous income" is not usually employed and must be seen as made up on revenue account of categories such as: fees and charges; sales; other income; rents. On the *capital* side, miscellaneous income comprises items such as: sale of fixed assets and other receipts. It is worth noting that the 1974/75 and 1982/83 issues of [QRL.41] contain, for England and Wales, a historical series for the composite category "miscellaneous income" with separate figures for revenue and capital.

### 8.1.3

To identify statistics of the various categories of miscellaneous income–and such statistics are far from complete–we require to trace them through to separate local authority accounts i.e. Rate Fund; Housing Revenue; Trading Services and Special Funds.

## 8.2 Rate Fund Services

### 8.2.1

The activities of the various Rate Fund Services (e.g. education, social services, libraries) generate local revenue in the form of items such as charges for school meals, tuition fees, charges for the services of home helps, public library fines and so on. These forms of income are of course, credited to the *Rate Fund Revenue Account* and appear in summary form under the various headings already referred to (sales, fees and charges, other income, income from local services). In England and Wales the statistics are published in [QRL.41] and [QRL.92] respectively, the source being the DOE/Welsh Office Revenue Outturn forms (RO) listed in Appendix 4 (specimens in Appendix 6). The Scottish figures appear in [QRL.77] with the data being collected by means of the set of Local Financial Returns listed in Appendix 4 (specimen in Appendix 6). Income from *interest* in England is shown only as a footnote to the Rate Fund Revenue Account in [QRL.41]; the source employed is either the appropriate RO Form or the DOE's Rate Fund Summary Forms (RSIB) shown in Appendix 8. There is no corresponding published statistic for Wales. In Scotland only, income from "interest on revenue balance" is separately identified; income by way of interest from other sources is not shown. The source of this statistic is the Scottish Office form LFR 10 (see list Appendix 4). With respect to the item "changes in balances", publication of the summary statistics in England, Wales and Scotland varies. The English outturn figure is not published; the Welsh figure is shown in [QRL.92] with the Rate Fund Summary Form RSIB (Appendix 8) as source. In Scotland the amount appears in [QRL.65] as "balances applied" and in [QRL.77] as a derived figure or residual, all other items of income and expenditure having been taken into account. Sources of [QRL.65] data are identified in the publication itself.

### 8.2.2

Of the "other income" categories listed in paragraph 8.1.1 the only one for which Rate Fund Revenue *estimates* are provided on a consistently defined basis throughout Great Britain, is "changes in balances" or "balances applied". Data for this item for individual authorities is published annually in CIPFA/SIS *Finance and General Statistics* [QRL.20]; in Scotland the corresponding data is provided by *Rating Review–Part 1* [QRL.64]. The England and Wales source is the Return of Expenditure and Rates (Appendix 7); the Scottish source is a CIPFA form not reproduced in this Review. Published estimates of Rate Fund Revenue income from interest earned are both incomplete and inconsistent as between England/Wales on the one hand and Scotland on the other. In the former case, estimates of "interest on revenue balances" for each authority are published in [QRL.20], the source is the Return of Expenditure and Rates (Appendix 7). In Scotland while the corresponding figure is not published, a footnote to the appropriate table in [QRL.64] shows income from interest on revenue balances in respect of part of Rate Fund Services only. The source used is a CIPFA form (see above).

**8.2.3**
So far as the Rate Fund *Capital Account* is concerned, the appropriate categories of income are: "sale of fixed assets originally charged to capital accounts" and "other capital receipts". In the England, Wales and Scotland summaries, the transactions included in the statistics relate of course to those between the local authority sector and all other sectors of the economy. Excluded from the figures published are therefore all inter-authority and inter-departmental income transactions. The kinds of transaction included are: receipts from sales of land, buildings or plant of outside bodies; payments made by Area Health Authorities for jointly funded capital schemes; assistance from the European Regional Development Fund or other quasi government body; repayment of staff car loans. In England and Wales the summary statistics are published in [QRL.41] and [QRL.92], the common source of the statistics being the set of Capital Outturn Forms (CO 1 to 20) listed in Appendix 4 (see also specimen form in Appendix 6 and accompanying instructions and definitions). In Scotland, the matching statistics are published in [QRL.77] with the figures being derived from the set of forms listed in Appendix 4 (LFR 1 to 10); a specimen form with covering instructions is shown in Appendix 6.

## 8.3 Housing Revenue Account

**8.3.1**
As housing statistics is the subject of a companion volume in this series of reviews (see Appendix 1), there is no need to elaborate further on sources. Suffice to say that the categories of "miscellaneous income" which appear in the Housing *Revenue* Account are: rents; changes in balances; other income (including interest). On the *capital* account they appear as: sale of fixed assets originally charged to capital; other capital receipts. In England and Wales the outturn statistics are published in [QRL.41] and [QRL.92] and in Scotland in [QRL.77]. *Estimates* of the Housing Revenue Accounts income are published by the Scottish Branch of CIPFA in [QRL.64].

## 8.4 Trading Services

**8.4.1**
By far, the major part of the income of Trading Services on *revenue account* fall within the category "miscellaneous income" as defined in this chapter. The classification used (sales; fees and charges; other income and interest receipts; changes in balances) are similar to those employed in the Rate Fund Services account and the qualifications which apply to that account pertain also to trading services (see paragraphs 8.2.1 and 8.2.2). Statistics of *actual* or *outturn* "miscellaneous income" are published by DOE and the Welsh Office in [QRL.41] and [QRL.92] for English and Welsh authorities respectively. However these statistics do not include data about the activities of passenger transport executives or the London Transport Executive since these bodies are regarded as public corporations for national accounting purposes. But the accounts of municipal bus undertakings, outside the

conurbations, are included. In Scotland, matching statistics are provided by the Scottish Office in [QRL.77].

**8.4.2**
The *sources* of the statistics in [QRL.41] and [QRL.92] are the DOE/Welsh Office revenue outturn TSR1 and TSR2 (see list in Appendix 4). The source of the Scottish statistics in [QRL.77] is the Scottish Office's Local Financial Return 22 (see list Appendix 4). In the case of Scotland only, water supply is a local authority Trading Service albeit partially supported by a specific water rate levied on ratepayers. The source of water supply "other income" data is Local Financial Return 21 (see list Appendix 4).

**8.4.3**
Trading Services *capital account* statistics use categories similar to those employed by Rate Fund capital account i.e. sale of fixed assets originally charged to capital; other capital receipts. Actual or outturn statistics are published by DOE/Welsh Office in [QRL.41] and [QRL.92] for England and Wales and by the Scottish Office in [QRL.77]. In England and Wales the sources of the figures are the DOE/Welsh Office Capital Outturn forms CO 20, CO 21 and CO 22 (see list Appendix 4); in Scotland the sources are the Scottish Office Local Financial Returns 21 and 22 (see list Appendix 4).

## 8.5 Special Funds

**8.5.1**
Local authorities maintain a number of special funds all of which, at least in England and Wales, are now regarded as statutory under Section 28 of the Local Government (Miscellaneous Provisions) Act, 1976. The individual names of the Special Funds in England and Wales are set out in Appendix 13. Understandably the vast majority of transactions recorded are within-sector transactions, that is to say they relate to items of income from other authorities or other funds within the same authority. As such they do not touch upon the subject of this chapter which deals almost exclusively, with other income to the local authority sector in England, Wales and Scotland. The only categories of "other income" to be considered here are items of "final income" as distinct from "transfer income". Final income of Special Funds is income received from outside bodies (e.g. private individuals, public corporations). In the publications in which *actual/outturn* statistics appear, [QRL.41] and [QRL.92], the relevant categories are: "interest receipts" and "other income". The sources of the data are DOE/Welsh Office forms SF1 and SF2 (see Appendix list 4). Although a separate return for capital and capital reserve funds is called for, these do not differ in their nature, from other Special Funds. The separate return is required because the capital funds demand a different sort of analysis.

**8.5.2**
In *Scotland*, the statistics of Special Fund "miscellaneous income" are published by the Scottish Office in [QRL.77]. The data source is the Scottish Office's Local Financial Return 23 (see Appendix 4). In a note to LFR 23 a distinction is drawn between statutory and non-statutory Special Funds, the latter being omitted from statistics publications. This distinction is in marked contrast to the situation in England and Wales where, as previously mentioned, Special Funds are regarded as statutory.

**8.5.3**
As already pointed out in paragraph 3.3.5, one Special Fund in Scotland which calls for separate mention is the Common Good Fund. It is unique in that it is the only Special Fund which maintains separate revenue and capital accounts. Summary data is published by the Scottish Office in [QRL.77], the source being form LFR 25 (see list in Appendix 4).

## 8.6 Superannuation Funds

**8.6.1**
As mentioned in paragraph 3.3.5, many authorities in Great Britain, run superannuation funds for their own employees (excluding school teachers, firemen and police) and for the staffs of other authorities. The authorities administering the funds are mainly London Boroughs and counties in England and Wales and the regions and islands councils in Scotland. From the point of view of the national accounts the superannuation funds are not part of the local authority sector; like other pension funds they are regarded as part of the "other financial institutions" sector of the economy (see paragraph 7.1.2). Nevertheless, in the publications [QRL.41], [QRL.92] and [QRL.77] they are regularly published as part of local government statistics. The source documents of the summary statistics in England and Wales are the DOE/Welsh Office returns SF3A/SF3B (see list in Appendix 4); in Scotland, the source is the Scottish Office's Local Financial Return 24 (see list in Appendix 4).

**8.6.2**
Certain statistics supplementary to the main superannuation Funds tabulation referred to above, call for explanation. First, "payments under the Pensions (Increase) Acts" made to superannuation Fund pensioners relate to that part of a pension which is financed wholly by the employing authority (e.g. the cost of index linking). It is therefore excluded from expenditure shown in the Funds tabulation. All three publications [QRL.41], [QRL.92] and [QRL.77], contain this supplementary statistic as a footnote, the sources of the data being the forms SF3 and LFR 24 listed in Appendix 4.

**8.6.3**
Further supplementary statistics in the form of footnotes to the Superannuation Fund tables refer to England and Wales publications only. In addition to the main Superannuation Funds described in paragraph 8.6.1 the police pension scheme and the firemen's pension scheme–both of which are unfunded and therefore entail the collection of contributions and the payment of benefits–are administered respectively by police and fire authorities. Summary statistics of payments made by these two schemes are published in [QRL.41] and [QRL.92] for England and Wales respectively. The source form is the SF3, see above.

**8.6.4**
Since 1982, CIPFA and SCT have together published annually *Local Authority Superannuation Fund–Investment Statistics 1977 to current year* [QRL.39]. The publication contains statistics of the performance of funds' investments, the direction of new investment and a wide range of data on the individual superannuation funds of many authorities in Great Britain. The CSO and two firms of stockbrokers are the sources of the background statistics contained in [QRL.39]; data on individual funds have been provided by individual authorities in annual surveys administered for CIPFA and SCT by Derbyshire County Council.

## 8.7 Northern Ireland

**8.7.1**
As in the rest of the United Kingdom, Northern Ireland local authorities before and after the radical change of October 1973, raise part of their income directly from rents, fees, fines and charges from those who use the services they provide whether these be abattoirs, swimming baths, burial grounds or whatever. Indeed in the case of trading services (of which only nine Gas Undertakings survive under the control of district councils) this type of current income is today and was before 1973, by far the most important.

**8.7.2**
*At present*, statistics of outturn "miscellaneous income" in summary form are published in *District Council Summary of Statement of Accounts* [QRL.15]; the sources are the Abstract of Accounts and the Gas Undertaking accounts (see paragraph 4.5.2). The publication adopts the convention in current or revenue accounts, of referring to the items as "miscellaneous income", in capital accounts the term used is "other income". Only in the "all accounts" summary is the item "decreases in balances" identified as a separate component. *Estimates* of "miscellaneous income" in respect of Rate Fund revenue only, are published and attributed to individual services in the Department of the Environment (Northern Ireland) *District Council Rate Statistics* [QRL.14]. The source of the data is the individual district council *via* its General Estimates of Rates (see paragraph 4.5.3). In [QRL.14] the statistics are of course tabulated under the heading "miscellaneous income".

**8.7.3**

*Prior to 1973*, statistics of *actual* income from "other sources" were provided in *Local Authority Financial Returns* [QRL.36] published by the Government of Northern Ireland Ministry of Development. In current or revenue accounts, the data was shown under the heading "miscellaneous receipts" whereas in capital accounts, they appeared under "other receipts". No finer subject breakdown of the figures is shown. *Estimates* of miscellaneous receipts for Rate Fund purposes only–although that term was not used in the publication–appeared in *Local Authority Rate Statistics* [QRL.38]. The estimates related to current or revenue accounts only. As in the case of [QRL.36] no finer breakdown of the figures by subject was made available.

CHAPTER 9

# RATEABLE VALUES AND PENNY RATE PRODUCTS

## 9.1 Rateable Values

### 9.1.1

The rateable value (RV) of a property forms one of the bases on which liability to pay local authority rates in this country, is founded. For full details see *Guide to Local Authority Finance* [B.23]. Throughout Great Britain, the RV is the rent at which a property might reasonably be expected to be let from year to year if the tenant were to bear the cost of repairs, insurance and maintenance. In England and Wales, RV's are determined periodically by valuation officers of the Board of Inland Revenue, who are independent of local authorities; the latest revaluation in England and Wales was effected in April 1973. In Scotland, valuation is carried out by Land Valuation Assessors who are employed by the Regional and Islands authorities, the latest Scottish revaluation dates from April 1985. It is the valuations produced by these authorities and published in valuation rolls or lists which form the foundation of rateable value statistics. (Note: In the RV context a property in England and Wales is referred to as a "hereditament"; in Scotland it is called a "subject".)

### 9.1.2

The statistics of RV's are usually analysed in two broad ways. First, the hereditaments or subjects are classified into the categories "domestic" and "non domestic" with the latter being subdivided into:

(a) commercial groupings such as shops, offices, hotels;

(b) an industrial category;

(c) public utilities such as gas and electricity undertakings;

(d) miscellaneous grouping.

Second, domestic properties are classified by value ranges and sometimes by type of ownership e.g. council dwellings and others. These analyses apply to the entity "actual rateable value" which is fairly self-explanatory and to the source of which we return below. In England and Wales only, "actual" rateable value of any authority has to be distinguished in some statistical tabulations from the quantity "rateable value for block grant purposes". Roughly speaking, the latter equates to actual rateable value minus a percentage of the RV's of those properties which remain empty on a particular date. The place of this quantity in the scheme of things is referred to in paragraph 9 of Appendix 12.

**9.1.3**
Of all the statistical entities dealt with in this review there is none more extensively publicised than rateable values in *England and Wales*; varying only in the range of analysis provided and the timing of publication which separate bodies publish. These are: the Board of Inland Revenue; the Department of the Environment; the Welsh Office; CIPFA/SIS; the Rating and Valuation Association; the Society of County Treasurers and Municipal Publications Limited. Of these the most authoritative is the Board of Inland Revenue whose Valuation Office is responsible for making property assessments not only on the occasion of periodic revaluations but from day to day as new properties come into existence and others go out. The Board publishes annually in *Inland Revenue Statistics* [QRL.31], the data of course being derived from the work of the Valuation Office.

**9.1.4**
The remaining bodies publish as follows:
(1) the DOE/Welsh Office in [QRL.41]–brief summaries for England and Wales separately;
(2) the Welsh Office in [QRL.92]–for all Welsh rating authorities;
(3) CIPFA/SIS in [QRL.20] and [QRL.58]–for all authorities in England and Wales in the former case and rating authorities in the latter;
(4) Rating and Valuation Association in [QRL.26]–data for all rating authorities;
(5) Society of County Treasurers in [QRL.21]–for English counties only;
(6) Municipal Publications Ltd., in [QRL.44]–for all authorities.

Up to and including the financial year 1975/76, the DOE/Welsh Office published "all authority" rateable values statistics in [QRL.63]; similar data up to and including the year 1978/79 was made available, in note form, by the DOE.

**9.1.5**
All publications referred to in the previous paragraph use the Inland Revenue, directly or indirectly as data source. The required information is provided by the Valuation Office's form VO 7140 (see Appendix 14) which is completed by that department on 1st April of each year for the area of each rating authority in England and Wales. In the case of CIPFA/SIS [QRL.58] the VO 7140 data is conveyed *via* the questionnaire Rate Collection Statistics and for [QRL.26] it is provided by means of questionnaires, neither of which is reproduced in this review. The *Municipal Year Book* [QRL.44] also uses the VO 7140 data which in the case of counties only is processed by the SCT for their publication. (Note: the content of all VO 7140 forms is now held in the CIPFA computer database, for authorities in England and Wales.)

**9.1.6**
In *Scotland* where the mechanics of valuation are quite different from those which prevail in England and Wales (see paragraph 9.1.1) publication also proceeds on

different lines. The most up to date and comprehensive publication of rateable values is provided by the Scottish branch of CIPFA in its *Rating Review–Part 1* [QRL.64]. Other bodies providing RV statistics are: the Scottish Office in [QRL.77]–for rating authorities; the Rating and Valuation Association in [QRL.26]–for districts and island councils; Municipal Publications in [QRL.44] for all authorities.

**9.1.7**

The basic sources of all the statistics mentioned in the previous paragraph are the Land Valuation Assessors' (see paragraph 9.1.1) who, for statistical purposes, complete forms RRV/ASS (see Appendix 14) and pass these to the Scottish Office which uses them as the source document for its [QRL.77]; it also makes them available to the compilers of [QRL.64]. The Assessors data having also been provided to individual districts, is in turn made available to the Rating and Valuation Association by means of a form not reproduced here. The Scottish branch of CIPFA in the course of compiling data for [QRL.64] makes it available to Municipal Publications Ltd., for inclusion in [QRL.44].

## 9.2 Penny Rate Product

**9.2.1**

An authoritys "penny rate product" which is an index of it's tax raising capa city is generally understood as the amount of cash which would be raised by the authority levying a rate of one penny per £ of its rateable value. As employed in local authority financial statistics however, the term is not used without qualification. At present–in *England and Wales*–the terms: (a) grant exclusive/excluding block grant and (b) grant inclusive/including block grant are variously employed. Again, prior to 1980, the term "inclusive penny rate product" which was then in use was defined quite differently from the "grant inclusive product" of today. Since the introduction of the block grant method of distributing most of RSG (see Section 6.4), the term "grant exclusive penny rate product" has referred to the amount collected locally from rates (inclusive of domestic rate relief and rate rebate grant) for every penny in its poundage, after allowing for costs and losses on rate collection. "Grant inclusive penny rate product" is the income which an authority receives for every penny it adds to its rate poundage plus the additional block grant which would be payable to the authority if its total expenditure was at a level of its Grant Related Expenditure (see Appendix 12) and its expenditure increased marginally. These definitions are set out in the Glossary and Notes to [QRL.92] which also contains the formula for calculating "inclusive penny rate product". The pre-1980 "inclusive penny rate product" was defined as "the amount which would be raised from rates and the resources and domestic elements of RSG by levying a rate of one penny in the £ after allowing for cost and losses of rate collection" (see paragraph 6.4.4).

**9.2.2**

Penny rate product statistics in *England and Wales* are published as follows:

(1) by CIPFA/SIS in [QRL.20]–both "grant inclusive and exclusive" for individual authorities;
(2) by CIPFA/SIS in [QRL.58]–"grant exclusive product" for individual authorities;
(3) by the Rating and Valuation Association in [QRL.26]–"grant exclusive product" for individual authorities;
(4) by Municipal Publications Ltd., in [QRL.44] both "grant exclusive and inclusive" for all authorities;
(5) by the Welsh Office in [QRL.92]–both "grant exclusive" and "grant inclusive" by county area;
(6) by the SCT for counties only in [QRL.21]–"exclusive penny rate product".

Up to and including the financial year 1975/76 the DOE/Welsh Office published "inclusive penny rate product" data for all authorities in their [QRL.63]; up to and including the year 1978/79 the figures were made available by DOE in note form.

**9.2.3**

The data source of the "grant inclusive" statistics in [QRL.20] is the Return of Expenditure and Rates (see Appendix 7) completed by individual authorities. "Grant Exclusive" data in the same publication is estimated by CIPFA following their preliminary calculation of each authority's block grant entitlement. For [QRL.58] data is assembled by means of the CIPFA/SIS form Rate Collection Statistics, not reproduced here. The Rating and Valuation Association collects data for [QRL.26] from a form not reproduced here, while for [QRL.44] Municipal Publications Ltd., now simply reproduces the CIPFA/SIS data already published in [QRL.20]. The Welsh Office source for its [QRL.92] data is the individual authority's own computation at the beginning of each financial year which is sent to the Welsh Office on a Block Grant Claim form (FL2). In respect of counties only in England and Wales, the Society of County Treasurers gathers data from councils by means of a questionnaire which has not been reproduced in this review.

**9.2.4**

In *Scotland* the relevant statistic is the "net penny rate product" which is defined by the Scottish branch of CIPFA in [QRL.64] as broadly the total amount which would have been produced by levying for each district/islands/regional authority a rate of one penny per £ of rateable value, after deducting the net loss suffered because of empty properties, reliefs, rebates and amounts written off as irrecoverable and the cost of collection of rates. The rate product so estimated disregards the reduction in respect of dwelling houses (see paragraph 5.3.5) which is at all events, reimbursed by central government. The statistics of "net penny rate product" are published as follows: (1) by the Scottish branch of CIPFA in [QRL.64]; (2) by the Rating and Valuation Association in [QRL.26].

**9.2.5**
The data for [QRL.64] is collected by the Scottish branch of CIPFA by means of a simple questionnaire, each authority making its own calculation of rate product in accordance with the Local Government (Rate Product) (Scotland) Regulations 1975 as amended. For publication [QRL.26] the Rating and Valuation Association uses its own form which is not reproduced in this review.

## 9.3 Northern Ireland

**9.3.1**
In the statistics of property valuation in Northern Ireland, both at present and prior to 1973, two slightly different concepts or definition of property values have their place. There is the *rateable value* of a property and its *net annual value*. Statistics of both are published by the authorities. The latter is defined as the average profit rent which a property would command if it were available for letting and on the assumption that all outgoings (rates, taxes, cost of repairs) were payable by the tenant. The rateable value of a property differs from its net annual value only to the extent that derating, exemptions or other statutory reliefs apply. As elsewhere in the UK rates are actually on rateable values.

**9.3.2**
Since 1977 statistics of rateable values (and net annual values) have been published annually in the DOE (NI) publication *District Council Rate Statistics* [QRL.14]. These relate to the valuation date of 30th September of the calendar year prior to the financial year for which expenditure estimates are published in [QRL.14]. The source of the figures is the Northern Ireland Commissioners of Valuations. These statistics are reproduced in [QRL.44]. Prior to 1973, corresponding statistics for the local authorities then in existence, were published by the Ministry of Development in *Local Authority Rate Statistics* [QRL.38]. In this case the rateable values related to 1st April i.e. the beginning of the financial year to which the estimates in [QRL.38] refer.

**9.3.3**
Matching *penny rate produce* statistics are published annually by the DOE (NI) in *District Council Rate Statistics* [QRL.14] for the period since 1973. The source of the data is the Rating Division of the DOE (NI). Prior to 1973 similar statistics were provided by the Ministry of Development in its *Local Authority Rate Statistics* [QRL.38]. In both cases, separate figures were tabulated for the produce of a penny rate in the £ on net annual valuation and rateable valuation, the latter figure being used to fix local rates. The difference between the two penny rate product figures is a measure of the extent of derating and forms the basis on which the "derating" part of the governments' General Grant to districts is calculated (see paragraph 6.5.1 above).

**9.3.4**

As in the remainder of the United Kingdom, there are periodic reviews of property valuation for rating purposes. These are referred to as General Revaluations. The latest–the Third General Revaluation–took effect from 1st April 1976.

## CHAPTER 10

# MANPOWER EARNINGS AND HOURS

### 10.1 Manpower

**10.1.1**
From 1952 to 1974 a detailed analysis by service, of aggregate local authority manpower in Great Britain was published annually in the Department of Employment's *Employment Gazette* [QRL.18] but see also *British Labour Statistics–Historical Abstract* 1866 to 1968 [B.13] and *British Labour Statistics Year Book* [B.14]. The statistics contain the numbers on the payrolls at mid June including those temporarily absent through sickness, holidays or other causes. The Department of Employment admits there is a marginal element of duplication in the figures as some part-time employees (e.g. teachers and lecturers at evening classes and technical colleges) are known to have had jobs with more than one local authority. The data was collected by the Department by means of a special return from individual authorities. But information about police force manpower was obtained from the Home Office (for England and Wales) and the SHHD for Scotland.

**10.1.2**
It is important to note that the Department of Employment's "Employees in Employment" series of statistics which related to activity throughout the British economy and which were analysed by SIC did not identify local government as an industry. The SIC category "local government service" (MLH 906) despite the title, was never all embracing since a large part of local authority personnel was classified outside MLH 906 i.e. in construction (MLH 500), in education services (MLH 972), in health (MLH 874) and in road passenger transport (MLH 702).

**10.1.3**
In 1975, at the request of central government, a manpower survey was instituted in England and Wales by the Local Authorities Conditions of Service Advisory Board (LAC SAB) in order to provide more regular information as a proxy for expenditure data which it was found impossible to collect within the timescale desired by central government. From 1976 a parallel survey was conducted in Scotland by the National Joint Councils for Local Authorities Services (Scottish Councils). This type of survey, which is voluntary and is still conducted quarterly, is now the main source of local authority manpower statistics throughout Great Britain. The survey is called *The Joint Manpower Watch (JMW)* and is compiled and issued in England and Wales by the Joint Manpower Watch Group on which both central government and

the local authority associations are represented. In Scotland, the JMW is compiled and issued on the authority of the Scottish Office and COSLA. The collection and tabulation of data from the authorities is carried out in England and Wales by the staff of LAC SAB and in Scotland by the NJC (Scottish Councils) staff.

**10.1.4**

Compiled alongside the JMW is the local authorities' part of the Department of Employment's *Quarterly Employment Survey* which in effect classifies the JMW data into MLH categories for the purpose of maintaining the Great Britain all industries employment statistics series, "Employees in Employment–Industrial Analysis" which has been published regularly since 1952. But note: as mentioned in paragraph 10.1.2, it is not possible to identify local authority manpower statistics from this published series because the local authority components of MLH's other then MLH 906 are aggregated with data from other employers. We return to this subject, including a description of the mechanics of collection, in paragraph 10.1.11.

**10.1.5**

*In England and Wales* JMW quarterly statistcs are published, for England and Wales respectively, by means of the DOE and Welsh Office press releases *Joint Manpower Watch Return–England* [QRL.32] and *Local Authority Manpower in Wales* [QRL.37]. These contain a 22 service or part service classification of the data–England and Wales summaries–and in addition totals for each authority with separate figures for the law and order services, also at authority level. In the case of the England summary only, an overall seasonal correction is estimated and although unpublished is commented upon in [QRL.32]–see also paragraph 10.1.8 and Scottish practice referred to in paragraph 10.1.7. The published figures may be revised in later quarters; the revision procedure is fully explained in paragraph (5) of "Notes of Guidance–December 1981" in Appendix 15. One other feature of the JMW statistics, with one or two "service" exceptions, is that they are analysed by:

(a) full-time staff;
(b) part-time staff;
(c) full-time equivalents.

The subject of converting part-time to full-time staff equivalents is dealt with in paragraph 10.1.9.

**10.1.6**

The separate England and Wales summary figures are later published in the LAC SAB Journal *Local Government Manpower* [QRL.42] and in [QRL.18] but it should be noted that in the latter–as the result of the aggregation of some law and order data–the statistics are grouped into 17 categories only. The June quarter JMW (summary) statistics are reproduced for England and Wales respectively in [QRL.41] and [QRL.92].

**10.1.7**

In *Scotland*, the JMW quarterly statistics are published first in a Scottish Office press release: *Joint Manpower Watch Survey Results* [QRL.33] and later in [QRL.18] and [QRL.77]. In [QRL.33] there are:

(a) national summaries (17 way and 41 way service breakdowns) not seasonally adjusted;
(b) a 17 way breakdown seasonally adjusted and
(c) individual authority totals.

[QRL.18] provides, as in England and Wales, the 17 way classification not seasonally corrected. In [QRL.77] a 23 way service classification plus a "job creation" group is published but not seasonally adjusted. As in the rest of Great Britain, the full-time, part-time, full-time equivalents classification of data is maintained. One final point, the Scottish statistics are not revised at a later date.

**10.1.8**

The data sources of the quarterly JMW statistics in England/Wales and Scotland respectively are the two sets of Employment by Local Authority Survey forms with accompanying notes shown in Appendix 15. These are completed quarterly (March, June, September and December) by each authority. In England and Wales non-response is virtually limited to a few small shire district councils; in Scotland there is a full response. In England and Wales, data for police personnel (all ranks, cadets, traffic wardens, civilian staff) is not collected via the JMW survey form but is supplied directly to LAC SAB by the Home Office. In Scotland, individual police authorities provide police manpower data which is checked by SHHD, the department of the Scottish Office responsible for the police. Seasonal adjustments to the statistics are calculated by DOE statisticians for England and, for Scotland, by the Scottish Office; both sets of statisticians using the X-11 variant of the Census Method II Seasonal Adjustment Program written by the US Bureau of the Census, Economic Research and Analysis Division, November 1968.

**10.1.9**

In England and Wales the conversion of *part-time* staff to *full-time equivalents* is effected by applying conversion factors which, although reviewed from time to time, have not been changed since 1975. The numerical values of these factors are printed as footnotes to the JMW tables in [QRL.18]. As will be seen from the JMW survey form in Appendix 15, authorities are not asked to apply conversion factors in making their returns; conversion is carried out by LAC SAB. In Scotland, conversion factors are revised annually and, as in the case of England and Wales, the numerical values are published as footnotes to the tabulation in [QRL.18]. The bases of revision are the annual surveys carried out by the NJC in Scotland. Although designed to serve a much wider purpose, the surveys yield data on the number of part-timers employed and the number of hours they actually work, from which conversion factors for various types of staff can be derived. The surveys use:

(a) Non Manual Salaries Survey (December);

(b) Building and Civil Engineering operatives–Numbers and Earnings Survey (April);

(c) Building and Civil Engineering Plumbers–Numbers and Earnings Survey (April).

**10.1.10**

One major difficulty attached to assembling the JMW statistics and a constantly recurring criticism of the value of the statistics for inter-authority comparison purposes, are both worth mentioning. Because of the variability of the departmental structures between councils, individual authorities often have difficulties in allocating their employees from their own departmental categories to the JMW service classification. (This is of course a problem which recurs throughout this review, especially where the staffs of central departments are concerned). With respect to inter-authority comparisons it is argued that these are invalid because authorities vary so widely in population structure, social characteristics, functions, seasonal employment needs and the extent to which they use private contractors or act as agents for other authorities. To the "allocation" point LAC SAB has replied that it is well aware of the difficulty but that JMW categories were devised in such a way as to minimise the problem. As to the "inter-authority comparison" issue, LAC SAB has made the point that JMW was not invented primarily as a management tool for local authorities but, as mentioned earlier, as a proxy for expenditure data required in connection with RSG considerations. The service classification used in the survey had its origin in the expenditure groupings used for RSG purposes but after extensive consultation with authorities a common service classification was adopted which represented the most acceptable compromise. We return to the whole subject of inter-authority comparisons in the final chapter of this review.

**10.1.11**

In paragraph 10.1.4, reference was made to the Department of Employment's Quarterly Employment Survey and to its severe limitations as a source of local authority manpower statistics. In the interests of clarity we return to the subject, since the collection of data for the Survey is bound up with the JMW survey. As will be seen in the England/Wales notes on completion of the JMW survey, reference is made to the Department of Employment's L42 form. This form is completed by authorities along with JMW survey forms and effectively reclassifies the JMW data into the SIC categories required by the Department of Employment for its publication of "Employees in Employment" in [QRL.18]. (Note: statistics of police personnel are omitted; the Home Office provides these figures–see paragraph 10.1.8). In Scotland, individual authorities do not complete forms L42. This return is completed centrally by the NJC (Scottish Councils) staff. The L42 survey is voluntary, but the Department of Employment has the authority to use a statutory alternative if necessary.

**10.1.12**
There is one difficulty attached to the classification of L42 data for publication purposes. As explained earlier, the L42 data is based on the 1960 SIC, whereas the publication in [QRL.18] rests on the 1980 SIC, the latter being more detailed in many of the service areas. The Department has some difficulty in allocating information from form L42 to the new classification, especially in those areas where local authorities are involved with private employers. We understand that the system is at present under review.

**10.1.13**
One separate category of local authority manpower calls for special mention: *registered disabled persons*. In common with large firms in industry and commerce, local authorities are required–under the Disabled Persons Employment Act 1944–to employ a certain quota of persons registered as disabled. Once per year, usually in January, the *Employment Gazette* [QRL.18] publishes, for each authority in Great Britain, statistics of the numbers of registered disabled staff employed and the percentage of all staff which these numbers represent. These figures have been published regularly since 1976. The source of the data is the Manpower Services annual enquiry of the quota position of all employers subject to quota; the enquiry relates to the position in June of each year and the publication of the local authority data in [QRL.18] is effected with the agreement of the employers in every case.

## 10.2 Earnings and Hours

**10.2.1**
The statistics of the *earnings* of local authority personnel lack the precision of definition of the JMW survey in that the local government boundary is more loosely drawn and the earning data itself is provided by sample. The earnings figures are collected by the Department of Employment under the authority of the Statistics of Trade Act, 1947 and published in the *New Earnings Survey* (NES) [QRL.46]. They are of two sorts: (a) average gross weekly earnings by sector i.e. local government sector; (b) analysis by collective agreement. Both relate to full-time personnel (male/female and manual/non-manual shown separately). The definitions of "local government" in the former case is such that personnel employed in construction and transport have been omitted (the full definition of the local government sector is set out in 1980 MLH terms in Section B of NES). But this omission has been rectified to some extent by the range of the second analysis i.e. by collective agreement. There, earnings are shown for the personnel covered by each of a series of collective agreements which apply to local authority services. The agreements relate respectively to police, municipal road passenger transport, teaching, the fire service, construction and various occupational groupings within local authorities generally (e.g. administrative, technical and clerical grades, manual workers). A detailed list of the agreements covered is set out in part B of the NES. In the analysis by collective agreement, a distinction is drawn in most cases between earnings in England and Wales on the one hand and Scotland on the other, as separate collective agreements apply.

**10.2.2**
The data source of the NES statistics is the questionnaire shown in Part B of [QRL.46]. Also contained in Part B are:
(a) a full description of the mechanics and techniques of the survey;
(b) a copy of the covering letters sent to employers completing the questionnaire;
(c) a glossary of terms and definitions used in the survey and
(d) a list of the major collective agreements affecting the tabulations.

**10.2.3**
Statistics of weekly *hours* worked by local government personnel published in the NES are analysed by the collective agreements listed in part B of the NES and referred to in paragraph 10.2.1 above. The figures relate to full time employees with male/female, manual/non-manual categories analysed separately. Also in most cases, there are separate figures for England/Wales and Scotland. The source of the data is of course the questionnaire contained in Part B of [QRL.46] and referred to in paragraph 10.2.2.

## 10.3 Northern Ireland

**10.3.1**
District council manpower statistics (summary figures only) were first provided in 1983 in the *Northern Ireland Annual Abstract of Statistics* [QRL.47]. Prior to 1983, that publication did identify a category "local government", within a table headed "public sector employment". However, in that case "local government" included teaching staff in all schools and non-teaching staff employed by the Education and Library Boards, bodies which are under the aegis of Northern Ireland central government. As in the rest of the United Kingdom, the Northern Ireland current and historical series "employees in employment" analysed by SIC orders, does not identify and indeed never has identified the local authority sector as a single industry and this is for the reasons set out in paragraph 10.1.1 above.

**10.3.2**
The source of the district council manpower statistics in [QRL.47] is the Northern Ireland Department of Economic Development's Quarterly Employment Enquiry–specimen copy not reproduced here. (Note: the published statistics in [QRL.47] relate to the June Enquiry). Although the Enquiry is referred to as a "sample enquiry", the fact is that it covers all pay points with 25 or more employees. Thus, so far as district councils are concerned, the coverage is complete.

**10.3.3**
It should be noted that statistics of district council employees' earnings in Northern Ireland are not published.

# CHAPTER 11

# ELECTIONS

## 11.1 Incidence of Elections

### 11.1.1

At the time of writing, *county councils* (metropolitan and non-metropolitan) in England and Wales are elected every fourth year, each electoral division within the county electing one councillor. The last election was held in May 1985, but note that metropolitan county councils were abolished in 1986 (see Section 2.1).

### 11.1.2

The incidence of *district council* elections in England and Wales varies as between metropolitan and non-metropolitan areas. In the former case, elections are held in May of each year except in those years in which county council elections take place. Each councillor is elected for a period of four years with one third of the council retiring at each election. In non metropolitan districts, councillors are also elected for four years but a council can choose whether to elect an entire council every four years or elect by thirds as in the case of the metropolitan districts. The Secretary of State for the Environment may if requested vary the system adopted by non-metropolitan councils but not more than once in ten years. The next whole election will be in 1986, 1987 and 1988.

### 11.1.3

Greater London Council elections take place every fourth year, the last being held in May 1981 (as in the case of Metropolitan counties, the GLC was abolished in 1986– see section 2.1). *London Borough* elections are also held every fourth year, the next being due in May 1990.

### 11.1.4

In Scotland, the *regional* and *islands councillors* are elected for a term of four years, elections being held in the month of May in every fourth year; the next election is due in 1990. Similarly, *district councillors* serve for a period of four years; the next election is due in 1988.

**11.1.5**
In addition to the ordinary elections referred to above, by-elections may be held at any time. This fact is worthy of special mention since it can have an impact on the statistics of the political composition of councils discussed in section 11.3.

## 11.2 Electoral Statistics

**11.2.1**
The statistics of local government *electors* in England and Wales ordered by county, district and at present Greater London (but see below) are published first in *OPCS Monitor–Reference EL* [QRL.48] and later in *Electoral Statistics* [QRL.17]. These publications now also contain corresponding Scottish data by district, islands council and region. Until 1985 the Scottish statistics appeared initially in *Parliamentary and Local Government Electors* [QRL.49] compiled by the Registrar General for Scotland and obtainable only on application to GRO, New Register House, Edinburgh. Summary data was published later in Section U of the *Annual Report of the Registrar General for Scotland* [QRL.5]. (Note: Publication of the former is now discontinued; it is not clear whether the statistics will continue to be provided by the latter.) A peculiarity of the presentation of the pre-1985 Scottish figures must be explained. The local government electorate is shown as consisting of two categories: (a) the parliamentary electorate; (b) local government electors marked "L" in the register of electors. This latter category, numerically very small, comprises those electors qualified only to vote at local government elections e.g. peers of the realm.

**11.2.2**
The sources of the published statistics throughout Great Britain are of course, inseparable from the electoral registration process, which is fully described in the introduction to [QRL.17]. (see also *The Electoral Registration Process in the United Kingdom* [B.19].) A copy of the recommended form of registration (Form A) is shown in [QRL.17]. Form RPF which OPCS uses to collect electoral data from local authority registration officers in England and Wales, is shown in Appendix 17. An approximately parallel process obtains in Scotland. (It should be noted that local registration officers are free to issue their own form of registration based on Form A.) A study into the accuracy of the electoral registers in Great Britain and by implication into the accuracy of the statistics produced there from, is described in *Electoral Registration in 1981* [B.18].

## 11.3 Election Results

**11.3.1**
Statistics of the political (party) composition of all councils in the United Kingdom are published annually in the *Municipal Year Book* [QRL.44]. The source of the statistics is the Editor's annual political composition questionnaires which is sent to every local authority in the United Kingdom. The data collected relates not only to

the outcome of ordinary or regular elections, of which there may not have been one during a year, but also to the result of any by-election which may have been held.

**11.3.2**

For election years only [QRL.5] contains tabulations analysing the regional/islands and district elections in Scotland. For contested elections the analysis covers: turnout of electorate, number of councillors returned, number of ballot papers in ballot boxes; it also details the number of councillors returned unopposed. The source of these statistics is the Statistical Return of Election Results completed by returning officers and sent to the Registrar General for Scotland (see Appendix 17). Up to and including 1979, similar statistics were published by OPCS in [QRL.17] for elections in England and Wales. This practice has now been discontinued.

## 11.4 Northern Ireland

**11.4.1**

Members of *district councils* in Northern Ireland are elected for a term of four years. The next elections are due to take place in May 1989. It is perhaps worth noting that members are elected by the single transferable vote method of proportional representation.

**11.4.2**

As in the rest of the United Kingdom the process of electoral registration is the basis of the statistics of district council electorates–see 11.2.2. The chief Electoral Officer for Northern Ireland is the registration officer for the whole of the province and the returning officer for all elections. He publishes annually the *Analysis of Register of Electors* [QRL.2] showing the electorates down to individual wards within districts. A useful secondary source of published statistics of electorates at district level, is the *Ulster Year Book* [B.35]. From 1985 the Northern Ireland figures are published in [QRL.17].

**11.4.3**

Statistics of election results (ordinary and by-elections) in the form of the political (party) composition of all district councils are published annually in [QRL.44]. As explained in 11.3.1 the source of the data is the Editor's political composition questionnaire which is not reproduced here.

CHAPTER 12

# PUBLIC PROTECTION

## 12.1 Police

### 12.1.1

It should be noted that a good deal of the ground covered in this section has already been reviewed in a companion volume (Crime–Volume XV), but some overlap is unavoidable. The subject matter of 12.1 is restricted for the most part, to those police statistics which are published at police force level and which therefore relate to most aspects of what is almost wholly a local authority service. With the exception of London, the police in *England and Wales* are under the control of individual or combined county authorities. In London, the police authorities are the City of London and the Home Office which controls the Metropolitan Police.

### 12.1.2

In the relevant statistics publications, the matters contained fall into the following broad categories:

(a) output/workload (e.g notified crimes and offences, breath tests, prevention of terrorism);
(b) input/resources (e.g manpower, transport, animals);
(c) input/finance;
(d) miscellancous (e.g earnings and hours of police, deaths in police custody).

In *England and Wales* publication is widely shared among:

(i) central government (Home Office, HM Chief Inspector of Constabulary, DOE, Welsh Office, Department of Employment);
(ii) CIPFA/SIS;
(iii) SCT;
(iv) LAC SAB.

### 12.1.3

With respect to the statistics of police force *output* or *workloads* the widest coverage is provided by the Home Office in *Criminal Statistics–England and Wales* [QRL.10] and especially *Criminal Statistics–England and Wales–Supplementary Tables–Volume 3* [QRL.11]. Many of the subjects ultimately included in [QRL.11] and others are contained in the quarterly or annual Home Office *Statistical Bulletins* [QRL.81] (e.g notifiable offences recorded by police, prevention of terrorism, breath tests). Fairly extensive coverage is also provided by CIPFA/SIS in *Police Statistics (Estimates)* [QRL.51].

**12.1.4**

The statistics of *input or resources* employed by the police are contained in:
(a) the *Report of HM Chief Inspector of Constabulary* [QRL.67];
(b) CIPFA/SIS *Police Statistics (Actuals)* [QRL.50] and the same body's [QRL.50];
(c) Joint Manpower Watch publications [QRL.37], [QRL.32] and the LAC SAB Journal [QRL.42];
(d) the DOE's [QRL.41] and the Welsh Office [QRL.92];
(e) SCT's [QRL.21] and the Department of Employments *Employment Gazette* [QRL.18].

With the exception of [QRL.51] the resource to which the statistics relate is manpower, the JMW and LAC SAB publications providing the most up to date figures–i.e at quarterly intervals. CIPFA/SIS in addition to publishing manpower statistics in great detail in [QRL.51] supply data on resources such as vehicles and animals employed in police work and also the number of police training establishments maintained.

**12.1.5**

All aspects of police finance–another aspect of resource statistics–are contained in: DOE's [QRL.41]; Welsh Office [QRL.92]; CIPFA/SIS [QRL.50], [QRL.51] and [QRL.20]; SCT's [QRL.21]. The statistical content of these publications varies from a national summary in the case of [QRL.41] to individual authority coverage in great detail for CIPFA/SIS [QRL.50] and [QRL.51]. With respect to the latter the status of the figures also varies in that the CIPFA/SIS publication [QRL.20] provides advance financial estimates whereas the latter publish actual or outturn data. It is worth noting that aggregate expenditure data, both current and capital, is published by CSO in [QRL.88]. However, since this data relates to the UK and therefore includes Northern Ireland in which the police service is not administered by local authorities, it falls strictly outside the boundaries of this review.

**12.1.6**

The items within the ill-defined miscellaneous category of police statistics, which includes matters such as the operation of Section 62 of the Criminal Law Act and deaths in police custody, are to be found in [QRL.67], [QRL.11] and [QRL.81]. Statistics of earnings and hours worked by police personnel are to be found in the Department of Employment's *New Earnings Survey* [QRL.46].

**12.1.7**

The *data sources* of DOE and Welsh Office publications discussed above (i.e. [QRL.41] and [QRL.92]) are the forms RO7 and C.O.7A/B (see list in Appendix 4); CIPFA/SIS publications [QRL.50] and [QRL.51] are questionnaires not reproduced, supplemented in the case of [QRL.50] by form RO7. For [QRL.20], the (estimated) expenditure data is provided by means of the RER form (see Appendix 7). The SCT's finance and manpower figures for [QRL.21] are obtained from counties *via* an SCT questionnaire in which data already provided for other purposes is entered. As explained in paragraph 10.1.8 the police manpower data for the JMW/LACSAB and Department of Employment publications [QRL.37], [QRL.32], [QRL.42] and [QRL.18] is supplied by the Home Office. With respect to the Home Office and HM Chief Inspector of Constabulary publications [QRL.10], [QRL.11], [QRL.81] and [QRL.67]: the data is collected from police forces by the Home Office Statistical Department. For more detailed information of these data sources the companion volume *Crime* (Volume XV) should be consulted. Details of the data source of [QRL.46] are set out at length in the publication itself.

**12.1.8**

In *Scotland* the police authorities are the regional councils, combinations thereof or combinations of regional and island councils. Statistical coverage is shared by:

a) HM Chief Inspector of Constabulary for Scotland;
b) the Scottish Office;
c) the Home Office;
d) JMW;
e) the Scottish Branch of CIPFA.

**12.1.9**

The statistics of *output/workload* are published in the annual *Report of HM Chief Inspector of Constabulary for Scotland* [QRL.68] and to a more limited extent in the Scottish Office (i.e SHHD) *Criminal Statistics Scotland* [QRL.12]. (It is understood that publication of [QRL.12] is being discontinued and that in future statistical bulletins are to be substituted. These will be obtainable on application to the Scottish Office Library, New St., Andrew's House, Edinburgh.) With respect to two specific "workload" subjects–prevention of terrorism and drugs offences–the Scottish statistics are included in Home Office *Statistical Bulletins* [QRL.81].

**12.1.10**

*Input/resources* statistics are contained most fully in [QRL.68] with manpower data being included each quarter in the Scottish Office press release, *Joint Manpower Watch Survey Results* [QRL.33], the Department of Employment's [QRL.18] and much later in *Scottish Local Government Financial Statistics* [QRL.77].

**12.1.11**

The most comprehensive coverage of police *financial* statistics is provided by the Scottish branch of CIPFA in *Rating Review–Part 2* [QRL.65] for actual expenditure, both revenue and capital, and *Rating Review–Part 1* [QRL.64] for estimated capital and revenue expenditure plus outstanding debt and loan charges. Financial statistics (actual/outturn) are also published in [QRL.77] and [QRL.68]. Scotland (summary) figures of planned expenditure for several years ahead are provided by the Scottish Office in *Public Expenditure to (date)* [QRL.55].

**12.1.12**
The set of *miscellaneous* statistics relating to the police service in Scotland (e.g complaints against the police, deaths of persons in custody) is wholly contained in the annual *Report of HM Chief Inspector of Constabulary for Scotland* [QRL.50]. Statistics of average earnings and hours worked by police personnel are included in [QRL.46].

**12.1.13**
The *data sources* for the statistics publications discussed above are as follows:
1) [QRL.68]–Form "A" series of documents (not reproduced), Form MKA (see below), LFR 4 series of forms (see list in Appendix 4), SHHD Police Division documents not reproduced;
2) [QRL.12]–Forms D/1, FP/A and FP/B not reproduced, forms MKM and MKA which together with completion instructions in the SHHD's "Criminal Statistics Handbook" are not suitable for reproduction as specimen documents;
3) [QRL.77]–on the financial side, LFR4 series of documents and on the manpower side the JMW forms in Appendix 15;
4) [QRL.33] figures are supplied each quarter to JMW in Scotland (see Appendix 15 and also paragraph 10.1);
5) [QRL.46]–the data sources are listed in the publication itself;
6) [QRL.64] data is provided *via* CIPFA questionnaire not reproduced;
7) [QRL.81] data is collected from police forces by the Home Office Statistical Department–for more detailed information on this source, see companion Volume XV;
8) [QRL.55] figures are planned aggregates arising directly from policy decisions made by the Secretary of State for Scotland.

## 12.2 Fire

**12.2.1**
At present the fire service in *England and Wales* is provided by the counties and the Greater London Council. As from 1st April 1986, in the metropolitan areas and Greater London new arrangements have been introduced (see section 2.1).

**12.2.2**
As in the case of the police service, publication of fire service statistics is widely spread. It is shared among:
a) central government (Home Office, HM Chief Inspector of Fire Services, DOE, Welsh Office, Department of Employment);
b) CIPFA/SIS;
c) LAC SAB.
The subjects included fall into four broad categories as follows:

i) output/workload (e.g fires, false alarm and special calls answered, fire prevention inspections);
ii) inputs/resources (e.g manpower, vehicles, fire appliances, stations);
iii) inputs/finance;
iv) miscellaneous (earnings, hours, causes of fire, sources of ignition).

**12.2.3**
The statistics of *output/workload* by fire authority area are provided by the Home Office in *Fire Statistics United Kingdom* [QRL.25]. Similarly disaggregated data in much less detail is provided by CIPFA/SIS in *Fire Statistics (Actuals)* [QRL.23] and *Fire Statistics (Estimates)* [QRL.24]. *The Report of HM Chief Inspector of Fire Services* [QRL.69] also contains information on workload, but not at individual authority level.

**12.2.4**
The most comprehensive coverage of *input resources* statistics is provided by CIPFA/SIS in [QRL.23] and [QRL.24]. One resource, manpower, receives special attention in that it is measured and recorded quarterly in [QRL.32] for England, [QRL.37] for Wales; [QRL.42] and [QRL.18] England and Wales. The figures are reproduced later, for England in [QRL.41] and [QRL.21] and for Wales, in [QRL.92]. The *Report of HM Chief Inspector* in [QRL.69] also includes manpower statistics at the level of the individual fire authority plus summaries of other resource statistics. *Input/finance* data by fire authority is provided by CIPFA/SIS in [QRL.23], [QRL.24] and [QRL.20]. An English summary only is later published by DOE in [QRL.41] and by the Welsh Office in [QRL.92] for each Welsh authority. Expenditure summaries also appear in [QRL.69] as do statistics of some of the costs of the fire research programme. Current expenditure data for individual English authorities is published in SCT's [QRL.21]. It is worth noting that the statistics of Fire Services expenditure on both current and capital account, which appear in [QRL.88] relate to the United Kingdom and therefore include Northern Ireland expenditure which falls outside the local authority boundary.

**12.2.5**
The *miscellaneous* statistics, which by definition fall out with those easily classified above, may be thought of as either output related or manpower related. The first group (e.g causes of fire, sources of ignition, casualties, risk category of area) are dealt with in [QRL.25] and [QRL.23], [QRL.24], [QRL.69]. The manpower related group (earnings and hours of personnel, pension schemes) is covered in [QRL.46], [QRL.23] and [QRL.24].

**12.2.6**
The *data sources* of the fire service statistics identified above are as follows:
1) the Home Office gathers its information from all authorities by means of the FDR forms and 44C (see Appendix 16); instructions for completion are

contained in the Home Office Booklet "Notes on the completion of Report of Fire form FDR1" which it is not possible to reproduce in the review;

2) the data for the *Annual Report of HM Chief Inspector of Fire Service* [QRL.69] is supplied *via* the Home Office Statistical Department which uses its sources described here; manpower data is of course obtained via JMW–see below;

3) the data for CIPFA/SIS [QRL.20] is collected by RER form (see Appendix 7); for [QRL.23] and [QRL.24] the figures are provided by questionnaires–Actuals and Estimates respectively–not reproduced;

4) the DOE and Welsh Office assemble the data for [QRL.41] and [QRL.92] by means of the forms RO9 (for revenue) and CO7A/B (for capital)–see Appendix 4;

5) the manpower statistics for [QRL.37], [QRL.32], [QRL.42] and [QRL.18] are collected from the JMW forms–see Appendix 15;

6) the earnings and hours data sources for [QRL.46] are fully set out in the New Earnings Survey Publication itself;

7) the SCT's data for [QRL.21] is obtained directly from authorities *via* questionnaire in which outturn figures already provided to DOE or JMW are entered.

**12.2.7**

In *Scotland* the regional councils, combinations thereof or combinations of regional and islands councils are the responsible fire authorities. As elsewhere in Great Britain the publication of fire service statistics is widely shared in this instance by:

a) central government (i.e Scottish Office, Home Office, HM Chief Inspector of Fire Services (Scotland), Department of Employment);

b) Joint Manpower Watch;

c) CIPFA–Scottish branch.

However, in the case of CIPFA the coverage is much narrower than that provided by CIPFA/SIS in England and Wales.

**12.2.8**

*Output/workload* statistics are provided by the Home Office in [QRL.25] and by central government in Scotland in the annual *Report of HM Chief Inspector of Fire Services (Scotland)* [QRL.70]. The latter also contains *input/resources* data in the form of establishment and manpower strengths. Quarterly manpower data is published also by JMW in [QRL.33], by the Department of Employment in [QRL.18] and later repeated by the Scottish Office in [QRL.77]. At the time of writing, statistics of other non-financial resources are not provided at national level in Scotland.

**12.2.9**

The statistics of *input/finance*, as in the case of the police service, are of three kinds: actual/outturn, estimates, planned. Actual or outturn figures are provided by the

Scottish Office in [QRL.77] and by the Scottish branch of CIPFA in [QRL.65]. Annual forward estimates of expenditure (current and capital) and outstanding debt are given in [QRL.64] while the Scottish Office in [QRL.55] publishes planned figures–summaries only–for several years ahead.

**12.2.10**
In the *miscellaneous* category HM Chief Inspector of Fire Services (Scotland) publishes data on fatalities [QRL.78]; the Home Office on casualties in [QRL.25]. The Department of Employment provides fire service earnings and hours figures in [QRL.46]. It should be noted that in [QRL.70] the Chief Inspector may include statistical tabulations on any aspect of fire service activities on which he cares to comment. The references above to the statistical content of [QRL.70] relate to recent publications.

**12.2.11**
The *data sources* of the Scottish fire service statistics are as follows:

(1) The Home Office collects data for [QRL.25] directly from Scottish fire authorities by means of FDR forms (see Appendix 16) (Note: form 44C is not used in Scotland). Instructions for completion are contained in the Home Office booklet "Notes on the completion of Report of Fire form FDR1" which it is not possible to reproduce in this review.

(2) HM Chief Inspector of Fire Services in Scotland gathers data for [QRL.70] *via* form FS(A) which is in 13 parts (see Appendix 16).

(3) JMW in Scotland collects figures for [QRL.33] and for the Department of Employment by means of the Scottish Local Authorities Negotiating Bodies form (see Appendix 15).

(4) The Scottish Office obtains its financial data from form LFR 4A and Supp 1 (see Appendix 4) and its manpower data from the Scottish JMW.

(5) the Scottish branch of CIPFA assembles its estimates data for [QRL.64] from its own form not reproduced; its [QRL.65] sources are set out in the publication itself.

(6) The earnings and hours statistics in [QRL.46] are displayed fully in the New Earnings Survey itself.

(7) The planned figures in [QRL.55] arise from policy decisions made by the Secretary of State for Scotland.

## 12.3 Probation

**12.3.1**
The Probation Service in *England and Wales* provides a wide range of services to the community, but its primary task is the supervision of offenders committed to its care under various orders of the courts notably the probation order, the supervision order (for juveniles) and the community service order. It also provides advice and reports to the criminal courts on persons charged with offences and to the civil courts in

domestic cases. By supplying probation officers seconded to Prison Department establishments and through visits and contacts of other probation officers, it provides a through–care service to persons serving custodial sentences and for a period after their release. Some of this supervision is "voluntary after-care and some of it is compulsory supervision on licence under parole etc.. The service is administered by 56 independent probation committees which employ the staff of their area probation services. All except the Inner London Probation Service and the City of London Probation Service, which are financed by the Receiver for the Metropolitan Police District, are financed by the local authorities for the areas; by the county councils in shire counties, by the metropolitan county councils in the metropolitan counties and by the Greater London Councils in outer London (but see note in section 2.1 re-position post-April 1986). However, these area services are not departments of the local authorities and 80% of the expenditure on the area services is supplied by Home Office specific probation service grant.

**12.3.2**

In Scotland, the corresponding range of "probation" services is dispensed by the Social Work departments of the regional and islands authorities, but statistical comparison with England and Wales is not possible (see Volume I in this series of reviews). It is worth noting that statistics of central government grant in support of this service are not made available in statistics publications.

**12.3.3**

Statistics publication in England and Wales is shared by the Home Office, DOE, Welsh Office, Joint Manpower Watch and CIPFA/SIS: in Scotland only the Scottish Office (i.e SHHD and SED) provides a few relevant statistics.

**12.3.4**

Home Office coverage is supplied by *Probation Statistics England and Wales* [QRL.53] which contains data on workload, manpower and finance. This publication also includes statistics of persons commencing probation, supervision under the Children and Young Persons Act 1969 and community service, each analysed by previous criminal record at commencement of the order, by probation areas. In [QRL.92] the Welsh Office publish summary details of expenditure and income for Wales. They also reproduce the JMW manpower statistics which are initially published by the JMW Group in [QRL.37] and [QRL.32]; LAC SAB also publish quarterly in [QRL.42]. CIPFA/SIS provides by far the widest coverage, by individual area probation service, in *Probation Statistics (Actuals)* [QRL.52] and *Probation Statistics (Estimates)* [QRL.54]. These contain not only financial statistics but manpower data and a series of case-load and quasi-case-load statistics. An attempt is also made to develop rough performance indicators in the form of net probation service expenditure related to the population at risk (i.e the 15/29 age group) or alternatively the number of persons proceeded against at Magistrates Courts.

**12.3.5**
Publication in Scotland is limited to work-load/case-load statistics produced by the Scottish Office (SHHD) in [QRL.12]–the number of probation orders imposed by the courts, and also for a short time by the Scottish Office (SED), in the *Social Work Statistical Bulletin* [QRL.80] which contained a series of case-load statistics. This bulletin has now been discontinued but recent and current statistics can be obtained on application to Social Work Services Group, Statistics Branch, Room 4B, 43 Jeffrey Street, Edinburgh EH1 1DH.

**12.3.6**
The *data sources* of the statistics publications discussed in paragraph 12.3.4 and 12.3.5 are as follows:
(a) [QRL.53]–collected by Home Office–covering documents not available since they would not be meaningful without the lengthy instructions which accompany them;
(b) [QRL.41] and [QRL.92]–see DOE/Welsh Office forms RO11 and CO7A/B listed in Appendix 4 and also JMW, see below;
(c) [QRL.37], [QRL.32] and [QRL.42]–see JMW forms (Appendix 15);
(d) [QRL.52] and [QRL.54] which are compiled by the Treasurer of an English County (see Appendix 5) are based on CIPFA/SIS questionnaires;
(e) [QRL.12] SHHD Form D/1 not reproduced here;
(f) [QRL.50] CIPFA/SIS questionnaire not available for reproduction here.

## 12.4 Administration of Justice and Miscellaneous Public Protection Items

**12.4.1**
In local authority statistics publications in England and Wales the definition of the term "administration of justice" varies widely not only as between publishing agencies but even within agencies. In CIPFA/SIS [QRL.20] and SCT's [QRL.21] the term applies to the sum of the following services: magistrates courts, other courts, probation, school crossing patrols, civil defence and registration of electors. In companion CIPFA/SIS publication, *Administration of Justice Statistics* [QRL.1] the term applies to magistrates and other courts only. Meanwhile the DOE and Welsh Office define the term as including courts and the probation service in aggregate. Joint Manpower Watch does not employ the term at all but identifies magistrates courts and the probation service separately. However, in this review the term embraces the following services which are examined separately: courts; school crossing patrols; civil defence; registration of electors. The statistics of the probation service have of course, been dealt with above.

**12.4.2**
In England and Wales, *magistrates courts* are administered by committees which, while strictly separate from the local authorities, nevertheless depend upon them to meet part of the expenditure incurred in their administration, but 80% of the

expenditure of this service is met by specific grant from the Home Office. The local authorities responsible are the non-metropolitan county councils, the metropolitan districts and in London, the Greater London Council and the City of London (see Section 2.1 for position in London from April 1986). *Coroner's Courts* are administered by all county councils (metropolitan and non-metropolitan) and the GLC. (See Section 2.1 for position in metropolitan areas and London from April 1986.)

**12.4.3**
The widest coverage of the statistics of the magistrates/coroners court service is provided by CIPFA/SIS in [QRL.1]. Subjects covered are: finance; estimated population or sub populations "at risk". (Note: at the time of writing, magistrates court statistics published for metropolitan areas relate to metropolitan county areas rather than to the individual metropolitan districts actually providing the service.) Financial summaries for England and Wales respectively are published in [QRL.41] and [QRL.92]; the former also provides summary manpower statistics for magistrates courts. The JMW statistics for magistrates courts are published in [QRL.32]–England; [QRL.37]–Wales and both in [QRL.42]. With respect to the work of coroners courts, it should be noted that in the Statistical Bulletin *Statistics of Deaths Reported to Coroners: England and Wales* [QRL.82] the Home Office, in addition to deaths reported, provides data on the number of post-mortems and inquests and on verdicts returned at inquests. Unpublished data by Coroners' areas is available from the Home Office on request.

**12.4.4**
*Civil Defence* or emergency planning is mainly the responsibility of the counties (metropolitan and non-metropolitan) and the GLC with districts and London Boroughs playing a subsidiary role. (See section 2.1 for position in metropolitan areas and London from April 1986.) Statistical coverage of the service encompasses finance only and is provided by DOE/Welsh Office in [QRL.41] for England and Wales separately and [QRL.92] for Wales only. The content is much more extensive in the latter case; in both, analysis goes down to class of authority level.

**12.4.5**
The *registration of electors* is the responsibility of the London Boroughs and the district councils in the rest of England and Wales. But note that in a few parts of England, shire counties carry out the work on an agency basis. Statistical coverage is limited to finance, class of authority summary in England and Wales separately, in the DOE publication [QRL.21] and in much greater detail for Wales in [QRL.92].

**12.4.6**
*School crossing patrols* are administered by the police authorities and the National Parks in England and Wales (see section 2.1 for position from April 1986).

Publication of the statistics of the service relate only to finance–specifically expenditure and income on revenue account since any capital expenditure or debt incurred is included in the police service budget. Coverage by class of authority in England and Wales separately is provided in [QRL.41] and, in greater detail for Welsh authorities in [QRL.92]. Manpower statistics for the service are not separately published having been included within the category "other services" (see Appendix 15).

**12.4.7**
Reference was made above to statistics publications which provide data for an aggregation of services called "administration of justice" (see paragraph 12.4.1 for precise definition). In [QRL.20] CIPFA/SIS publish revenue expenditure estimates for each authority in England and Wales; in [QRL.21] the Society of County Treasurers publish revenue expenditure outturn statistics for all English and Welsh counties.

**12.4.8**
The *data sources* for the statistics described in this section are as follows. For [QRL.1] CIPFA/SIS collect data by means of a questionnaire which it has not been possible to reproduce here. The questionnaire is however not all inclusive but the publication itself contains notes on the sources of items excluded. The financial statistics in [QRL.41] and [QRL.92] are obtained from the annual outturn form (see list in Appendix 4); manpower statistics in [QRL.41] and [QRL.92] are simply the JMW tabulations, see below. The statistics in [QRL.42], [QRL.37] and [QRL.32] are assembled *via* the JMW questionnaire (see Appendix 15). The expenditure estimates in [QRL.20] are obtained from the RER forms (see Appendix 7) by CIPFA/SIS. For [QRL.21] the data is collected from the counties by means of an SCT questionnaire in which authorities reproduce the financial outturn figures already provided to DOE/Welsh Office (see above) and the JMW manpower data. The source documents for [QRL.82] are not available for reproduction.

**12.4.9**
In local government statistics publications in *Scotland*, the term "administration of justice" is used only by the Scottish Office to apply to district courts. The term is similarly employed by other statistics publishing agencies (i.e the Scottish Branch of CIPFA and Joint Manpower Watch). In addition to the district courts statistics, this section encompasses also civil defence (or emergency planning) and school crossing patrols. Unlike the situation in England and Wales, the statistics of electoral registration cannot be separately identified since for statistical purposes, the service is merged with that of the Land Valuation Assessor (see paragraph 9.1.1).

**12.4.10**
In Scotland the *district courts* are administered by the district and islands councils. They are financed as standard local authority services, except that they retain 10% of

fixed penalties and the larger part of the income received from fines imposed by the courts. (Note: there are no coroners' courts in Scotland.) Outturn financial statistics (Scotland summaries) are published by the Scottish Office in [QRL.77] and by CIPFA in Scotland in [QRL.65]; forward estimates are also published by the Scottish branch of CIPFA in [QRL.64] while planned expenditure for several years ahead is supplied by the Scottish Office in [QRL.55]. Quarterly manpower data is published by JMW in Scotland in the Scottish Office/COSLA press release [QRL.33] and by the Department of Employment in [QRL.18]. The statistics are reproduced much later in [QRL.77]. At the time of writing, other statistics on the work of the district courts in Scotland are not published. However, data which is collected centrally on particular aspects of the courts work may be provided on request to: Secretary of Commissions, St. Andrew's House, Edinburgh. It is understood that the SHHD is currently considering the possible publication of statistics comparing sentences imposed by individual courts.

**12.4.11**

*Civil defence* (referred to by the Scottish branch of CIPFA as "emergency planning") is the responsibility of the regional and islands authorities in Scotland. The content of the published statistics of this service is restricted to finance. In [QRL.77] and [QRL.64] there are Scotland summaries of respectively, outturn and estimated expenditure: [QRL.55] contains expenditure outturn data previously published and planned figures for several years ahead.

**12.4.12**

The provision of *school crossing patrols* is the responsibility of the regions and islands councils. Only financial statistics for this service are published nationally. Scotland level expenditure figures are provided in [QRL.77] for outturn expenditure and [QRL.64] for estimated expenditure. The Scottish branch of CIPFA now publish outturn expenditure data for individual authorities in [QRL.65].

**12.4.13**

The *data sources* of the publications discussed in paragraph 12.4.10 to 12.4.12 are as follows: the financial statistics in [QRL.77] in respect of district courts, civil defence and school crossing patrols are gathered from Scottish Office forms LFR 4B, LFR 9A and LFR 4A respectively (see list in Appendix 4). The manpower statistics in [QRL.77] are copied straight from JMW (see below). The estimates in [QRL.64] are obtained from the authorities by means of a CIPFA form not reproduced here. The publication [QRL.65] lists its own data sources and [QRL.55] of course, contains only planned figures drawn up by the Secretary of State for Scotland. Manpower statistics in [QRL.33] and [QRL.18] are assembled by the JMW in Scotland (see Appendix 15).

## 12.5 Consumer Protection (including Trading Standards)

**12.5.1**
This service is part advisory/part law enforcement. The *Municipal Year Book* [QRL.44] points out that while the service provided varies between authorities, the work of consumer protection/trading standards departments falls into the following five broad areas of responsibility:
a) metrology and quantity control;
b) controls over quality and performance of goods and services;
c) safety and licencing controls;
d) controls over consumer credit and price information;
e) information and advice to consumers and the trade.
Details of the main legislation administered by each type of authority are also set out in [QRL.44].

**12.5.2**
In England and Wales the service is in the hands of the counties outside London, where it is administered by individual boroughs or consortia of these. However in England, district councils incur expenditure on this service in that they would appear to discharge some responsibilities on an agency basis. In Scotland the regional and islands councils are the responsible authorities.

**12.5.3**
In *England and Wales*, the widest statistics coverage of this service is provided by the Society of County Treasurers in *Trading Standards Statistics (Estimates)* [QRL.86] and *Trading Standards Statistics (Actuals)* [QRL.86]. While the subject coverage of the latter is almost wholly financial, the former in addition to financial data, contains manpower and workload figures. The statistics relate to individual authorities. Financial (outturn) summaries of varying levels of disaggregation are published by the DOE and Welsh Office in [QRL.41] for separate figures for England and Wales, and by individual authority in Wales in [QRL.92]. For each county council in England and Wales the SCT also provides data on revenue expenditure per head in [QRL.21].

**12.5.4**
In *Scotland* summary financial statistics at Scotland level are provided by the Scottish Office in [QRL.77] for outturn figures and by the Scottish branch of CIPFA in [QRL.64] for forward estimates. Outturn expenditure data by individual authority is now provided in [QRL.65] by the Scottish branch of CIPFA. Quarterly manpower summaries are published in [QRL.33] by the JMW in Scotland. A wealth of relevant statistics, although not standardised, is published in the annual reports of the responsible authorities in Scotland. A set of these reports is held by the Tayside Regional Council's Director of Trading Standards, 1 Riverside Drive, Dundee DD1 4DB, to whom application can be made for further details.

**12.5.5**
The data sources of the publications discussed in paragraph 12.5.3 and 12.5.4 are as follows. For [QRL.41] and [QRL.92] the data is gathered from DOE/Welsh Office forms RO12 and CO12A/B (see lists in Appendix 4). The SCT assembles its data for [QRL.86] and [QRL.85] by questionnaire to individual authorities which simply reproduce the RO12 figures already provided to central government. In Scotland, the information for [QRL.77] is obtained from authorities by means of the Scottish Office form LFR 9 (see list in Appendix 4). The Scottish branch of CIPFA obtains its estimates data from its own form, not reproduced here, while, for its [QRL.65] the data source is set out in the publication itself. Manpower figures for [QRL.33] are collected quarterly for the JMW by means of the NJC (Scottish Councils) survey form (see Appendix 15).

## 12.6 Northern Ireland

**12.6.1**
None of the public protection services considered in this chapter is the responsibility of the district councils in Northern Ireland.

CHAPTER 13

# TRANSPORT

## 13.1 Highways and Transportation (England and Wales)

### 13.1.1

This heading first of all, encompasses the following (Rate Fund) local transport functions:

a) the construction, maintenance and lighting of highways;

b) the provision of parking for vehicles;

c) road safety;

d) financial support for local passenger transport;

e) the provision of concessionary fares;

f) administration of these functions.

In addition to the Rate Fund services the heading also includes those transport Trading Services (see section 8.4) which are carried on by some authorities (i.e passenger transport; fishing harbours; ports and piers; tolled bridges, roads, ferries and tunnels). It should be noted that aerodromes are considered separately in section 13.3.

### 13.1.2

Outside London, the county councils are responsible for the highways but may allow districts to carry out this work on an agency basis, for urban roads. Road safety, including traffic management, is also a county function. Certain minor highway functions such as street naming, footway lighting and pedestrianisation are carried out by districts. Revenue support for buses and railways, concessionary fare schemes, the provision of vehicle parking and the construction and maintenance of harbours, docks, toll bridges are shared between counties and districts. In London, the boroughs are responsible for highways and most other transportation services set out above. The exceptions are: the construction and maintenance of principal roads, traffic management, the construction of piers, tunnels and bridges. Until April 1986, these were the responsibility of the GLC which could however, delegate principal road maintenance to the boroughs. For changes from April 1986 onward, see section 2.1 above. Transport Trading Services may be carried out by any of the authorities above but are more usually carried out by lower tier authorities. Note that road accident reporting is a police authority responsibility.

**13.1.3**

The statistics of highways and transportation in England and Wales are published by central government (DOE, Welsh Office, Department of Transport) and two local government bodies (CIPFA/SIS and SCT). The subjects covered are: finance and work load related matters (road lengths to be maintained; licensed vehicles; road accidents). Additionally, the Joint Manpower Watch in [QRL.42], [QRL.37] and [QRL.32] also publishes a "transport services" category which however refers in the main, only to manpower employed by local authorities' Trading Services. Manpower engaged in the (Rate Fund) highways and transportation services are partly contained in transport, construction and a number of other categories.

**13.1.4**

The statistics of the *finance* of the Rate Fund items of highways and transportation are provided at various levels of disaggregation in the following central government publications: DOE/Welsh Office [QRL.41] and [QRL.92]; Department of Transport's *Transport Statistics–Great Britain* [QRL.87] which also contains data relating to Scotland. However the most extensive coverage on the Rate Fund financial side is supplied by CIPFA/SIS in *Highways and Transportation Statistics–Estimates* [QRL.30] and *Highways and Transportation Statistics–Actuals* [QRL.29]. Greatly condensed expenditure estimates for each authority, are published by CIPFA/SIS in [QRL.20]; for counties only, condensed figures of actual expenditure are provided by SCT in [QRL.21]. With respect to public transport services which are run as Trading Services, detailed national summary data is provided in [QRL.41] for England and [QRL.92] for Wales.

**13.1.5**

(While the remaining subjects included in highways and transportation statistics have been classified as workload-related, in this review, it should be understood that the purpose of collecting these statistics is not primarily one of attempting to measure work load, albeit imperfectly. However, looked at from the point of view of the growing tendency in local government to measure performance (see paragraph 17.1.2), it is a useful grouping.) Road lengths within the areas of highway authorities are published in [QRL.30], [QRL.29], [QRL.87], [QRL.21] and *Statistics of Road Lengths in Wales* [QRL.83]; the number of vehicles licensed, by highway authority area, is contained in [QRL.87]; statistics of road accidents resulting in injuries for all areas in Great Britain are supplied in *Road Accidents–Great Britain* [QRL.73]. Further road accident statistics in English and Welsh local authority areas are published respectively in *Road Accident Statistics–English Regions* [QRL.74] and *Road Accidents: Wales* [QRL.75].

**13.1.6**

The *data sources* of the statistics identified above are as follows: [QRL.41] and [QRL.92] Rate Fund revenue account figures are obtained from forms RO18 and RO19 (see list in Appendix 4) with capital items coming from forms CO18 and CO19 also listed in Appendix 4. The Trading Services data is assembled *via* forms TSR1 and TSR2 (see Appendix 4 list) on revenue account and CO20, 21 and 22 on capital account (see list Appendix 4). On the financial side [QRL.87] data is collected *via* DOE/Welsh Office using either the Rate Fund forms listed above or in the case of Grants, directly from DOE Finance Division. Road length figures are obtained from highways authorities by means of an annual return, not reproduced here. Vehicles licensed data is collected from the Driver and Vehicle Licensing Centre (DVLC) at Swansea which provides to the Department of Transport/Welsh Office a computer print out of vehicles licensed in each county. Data for the CIPFA/SIS [QRL.30] and [QRL.29] publications is obtained from four sets of questionnaires, two relating to counties–estimates and actuals separately–and one each to English/Welsh districts and London Boroughs respectively, estimates only in each case. It has not been possible to reproduce these substantial questionnaires. [QRL.20] data comes to CIPFA/SIS from the RER forms (see Appendix 7). SCT obtains its financial data from counties which simply reproduce the Rate Fund figures as above; road lengths are also collected from counties which repeat the figures previously sent to the Department of Transport/Welsh Office. The source of [QRL.83] is of course similar to that described above for [QRL.87]. The statistics of road accidents resulting-in-injuries in [QRL.73], [QRL.74] and [QRL.75] are compiled from the basic Statistics 19 form which is completed by the police attending the scene of an accident. The resulting figures are usually committed to computer by the police authority in question and forwarded by tape to the Department of Transport/Welsh Office. A specimen copy of Statistics 19 is contained in each of the statistics publications mentioned.

**13.1.7**
It is worth noting that statistics of car ownership or cars available to households within counties (a further workload-related subject) are published in the ten yearly population census volumes.

## 13.2 Highways and Transportation (Scotland)

**13.2.1**
Although the terms "highways and transportation" and "local transport" are not used in the literature of local authority statistics in Scotland, a matching set of services to that identified in paragraph 13.1.1 for England and Wales, is provided by the region and islands councils. In Scotland the collective term used by the Scottish Office is "roads and transport" but CIPFA in Scotland does not employ the term preferring, as in England and Wales, to divide the services into three categories:

a) roads and lighting;
b) transport or transport subsidies;
c) concessionary fares.

As in the rest of Great Britain, some transport services, mainly bus passenger transport, are Trading Services, the statistics of which are separated from those of roads and transport financed from the Rate Fund.

**13.2.2**
The statistics of transport are published mainly by the Scottish Office and the Department of Transport, but the Scottish Branch of CIPFA also provides some figures. As in England and Wales, subject coverage may be divided into: finance and work load related matters. It is true that the JMW in Scotland publishes manpower data on "transport services" and "roads/highways". However, neither of these categories matches the services or components thereof identified above. The fact is that manpower employed in roads and lighting may be included in any one of four JMW categories (i.e construction, technical services, roads/highways, direct works); the transport services group includes not only the transport Trading Services category but also school transport and the Strathclyde PTE which falls outside the local authority financial sector.

**13.2.3**
On the *financial* side, the Scottish Office publishes Rate Fund outturn statistics in *Scottish Local Government Financial Statistics* [QRL.77] and *Scottish Transport Statistics* [QRL.78]; outturn expenditure data by region in Scotland, is also provided by the Department of Transport in [QRL.87]. Planned expenditure for several years ahead is published by the Scottish Office in *Public Expenditure to...(Years)* [QRL.55]. Outturn financial statistics–much earlier and in greater detail than in [QRL.77]–are now published by the Scottish Branch of CIPFA in *Rating Review–Part 2* [QRL.65]; estimated expenditure, one year ahead, is published by the same body in *Rating Review–Part 1* [QRL.64].

**13.2.4**
The statistics of *work load related* matters in Scotland are published as follows. Road lengths and vehicles licensed data are most widely covered by the Scottish Office's [QRL.78]; the Department of Transport's [QRL.87] also contains figures on these subjects. Accidents and casualties statistics, by region, are published in [QRL.73] produced by the Department of Transport. The Scottish Office's [QRL.78] also includes road accidents data. Some decennial population census data on car availability by regional/islands area is also published in [QRL.78].

**13.2.5**
The *data sources* of outturn finance in [QRL.77], [QRL.78] and [QRL.87] are the Scottish Office forms LFR5 and LFR22 (see list in Appendix 4). But note that expenditure on motorways and trunk roads by regional/islands councils (although central government responsibilities) are published in [QRL.78] on the basis of scheme by scheme accounts submitted by the regions and islands councils acting as agents for the Scottish Office. Planned expenditure figures in [QRL.55] are of course policy matters provided by the Secretary of State for Scotland. The sources of the outturn data in [QRL.65] are listed in the publication itself; the [QRL.64] estimates are provided by a Scottish Branch of CIPFA form not reproduced here. Trading Services financial summaries in [QRL.77] are assembled from LFR22 (see Appendix 4 list).

Statistics of public road lengths in [QRL.78] and [QRL.87] are collected from local highways authorities by means of the return "Road Lengths in kilometres as at the 1st April". Data on vehicles licensed is obtained from DVLC at Swansea (see paragraph 13.1.6). Road accidents figures are collected from Statistics 19 forms a specimen copy of which is contained in [QRL.73]–see also paragraph 13.1.6. Car availability data is obtained from the census of population (see paragraph 13.1.7).

## 13.3 Ports

### 13.3.1

Approximately 30 ports in Great Britain are owned by local authorities. Separate statistics relating to their operation have been dealt with in a companion volume in this series of reviews (see *Ports and Inland Waterways* Volume X).

## 13.4 Airports

### 13.4.1

A number of airports in Great Britain are administered by local authorities. Statistics of their operation having been published, they are treated here in a separate section although some overlap with sections 13.1 and 13.2 is entailed.

### 13.4.2

Authorities responsible for airports administration in the case of England and Wales, are either counties, districts or a combination of both. In Scotland, one of the regional councils runs the only local authority airport there.

### 13.4.3

The statistics of local authority airports in Great Britain are published by CIPFA/SIS in *Local Authority Airports–Accounts and Statistics* [QRL.34] and *Local Authority Airports–Accounts and Statistics (Estimates)* [QRL.35] (note: the latter has now been discontinued). The subjects covered in these publications are: finance; output (e.g passengers and cargo handled, aircraft movements, scheduled services); input (e.g manpower, runways). On the financial side there are both revenue and capital accounts and a balance sheet. The CIPFA/SIS compilers (see Appendix 5) have drawn attention to the difficulty encountered in preparing comparable revenue accounts for airports which provide different services, have varying accounting procedures according to objectives, scale of operation and whether operated by a single authority or joint committee.

### 13.4.4

The data sources of [QRL.34] and [QRL.35] are CIPFA/SIS questionnaires circulated and processed using the compiler system referred to in Appendix 5. It has not been possible to provide specimens of the questionnaire in this review.

## 13.5 Northern Ireland

### 13.5.1

The only transport related service provided by local authorities in Northern Ireland is the operation of *harbours* by several district councils. Financial statistics are published in *District Council Rate Statistics* [QRL.14] which contains not only a national summary but estimates of expenditure, income and net cost (i.e cost to the ratepayer) for one year ahead for each district council concerned. The *data source* is the General Estimates of Rates (see paragraph 4.5.5).

CHAPTER 14

# ENVIRONMENTAL SERVICES

## 14.1 Definition and Scope

### 14.1.1

Many aspects of local authority environmental services have already been covered in companion volumes (see Volume III *Housing*; Volume IV *Leisure and Tourism*; Volume VIII *Land Use* and *Town and Country Planning*). The scope of this chapter is therefore limited to an examination of those environmental services devoted to the protection of the citizen from a wide range of health risks such as atmospheric pollution, waste hazards, unhygienic food, water pollution, occupational hazards and so on. In the event, providing a reasonably clear and convincing guide to relevant statistical sources in this area, has proved far from simple. The difficulty is that in the literature the term "environmental health" is inconsistently defined. However, the major inconsistencies are confined to England and Wales and, with minor exceptions, do not apply to the following services:

a) waste collection;

b) waste disposal;

c) cemeteries and crematoria.

Since there are differences in functions and terminology as between England and Wales on the one hand and Scotland on the other, consideration of the statistical sources in the two areas has been separated below.

### 14.1.2

What is provided in this chapter is, for England and Wales, an examination of the statistical sources of (i) a set of fairly clearly identified services (waste collection and disposal, the administration of cemeteries and crematoria) and (ii) a collection of services (environmental health) the definition of which varies with the providing agency (CIPFA/SIS; DOE; Welsh Office; JMW). For Scotland, five distinct services are identified (cleansing, environmental health, sewerage, public water, burial grounds and crematoria) the statistics definitions of which are relatively coherent. In respect of Northern Ireland, the district-council-provided services falling within the ambit of this chapter are: health inspection, clean air, public conveniences, burial grounds and crematoria, refuse collection and disposal, street cleansing and abattoirs.

## 14.2 Environmental Health (England and Wales)

### 14.2.1
In England almost all of these services are provided by districts (metropolitan and non-metropolitan) plus the London Boroughs. The services are provided by the districts in Wales.

### 14.2.2
Statistics of outturn expenditure and income both revenue and capital, are published by the DOE and Welsh Office in, respectively, *Local Government Financial Statistics–England and Wales* [QRL.41] and *Welsh Local Government Financial Statistics* [QRL.92]. The definition of the service employed in these publications is set out in DOE/Welsh Office form RO14 (see list in Appendix 4). It comprises: public conveniences; prevention of air pollution; cesspool emptying; baths and public launderies (other than swimming baths); services under the Public Health Hygiene Acts; urban and domestic rodent extermination; street sweeping and cleansing not chargeable to highways. Estimates of revenue expenditure for this range of services are published by CIPFA/SIS in *Finance and General Statistics* [QRL.20].

### 14.2.3
The data sources of [QRL.41] and [QRL.92] are the revenue (RO14) and capital (CO14) forms listed in Appendix 4. Data for the estimates published in [QRL.20] are provided on form Return of Expenditure and Rates (RER) a specimen copy of which is shown in Appendix 7.

### 14.2.4
Manpower figures for Environmental Health are made available by the Joint Manpower Watch survey (see section 10.1 above). The publications concerned are the Joint Manpower Watch Group quarterly *Joint Manpower Watch Return–England* [QRL.32], for English authorities and the Group's *Local Authority Manpower in Wales* [QRL.37]. The statistics are later published in the LAC SAB Journal *Local Government Manpower* [QRL.42] and in the Department of Employment's *Employment Gazette* [QRL.18]. In this context the definition of the Environmental Health service is set out in the Quarterly Joint Manpower Watch–Notes of Guidance shown in Appendix 15. It consists of public health; meat inspection; prevention of air pollution; public conveniences; port health; any other environmental health work undertaken on an agency basis (e.g for a water authority). It should be noted that the June figures are reproduced in [QRL.41] and [QRL.92].

### 14.2.5
The data source of the manpower statistics is the LAC SAB Quarterly Joint Manpower Watch return a specimen of which is shown in Appendix 15 with accompanying Notes of Guidance (for fuller treatment see section 10.1).

**14.2.6**

In 1982 the CIPFA/SIS Environmental Health Working Party (see Appendix 5) produced for statistical purposes, a definition of what it called the "core functions" of the service. This definition was developed and published in *Environmental Health Core Function Statistics–Report of a Pilot Survey* [B.20]. Following from the survey the CIPFA/SIS statistics booklet *Environmental Health Statistics–Actuals* [QRL.19] has now been published. The first issue relates to the year 1982/83 and contains a readily available definition of the "core functions". They consist of: monitoring of housing conditions; pollution control; occupational health and safety; Shop Act administration; food hygiene and Food and Drugs Acts; public health; animal health; port health; health education.

**14.2.7**

The subject coverage in [QRL.19] is quite extensive and includes for each authority the following:

a) area characteristics (analysis of hereditaments);
b) staff (analysis by precise function);
c) staff time analysis;
d) enforcement actions, inspections and visits of various sorts;
e) finance.

A specimen copy of the data source questionnaire is shown in Appendix 18 together with notes for guidance in making the return. The completed questionnaires are scrutinised by the Institution of Environmental Health Offices with assistance from one district council in England.

**14.2.8**

When the Health Core Function Pilot Survey was undertaken, it was recognised that, having regard to the wide variation in the scope of environmental health functions carried out by authorities and given the wide range of authority sizes, statistical comparisons would only be valid and useful provided the clustering of like authorities was effected. In the event, the clustering, described in [B.20], was found not to be successful and the ordering of data in [QRL.19] is not therefore based on clustering. We return to the subject of statistical clustering and its application to local authority statistics in section 17.1 below.

**14.2.9**

A useful publication which provides national statistics on environmental protection is the DOE's *Digest of Environmental Protection and Water Statistics* [B.17]. It has been published annually since 1978 and relates for the most part to the United Kingdom. While its scope extends well beyond that of the local authority environmental services covered in this chapter, it does in many cases, fit the latter into the broader context of environmental protection in general.

## 14.3 Refuse Collection (England and Wales)

**14.3.1**
The door to door collection of refuse from houses and business premises in England and Wales is the responsibility of the district councils (metropolitan and non-metropolitan) and the London Boroughs.

**14.3.2**
Statistics of annual outturn total expenditure (revenue and capital) are published in [QRL.41] and [QRL.92] by DOE and the Welsh Office respectively. Advance estimates of revenue expenditure by individual authority are provided by CIPFA/SIS in [QRL.20]. The data source forms for outturn statistics are the DOE/Welsh Office forms (RO 14) for revenue figures and (CO 14) for capital, listed in Appendix 4. The CIPFA/SIS revenue estimates are collected by means of the Return of Expenditure and Rates (RER), see Appendix 7. It should be noted that in [QRL.20] statistics for waste collection and disposal are combined. Since at present, the two functions are carried out by different types of authority and since data is published at the individual council level, it is quite easy to identify expenditure on the separate services.

**14.3.3**
In the quarterly manpower statistics published in [QRL.32], [QRL.37], [QRL.42] and [QRL.18] the two functions, collection and disposal, are aggregated. It is therefore impossible, at national summary level, to distinguish one service from the other. This of course, is also true of the June quarter manpower statistics reproduced later in [QRL.41] and [QRL.92].

**14.3.4**
The data source of the manpower figures in [QRL.32], [QRL.37], [QRL.42] and [QRL.18] is the LAC SAB Quarterly Joint Manpower Watch return shown in Appendix 15 (see section 10.1).

**14.3.5**
The widest coverage of waste collection statistics is provided at individual council level by CIPFA/SIS in *Waste Collection Statistics–Actuals* [QRL.89]. The principal non-financial subjects included are: output (e.g waste collected, waste reclaimed); resources employed (e.g manpower, vehicles, bulk storage containers, bottle banks). Financial coverage includes:

a) revenue expenditure (analysed by economic category);
b) current income (from charges and sales of reclaimed waste);
c) capital expenditure and means of financing;
d) costs of collection;
e) collection charges.

**14.3.6**
The data source of [QRL.89] is a CIPFA/SIS questionnaire which it has not been possible to reproduce.

**14.3.7**
The compilers (see Appendix 5) report a number of substantial difficulties met in producing waste collection statistics. First, because of the widely varying methods of charging for the collection of bulky and commercial waste, it is impossible to provide summary statistics of these items. In the 1982/83 edition of [QRL.89], appendices containing detailed information on variations run to five pages. This type of information is omitted from later issues but can be obtained on request from CIPFA. Second, an obvious and recurring problem to many authorities, relates to the difficulty of allocating central administration charges over the various accountancy headings.

**14.3.8**
A third difficulty arises from the growing practice of privatisation i.e the use by authorities of private contractors to collect all or part of the waste generated within their areas. Whereas the published statistics in [QRL.89] include the volume of waste collected by and on behalf of an authority, data on resources (vehicles and operatives) employed by private contractors are omitted. Also, an incomplete statistical picture is painted in the case in which an authority changes over to the use of private contractors in the course of a year since the published figures of resources relate to the position at a single point in time (31st March). Clearly, the statistics consequences of privatisation, so far unresolved, are to reduce the possibility of valid inter-authority comparison which to managers and elected members at least, is perhaps the whole point of collecting them. While it is of course difficult to obtain certain information from contractors (waste collection for a given authority may be only one of many contracts) some information on contractors vehicles and manpower was requested for the financial year 1983/84 and subsequent years.

## 14.4 Refuse Disposal (England and Wales)

**14.4.1**
At the time of writing, the local authorities responsible for waste disposal as distinct from the door to door collection of waste, are the English counties (metropolitan and non-metropolitan), the GLC and the districts in Wales (but see section 2.1 for post-April 1986 position).

**14.4.2**
As in the case of waste collection, statistics totals for the waste disposal service are published as follows:

a) outturn revenue and capital income and expenditure [QRL.41] for England, [QRL.92] for Wales;

b) estimated revenue expenditure for English and Welsh authorities [QRL.20].

The qualifications identified in paragraphs 14.3.2 and 14.3.3 also refer to the statistics of waste disposal. The data source documents referred to in paragraph 14.3.2 for financial statistics also apply.

**14.4.3**

By far, the widest range of statistics of waste disposal is published for each authority in the CIPFA/SIS booklets *Waste Disposal Statistics–Actuals* [QRL.90] and *Waste Disposal Statistics–Estimates* [QRL.91]. In addition to detailed financial and manpower statistics, those publications contain data on output (e.g waste disposed of and method of treatment) and performance (e.g expenditure per tonne, income per tonne of waste reclaimed). The sources of the data are CIPFA/SIS questionnaires which it has not been possible to reproduce in this review. (Note: the content of [QRL.91] is now being narrowed in scope.)

**14.4.4**

The compilers of the statistics (see Appendix 5) have drawn attention to a number of difficulties which they have met and which substantially affect interpretation. First, there is the recurring problem of defining the boundary between administrative and operational costs. Next, there is the difficulty of apportioning operational costs over methods of treatment and final disposal in cases in which waste is treated by more than one method. Problems of data consistency and appropriateness arise from the fact that expenditure data and tonnage information originate in different parts of an authority. In addition:

a) since less than half of the tonnage of waste is actually weighed;

b) since there are discrepancies in waste volumes in situations where waste arises in one WDA area and is disposed of by another WDA;

c) since there are discrepanices between tonnages of waste from collecting authorities, as recorded in waste collection statistics, and those provided by the waste disposal services; those performance measures in the form of unit costs based on tonnages need to be treated with caution.

It is understood that some users of the statistics including the County Surveyors Society, have criticised as misleading the use of an overall cost per tonne as the prime performance measure applied to this service. Conscious of these problems and difficulties the SIS Environmental Health Working Party (see Appendix 5) has redesigned the "Estimates" questionnaire in an effort to improve inter-authority comparisons in the matter of unit costs of particular methods of treatment and to clarify the distinction between "primary processing and treatment" data on one hand and "final disposal" data on the other.

## 14.5 Cemeteries and Crematoria (England and Wales)

**14.5.1**
The provision and administration of cemeteries is the responsibility of the district councils and the London Boroughs. Crematoria are also the responsibility of these authorities but understandably not every council provides a crematorium within its boundaries. Some crematoria are provided and administered by Joint Boards or Joint Committees of authorities.

**14.5.2**
Statistics of outturn, revenue and capital income and expenditure for cemeteries and crematoria combined are published by the DOE/Welsh Office respectively in [QRL.41] and [QRL.92]. Advance estimates of revenue expenditure for individual authorities throughout England and Wales are published by CIPFA/SIS in [QRL.20]. But it is worth noting that, manpower statistics for cemeteries and crematoria are not provided by Joint Manpower Watch. The manpower data for these services is included in the broad category "miscellaneous services" in the JMW tabulations.

**14.5.3**
Data sources of the [QRL.41] and [QRL.92] outturn statistics are the revenue (RO14) and capital (CO14) forms listed in Appendix 4. The data for the CIPFA/SIS estimates in [QRL.20] is obtained from the Return of Expenditure and Rates (RER) a specimen of which is shown in Appendix 7.

**14.5.4**
An extensive coverage of the statistics of cemeteries and crematoria–at the level of the individual council–is provided by CIPFA/SIS in *Cemeteries and Crematoria–Actuals* [QRL.9]. Publication of the statistics of crematoria predates that of cemeteries by over 20 years. The first publication of cemeteries statistics relates to the financial year 1980/81 and the Joint publication with crematoria statistics dates from the 1981/82 issue of [QRL.9]. With respect to cemeteries, the subjects included are:
a) resources (staff, number of crematoria);
b) activities or activity related matters (interments, new graves excavated, grave spaces maintained, deaths in the area);
c) finance (income, expenditure, range of charges imposed).

The crematoria figures not only relate to the range just referred to but note that manpower figures are not provided; they also include estimates of surpluses or deficits on the operation of the crematoria and identify the means of financing deficits (by precept, rate fund contribution, application of former surplus).

**14.5.5**
The primary data sources of [QRL.9] are CIPFA/SIS questionnaires which it has not been possible to reproduce in this review. The compilers (see Appendix 5) report that

in the case of cemeteries returns, response is fairly good with some 10% of authorities outside London not being accounted for. Response in the case of the crematoria statistics is said to be excellent. There is the continuing and widespread difficulty of apportioning central administration charges (in the case of crematoria) and the compilers confess that in some instances, "a notional figure is inserted" in order to ensure compatability between authorities.

### 14.6 Scotland

**14.6.1**
In Scotland, the environmental services not only include those described above for England and Wales–though some of the terminology differs, but also encompass water supply and sewerage. The latter services are the responsibility of the regions and islands councils, the remainder (cleansing, environmental health and burial grounds and crematoria) are provided by the district and islands councils.

**14.6.2**
In the publication of Scottish local authority financial statistics the terms *refuse collection and disposal* and *cleansing* are used by different agencies to apply to the same set of service i.e (a) refuse collection; (b) refuse disposal; (c) street cleaning. The former term is employed by the Scottish Office in the *Scottish Local Government Financial Statistics* [QRL.77] the latter by the Scottish branch of CIPFA in *Rating Review–Part 1* [QRL.64] and *Rating Review–Part 2* [QRL.65], although in the case of [QRL.65] data for each of the three elements of the set of services is published. In [QRL.77] statistics of (outturn) summary expenditure and income on revenue and capital accounts are published; in [QRL.65] outturn data in somewhat greater detail and much earlier in time is provided. Forward estimates of net revenue expenditure, capital expenditure and loan charges are published in [QRL.64] which also contains statistics of capital debt outstanding at the end of the previous financial year.

**14.6.3**
Quarterly manpower statistics relating to this set of services are published by the Joint Manpower Watch in Scotland in *Joint Manpower Watch Survey Results* [QRL.33] and in the Department of Employment's [QRL.18].

**14.6.4**
The source of the Scottish Office financial statistics in [QRL.77] is the form LFR 6 listed in Appendix 4 (this form relates both to revenue and capital). Data for the CIPFA estimates is collected *via* a CIPFA return which has not been reproduced here. The data source forms for [QRL.65] are clearly set out in the publication itself. JMW data for [QRL.33] and [QRL.18] is of course gathered by means of the Scottish Local Authorities Negotiating Bodies forms, specimens of which are shown in Appendix 15.

**14.6.5**
The term *environmental health* is used by the Scottish Branch of CIPFA and in the Scottish JMW statistics publications. However in the Scottish Office statistics publication [QRL.77] the following service components are separately identified and in sum equate approximately to the CIPFA "environmental health" set:
a) clean air;
b) public conveniences, baths and wash houses;
c) other public health.
The JMW definition of "environmental health" which is set out in the Notes on completing the JMW form shown in Appendix 15 relates approximately to the same set of services but may include (see paragraph below) manpower employed by the burial grounds and crematoria service.

**14.6.6**
Statistics of Scotland (summary) outturn expenditure and income, both revenue and capital, for the component services identified in paragraph 14.6.5–are published by the Scottish Office in [QRL.77]. Somewhat earlier in time a much wider range of outturn financial statistics is published by the Scottish branch of CIPFA in [QRL.65]. Unlike [QRL.77] these figures relate not to the components of environmental health but to the total service itself. In [QRL.64] advance estimates of revenue and capital expenditure plus capital debt and loan charges for environmental health are provided by the Scottish branch of CIPFA. With respect to manpower statistics, a quarterly series is published by JMW in Scotland in [QRL.33] and reproduced later in [QRL.77].

**14.6.7**
Form LFR 6, which is listed in Appendix 4, is the primary source of the revenue and capital statistics in [QRL.77]. The CIPFA (Scottish Branch) sources for [QRL.64] are CIPFA forms not reproduced here. With respect to the financial statistics in [QRL.65] the source forms are listed in detail in the publication itself. Scottish JMW data for [QRL.33] is collected by means of the Scottish Local Authorities Negotiating Bodies forms (see Appendix 15).

**14.6.8**
In statistics publications the terms "sewers and drains" and *sewerage* are used respectively by (i) the Scottish Office and the Scottish branch of CIPFA and (ii) Joint Manpower Watch to denote the same service. Statistics of various financial outturns are published by the Scottish Office in [QRL.77]. However, rather earlier a much wider range of outturn financial statistics is provided by the Scottish branch of CIPFA in [QRL.65]. Annual forward estimates of net revenue and capital expenditure, loan charges and capital debt are published by the same body in [QRL.64]. Manpower statistics relating to this service are published quarterly by JMW in [QRL.33]. Those figures are later reproduced in [QRL.77].

**14.6.9**

There are several accounting features of the *water supply* service in Scotland which may make it difficult for the layman to comprehend its financial statistics. While the service is a Trading Service it also has a special rate (domestic water rate) levied in its favour (see [QRL.65]). In addition its operating deficit is met by Rate Fund contribution which oddly enough is referred to by the Scottish Office as the "public water rate", although it is not a "rate" in the sense in which that term is normally used in local government finance. Furthermore, in Rate Fund and Trading Service revenue accounts the term "running expenses" in relation to water supply is used to refer to two quite different entities (see below).

**14.6.10**

The Scottish Office publication [QRL.77] provides a range of water supply outturn financial statistics (gross revenue and capital income and expenditure) within a Trading Services context. Also, within the Rate Fund context a single item "running expenses" is shown against "Public Water". This figure represents the total Rate Fund contribution to the water supply operating deficit but note that it does not precisely equate with the revenue contributions which appear against water supply in the Trading Services revenue and capital accounts. Note also that the term "running expenses" in the water supply (Trading Services) revenue account is employed in its normal accountancy sense.

**14.6.11**

The CIPFA (Scottish Branch) publication [QRL.65] publishes outturn expenditure statistics for water supply which refer to Rate Fund expenditure only i.e the operating deficit which (see paragraph 14.6.9) is designated the "public water rate" or simple "public water". Forward estimates of this figure (for each regional/islands authority) plus estimates of capital expenditure, loan charges and capital debt are contained in [QRL.64].

**14.6.12**

The data sources for the [QRL.77] statistics referred in 14.6.4 are the LFR 21 for Trading Services figure (see list in Appendix 4) and the form LFR 6 for the Rate Fund figures (see also Appendix 4). The [QRL.64] data is collected *via* a CIPFA form for regions, not reproduced here. Sources of data for [QRL.65] are listed in detail in the publication itself and the JMW statistics are collected by means of the JMW forms shown in Appendix 15.

**14.6.13**

Domestic water rate poundages for regional/islands authorities are published in [QRL.77] and [QRL.64]. The source of the latter is a simple questionnaire completed by rating authorities and forwarded to the Scottish Branch of CIPFA. The data sources of [QRL.77] are the individual authorities Rate Demand Notes sent to the Scottish Office in the early part of the year to which they refer.

**14.6.14**
Manpower statistics of the water supply service are provided by the Joint Manpower Watch in [QRL.33], data for which is gathered by means of the Scottish Local Authority Negotiating Bodies forms displayed, with explanatory notes in Appendix 15. The figures are reproduced annually in [QRL.77].

**14.6.15**
Statistics of annual consumption of water and available water resources for each region and islands council area are published by the Scottish Office in the *Scottish Abstract of Statistics* [QRL.76]. The data source for water consumption is the Scottish Office form WR 14/R (not reproduced in this review) which is completed by the regional and island authorities. Water resources data (i.e the number and yield of sources such as reservoirs, lochs, rivers, bore holes, springs) is collected *via* a Scottish Office computer printout–Water Resources Surveys–which relates to individual supply points identified by name and National Grid Reference (NGR), the printouts having been originally completed for a base year for which surveys were carried out and updated annually *via* the regions and islands authorities. The printouts are returned annually by the Scottish Office, to the water authorities which either confirm the data or make manual alterations before returning them for inclusion in [QRL.76].

**14.6.16**
Statistical coverage of the *burial grounds and cremation* service in Scotland is almost wholly financial. Annual outturn statistics of (summary) income and expenditure are published by the Scottish Office in [QRL.77]. In addition outturn data of net revenue and loan charges are now provided by CIPFA (Scottish branch) in [QRL.65] while, the same body also publishes in [QRL.64] advance estimates of capital and revenue expenditure plus loan charges and capital debt. In the case of crematoria only the CIPFA/SIS publication [QRL.9] relates to Scotland (see paragraph 14.5.4).

**14.6.17**
The primary data sources of the statistics in [QRL.77] is the Scottish Office form LFR 9 (see list in Appendix 4); the publication [QRL.65] contains full details of its own original sources. The Scottish Branch of CIPFA collects data for its estimates in [QRL.64] by means of its own form (specimen not reproduced). For the crematoria statistics published by CIPFA/SIS in [QRL.64], the source questionnaire and completion notes are not reproduced here (see also paragraphs 14.5.4 and 14.5.5).

**14.6.18**
It should be noted that, unlike the position of other services considered in this section, the JMW does not publish manpower statistics for burial grounds and crematoria. It is understood that figures for manpower employed in this service may be included with "environmental health" or "leisure and recreation" depending on the place which the service occupies within the organisation of a district council.

## 14.7 Northern Ireland

**14.7.1**
The environmental services which fall within the scope of this chapter and which are provided by the district councils in Northern Ireland are as follows: health inspection; clean air; public conveniences; burial grounds and crematoria; refuse collection and disposal; street cleansing; abattoirs. For each of these services, *estimates* are provided for:
a) expenditure;
b) income from specific government grants and from miscellaneous sources;
c) net cost after allowing for item b.
The estimates are published for each district council in *District Council Rate Statistics* [QRL.14] the data source being the General Estimate of Rates (see paragraph 4.5.3). Statistics of *actual* expenditure on both revenue and capital account (Northern Ireland Summary only) are provided by the Department of the Environment for Northern Ireland in *District Council Summary of Statement of Accounts* [QRL.15]. The source of the summary figures is the Abstract of Accounts completed by district councils–see paragraph 4.5.2.

**14.7.2**
*Prior to 1973* the relevant environmental services provided by the (then) local authorities (i.e county councils, county borough councils, borough councils, urban district councils, rural district councils) were identified in statistics publications under the following headings:
a) health committees;
b) Drainage Acts;
c) water supplies;
d) sewerage services;
e) Urban Drainage Act;
f) burial grounds;
g) refuse collection and disposal;
h) street cleaning;
i) public conveniences;
j) abattoirs;
k) clean air;
l) other sanitary services.
For each of these the (then) Ministry of Development published advance estimates of receipts and payments in *Local Authority Rate Statistics* [QRL.38]. Statistics of actual receipts and payments on both revenue and capital account were provided by the same source in *Local Authority Financial Returns* [QRL.36]. In the case of both publications, statistics were published not only in summary but for individual authorities.

CHAPTER 15

# MISCELLANEOUS

## 15.1 Museums, Art Galleries and Library Services

### 15.1.1

In *England and Wales* local authority museums and art galleries may be provided by a county, a district, a London Borough, the City of London or the GLC (see the *Municipal Year Book* [QRL.44] for fuller details and see section 2.1 above for position after April 1986). The authorities responsible for the public libraries service in England are the metropolitan districts, the non-metropolitan counties, the London Boroughs and the City of London. In Wales, the counties are responsible but at present, four of the district councils have been designated library authorities. It is understood that the long term responsibility for libraries in Wales is now under review.

### 15.1.2

Published statistics relating to this group of services are in some publications combined but in others appear as two distinct services: a) museums and art galleries; b) libraries. The bodies responsible for publication are the DOE/Welsh Office, CIPFA/SIS, SCT, Municipal Publications Ltd. and joint Manpower Watch; subjects covered are finance, manpower and in the case of libraries only a wide range of input, output, work load and miscellaneous items.

### 15.1.3

In *Local Government Financial Statistics–England and Wales* [QRL.11] and *Welsh Local Government Financial Statistics* [QRL.92] the DOE and Welsh Office respectively publish financial outturn for the two component services separately and manpower data for the two combined. CIPFA/SIS in *Finance and General Statistics* [QRL.20] provide expenditure estimates for individual authorities for the combined services while outturn financial statistics are produced by SCT in *Financial and General Statistics of County Councils* [QRL.21] with separate figures for the two services; book stock and staff statistics are also included for libraries. Combined manpower (summary) statistics are published for England in the press release *Joint Manpower Watch Return–England* [QRL.32] and for Wales in *Local Authority Manpower in Wales* [QRL.37]. Later these statistics are published in more readily accessible form in *Local Government Manpower* [QRL.42] and the *Employment Gazette* [QRL.18].

**15.1.4**

As to coverage of the library service itself: this is most extensively provided by CIPFA/SIS in *Public Libraries Statistics–Actuals* [QRL.56] and *Public Libraries Statistics–Estimates* [QRL.57]. The first of these relates to library authorities throughout Great Britain and indeed the United Kingdom (but see below for treatment of Northern Ireland). The second [QRL.57] however, relates to authorities in England and Wales only. [QRL.56] contains not only a wide range of financial statistics but also includes figures on manpower, service or delivery points, agency services provided, books and other materials available to library users. In [QRL.57] which is much narrower in scope, the main subjects covered are: estimated expenditure (in some detail), staffing, establishments, service points. It is expected that a substantial revision of these statistics may be undertaken shortly as an outcome of the regular collaboration between the Advisory Committee for Public Library Statistics (which meets under DES auspices) and the CIPFA/SIS Leisure Volume Working Party (see Appendix 5) which has ultimate responsibility for the contents of [QRL.56] and [QRL.57].

**15.1.5**

Oddly enough, neither of the CIPFA/SIS publications contains statistics of one of the major output measures in the public libraries service (i.e issues from lending libraries) although CIPFA collects the necessary data. However the statistics are held in the CIPFA computer database and are made available automatically to database subscribers; but they can be purchased by non-subscribers. Also, issues statistics (i.e books and records separately) are published for each library authority in the United Kingdom in the *Municipal Year Book* [QRL.44].

**15.1.6**

The *data sources* of the publications identified above are as follows: [QRL.11] and [QRL.92] obtain their financial data from the DOE/Welsh Office forms RO2 and CO2 listed in Appendix 4; manpower figures are obtained *via* the JMW (see below). The [QRL.24] expenditure estimates are collected from individual authorities by the RER forms (see Appendix 7) while SCT obtain from counties, a note of RO2 entries already forwarded to DOE/Welsh Office. Manpower data in [QRL.37], [QRL.32], [QRL.73] and [QRL.18] is provided by the Joint Manpower Watch (see Appendix 15). The material for the CIPFA/SIS publications [QRL.56] and [QRL.57] is collected by questionnaire (not reproduced here) while Municipal Publications Ltd., gather data for [QRL.44] also by questionnaire not reproduced in this review.

**15.1.7**

In *Scotland*, museums and galleries are owned by every type of authority; public libraries are the responsibility of districts in six of the regions, of three regional councils and of the island councils in the rest of the country. Statistics are published by the Scottish Office, the Scottish Branch of CIPFA and Joint Manpower Watch. As in the case of England and Wales, the statistics sometimes relate to the combined services ((a) museums and galleries; (b) libraries) and occasionally to the services separately. Additionally, library statistics only are provided by CIPFA/SIS and Municipal Publications Ltd.

**15.1.8**
The Scottish Office publication *Scottish Local Government Financial Statistics* [QRL.77] provides outturn financial statistics for the combined services; manpower figures also relate to these. Earlier, the financial outturn data for districts only, is now published by the Scottish Branch of CIPFA in *Rating Review–Part 2* [QRL.65] with matching forward estimates in *Rating Review–Part 1* [QRL.64]. Planned expenditure figures for a number of years ahead, for the two services separately, are published in *Public Expenditure to...(Year)* [QRL.55]. A manpower summary for the combined services is published quarterly in *Joint Manpower Watch Survey Results* [QRL.33] and repeated in [QRL.18]. Extensive statistical coverage of the library service alone is provided by CIPFA/SIS in [QRL.56]–see paragraph 15.1.4. The subject of lending library issue of books and records is covered in [QRL.44] and can be made available in some detail from the CIPFA database–see paragraph 15.1.5.

**15.1.9**
The *data sources* of these publications are as follows. Financial outturn data for [QRL.77] is collected by the Scottish Office form LFR2 (see Appendix 4 list); manpower figures are obtained from JMW (see below). The sources of [QRL.65] are identified in that publication itself, while for [QRL.64] figures are gathered by the Scottish Branch of CIPFA by means of a questionnaire not reproduced here. The [QRL.55] figures, being planned figures, are supplied by the Secretary of State for Scotland. Manpower data for [QRL.33] and [QRL.18] is obtained by the JMW questionnaire (see Appendix 15). Data for [QRL.56] is obtained by a CIPFA/SIS questionnaire not reproduced here; Municipal Publications Ltd., gather data for [QRL.44] by means of their own questionnaire which is not shown in this review.

## 15.2 Direct Labour Organisations

**15.2.1**
Local authorities in Great Britian usually employ work forces to carry out at least part of the new construction and maintenance of their buildings and roads and, in the case of Scotland only, water supply facilities and sewerage. Those work forces are known as direct labour organisations (DLO's). National statistics relating to the activities of DLO's were published (for the first time ever), for the financial year 1981/82. The content and form of these statistics has been determined by the nature of recent legislation which relates to their formal status and accounting arrangements. In terms of Part III of the Local Government Planning and Land Act, 1980 DLO's of authorities in England and Wales were to be treated as quasi-trading organisations and authorities were required to maintain a balance sheet and revenue account, to prepare a statement of the rate of return and state of affairs of the

DLO's. Similar legislation applied to Scotland and took effect one year later, in April 1982. The legislation applied not only to local authorities as defined in this review but covered joint committees and development bodies including New Town Development Corporations. (Note: Authorities employing fewer than 30 in a DLO were exempt from the need to keep records as above.)

**15.2.2**

CIPFA/SIS now publish statistics relating to the operation of DLO's in *Direct Labour Organisations Statistics–Actuals* [QRL.13]. The first issue (1981/82) referred to authorities in England and Wales; from 1982/83 onwards Scottish authorities were included. The non financial subjects in [QRL.13] are: manpower, proportion of an authority's construction and maintenance work carried out by DLO's, method of charging for DLO work (e.g lump sum, target hours, day work). On the financial side the statistics relate to: revenue account; DLO appropriation account (i.e balance at beginning and end of year plus contribution of DLO to or from rate fund and reserve fund); calculation of rate of return on capital employed. It is worth noting that by statute, "rate of return" calculations require to be made for each of 4 separate categories of work i.e general highways work, major works of new construction, minor works of new construction, general maintenance works. In Scotland there is a fifth category (water and sewerage works) which for statistical purpose is to be combined with "general highways work" in [QRL.13]. This publication has a useful introduction, commentary and copious notes which collectively contain definitions of terms, references to the appropriate legislation and elaboration of concepts used in the statistics presentation.

**15.2.3**

The booklet [QRL.13] is compiled by staff at CIPFA headquarters without assistance from a "field" compiler. A specimen copy of the source questionnaire which includes notes for guidance on completion together with a covering letter are shown in Appendix 19. On receipt from authorities the data in questionnaires is fed to a computer, verified, manipulated and output to camera-ready sheets to be printed. CIPFA also prepare commentary notes and tables. However it should be noted that as in all CIPFA/SIS publications, a working party of officers from a number of authorities, in this case the Direct Works Accounting Working Party, advise as to content and presentation of the statistics.

**15.2.3**

Understandably compilers report certain difficulties encountered in producing this new set of statistics. They include: the tendency of some authorities to combine figures for two or more of the work categories referred to in paragraph 15.2.2; the inability of some authorities to provide data on "indirect employment"; the difficulty of providing financial data on current as distinct from historical cost. Many of the more detailed problems and "special cases" are listed in a valuable set of notes in the first issue of [QRL.13].

**15.2.4**
In Scotland, a national summary of manpower engaged in "direct works" is published quarterly in [QRL.33], the data source being the JMW questionnaire (see Appendix 15).

## 15.3 Smallholdings

**15.3.1**
In England and Wales, county councils and the Greater London Council own and administer agricultural smallholdings which they rent to tenant farmers wishing to start up in agriculture. The holdings were first acquired and administered following the first world war with a view to rehabilitating ex-servicemen in agriculture. The management of the service is now undertaken in accordance with the provisions of the Agriculture Act 1970. In addition to letting the holdings as they become vacant, the councils review rents, design and build new properties on the holdings, maintain existing properties and provide agricultural advice to tenants. These activities are carried out by a council official usually the County Land Agent, to whom reference is made in the data source document referred to in paragraph 15.3.3 (the actual title of the responsible officer varies between authorities). Note section 2.1 for position in London from April 1986.

**15.3.2**
The most comprehensive statistical coverage of smallholdings is provided annually by the Society of County Treasurers (SCT) in a booklet *Smallholding Statistics* [QRL.79]. Beginning with the 1982/83 issue, the format of the booklet was revised by the SCT in conjunction with the County Land Agents and Valuer's Association. The statistical coverage in [QRL.79] emphasises the financial aspects of administering the smallholdings and is largely designed to bring out the extent to which the service generates: (i) an *operational* surplus or deficit, and (ii) a *net* surplus or deficit after taking account of income from MAFF grants, recovery of fees and charges etc. and expenditure on administration plus items such as legal and professional charges. The booklet also contains a number of other financial statistics (e.g loan debt, analysis of administration costs, debt charges, average rent per hectare). On the non-financial side, the statistics cover: (a) staff employed by the council in managing the smallholdings; (b) details of the holdings themselves (e.g distribution by size, number of tenants, number of fully or partially equipped holdings).

**15.3.3**
The SCT *collects data* for [QRL.79] using a compiler system somewhat similar to that practised by CIPFA/SIS. In this case the compiler is the County Treasurer of an English shire county. The data is collected from authorities *via* the questionnaire "Society of County Treasurers–Smallholding Statistics" which it has not been possible to reproduce. This statistical return overlaps to some extent with the Return of the Annual Report on Smallholdings to the Minister of Agriculture, Fisheries and

Food (MAFF). The latter arises from MAFF responsibilities under the Agricultural Act, 1970 mentioned in paragraph 15.3.1

**15.3.4**
National financial (outturn) summaries of the smallholdings service are published annually by DOE/Welsh Office in [QRL.11] and [QRL.92] for England and Wales respectively. In addition, in [QRL.21] the SCT provides total (outturn) expenditure and smallholding area figures for each authority. The *data sources* for [QRL.11] and [QRL.92] are the revenue (RO12) and capital (CO12) forms listed in Appendix 4. Financial data for [QRL.21] is obtained by SCT by questionnaire in which authorities provide RO12 figures. Area data is obtained *via* the SCT questionnaire referred to in paragraph 15.3.3.

## 15.4 Surveys

**15.4.1**
Government departments are required to notify the Survey Control Unit of the Central Statistical Office of any surveys of local authorities to be carried out by them or their agencies. A list of the surveys notified is published quarterly in *Statistical News*. A computer listing of surveys notified over the past twelve years can be obtained from the Survey Control Unit, Central Statistical Office, Great George Street, London SW1P 3AQ (telephone 01-233-5018).

## 15.5 Northern Ireland

**15.5.1**
Statistics of estimated expenditure by district councils on *museums and galleries* are published annually in *District Council Rate Statistics* [QRL.14]; the data source is the General Estimate of Rates completed by each district council (see paragraph 4.5.3).

**15.5.2**
Although *public libraries* are not administered by district councils in Northern Ireland the CIPFA/SIS publication [QRL.56] includes a full range of statistics relating to the operation of public libraries by the five Education and Library Boards in the province; details of coverage and sources are set out in paragraphs 15.1.4 and 15.1.6 above. Statistics of issues from lending libraries are contained in [QRL.44] and also held in the CIPFA computer database (see paragraph 15.1.5 and 15.1.6).

CHAPTER 16

# COMPLAINTS

## 16.1 Great Britain

**16.1.1**
Under the Local Government Act 1974 and the Local Government (Scotland) Act 1975, Commissions for Local Administration (Local Ombudsmen) were set up for England and Wales and Scotland respectively. The Commissions investigate complaints made by members of the public of injustice resulting from maladministration in local government. The Commissions may consider complaints against committees, elected members and officers not only of all those bodies considered in this review but also joint boards of local authorities, including National Park boards. A complaint needs to be made in writing to an elected member of the authority complained against and can only be referred to the appropriate Commission after first having been made to the authority in question, the latter having had a reasonable time in which to reply to the complaint. Commissions in England and Wales have no power to investigate complaints about any action taken prior to 1st April 1974; in Scotland the relevant date is 16th May, 1975. Fuller details about powers and activities of the Commissions are contained in handy reference form in the *Municipal Year Book* [QRL.44].

**16.1.2**
The three Commissions publish annual reports, each containing statistics relating to their work. The report of the English Commission is entitled: *Your Local Ombudsman* [QRL.93]. The Welsh Commission's report is called: *Annual Report of the Commissioner for Local Administration in Wales for Year Ended 31st March* [QRL.3]. The Scottish report is entitled: *Report of the Commissioner for Local Administration in Scotland for Year Ended 31st March* [QRL.66]. The sources of the statistics in the reports are of course the Commissions themselves which simply record and classify the complaints received.

**16.1.3**
The statistical tables in the reports, with accompanying texts, analyse the complaints by type of authority complained against, subject of complaint (e.g housing, planning, education) and outcome of complaint i.e not accepted, withdrawn, settled locally, investigation report issued. Complaints not accepted for investigation (complaints where investigation was "terminated" in the case of England) are further analysed by reason for non-acceptance.

## 16.2 Northern Ireland

### 16.2.1

Northern Ireland has a Commissioner for Complaints appointed under the Commissioner for Complaints Act (Northern Ireland) 1969 as modified by the Northern Ireland Act 1974. The functions of the Commissioner are approximately similar to those of the Great Britain Commissions for Administration. However there is an important difference in the scope of his jurisdiction. He considers complaints not only against the district councils but against government departments, public bodies such as for example the Education and Library Boards, the Health and Personal Social Service Boards and the Northern Ireland Housing Executive. Further details of the work of the Commissioner are to be found in [QRL.44].

### 16.2.2

The Commissioner publishes statistics of complaints against the district councils analysed on almost the same basis as that prevailing in the rest of the United Kingdom. The only difference is that, from the analysis of complaints which were rejected or discontinued after preliminary investigation, it is not possible to identify local authority complaints since the data is aggregated to include "local bodies" (i.e local authorities), government departments and other public bodies. The statistics are published in a House of Commons paper: *Annual Report of the Northern Ireland Commissioner for Complaints* [QRL.4]. The source of the statistics is the Commissioner whose staff simply record and process the complaints received.

## CHAPTER 17

# CONCLUSION

### 17.1 Recent Developments

#### 17.1.1

The late 1960's and early 1970's saw a profound change in attitude towards the use of statistics within local government. This change in the first instance, related to non-financial statistics and did not initially have an impact on the publication of statistics nationally. Hitherto, as *Local Government Statistics* [B.1] shows, a wide range of mainly financial statistics was available at national level and there was the tacit assumption within authorities that these could be employed, and doubtless were employed in many cases, for the purpose of inter-authority comparison. However, beginning roughly with the London Government Act of 1963 which laid down that the GLC was to establish a Central Intelligence Unit, and followed by a series of new statistical demands in the planning field (e.g the conurbation land use transportation studies and later the new structure plans required by the Town and Country Planning Act 1971), emphasis began to shift towards the collection of statistics of demand or more accurately the assessment of need for various community services. In 1967 the *Maud Committee on the Management of Local Government Volume I* [B.15] had drawn attention to the unsatisfactory nature of the piece-meal means by which local government itself collected, analysed and disseminated statistical information about the way it carried out its functions. A much closer liaison between central government statisticians and those working on statistics within local authorities was signalled by the establishment of the National and Local Government Statistical Liaison Committee (NLGSLC) in 1970, see [B.10]. The reorganisation of local government, both as it approached and as it was effected, led to an intensification of interest in statistics for both management and planning within the authorities themselves. Research and Intelligence Units, approximately in the style of the GLC's Intelligence Unit, were set up by many of the larger authorities. This development in turn spawned: (a) the publication of many *local* statistical bulletins, abstracts and digests (a subject which is strictly outside the terms of reference of this review); (b) the setting up of the Local Authority Research and Intelligence Association (LARIA) which not only sponsors conferences and the publication of research papers but has also led to the exchange of views on, among other things, the use of statistics for management, planning and policy formation purposes.

**17.1.2**

The demands of central government from the middle 1970's but more especially from the formation of the Conservative administration in 1979, have begun to transform the local authority scene still further, both with respect to financial and non-financial statistics. The publication for the first time of appropriate manpower statistics as a proxy for financial data, stemming from central government's desire to monitor local authority expenditure, has already been mentioned in chapter 10 and of course preceded the Conservative government. However, two major products of that government's policies have powerfully affected both the publication and content of local government statistics. First in England and Wales, the new method of distributing the block grant part of RSG (described fully in Chapter 6) has now led to the publication of a vast mass of statistics on the needs of individual areas for services. (There has been an approximately parallel development in Scotland–see Chapter 6 and Appendix 12–but matching Scottish data is not published.) Second, there was government insistence, embodied in the Local Government Planning and Land Act 1980, that authorities must publish much more information about their performance or discharge of their function, a development which has resulted in the growing production of comparative statistics, and the identification of statistical comparators (see below).

**17.1.3**

The further impact of the developments outlined above has been considerable in terms of changes in institutional arrangements and statistical publication. Already referred to is the establishment of CIPFA/SIS, see 3.2.2, the origin, functions and activities of which have been fully described in *The Statistical Information Service* [B.11] and the large and varied output of which permeates this review. Next, in line with the Local Government Planning and Land Act 1980, the DOE/Welsh Office have published *Local Authority Annual Reports* [B.26] and the Scottish Office has produced *Publication of Financial and Other Information by Local Authorities* [B.31] both of which are codes of practice which prescribe the content of Annual Reports by authorities and recommend comparative statistics which authorities ought to publish as performance indicators e.g they were asked to compare their own outturn figures with those for the average for the same class of authority and with those of other authorities having similar characteristics. (Note: the Scottish publication was less specific, preferring to emphasise comparison between planned and outturn expenditures.) While the Annual Reports themselves as purely statistical publications, must be regarded as lying outside the scope of this review, the national statistics publication effects need to be mentioned. With the aim of providing suitable *comparative* statistics, both financial and non-financial, to meet the demands of the England and Wales codes of practice, CIPFA/SIS first published in 1982, the booklet *Local Government–Comparative Statistics* [QRL.46]. The background and rationale of this publication were set out in the *Statistical News* article "Developments in Local Authority Comparative Statistics" [B.5]. The booklet [QRL.46] contains a very useful introduction to the indicators, to the ways in which they may be used and to relevant background studies.

**17.1.4**
There were two further developments in national publication which flowed from the need to publish comparative statistics *via* Annual Reports. First, the Scottish Branch of CIPFA has now produced *Rating Review–Part 2* [QRL.65] to cater for actual expenditure details as a reliable source of inter-authority comparative statistics. The subject matter of this publication is under review and it is expected that the contents will be expanded beyond the purely financial. The second development worthy of mention is the inclusion from 1983 by the SCT in [QRL.21] of summary comparisons of the *actual* income and expenditure of county councils with *estimates* previously published by CIPFA/SIS in [QRL.20]. This type of comparison (plan versus achievement) is one which is commended in both codes of practice referred to in paragraph 17.1.3.

**17.1.5**
In a survey of recent developments in local authority statistics, most of which have been covered in this review, Roger Sellwood and Chris Griffin drew attention to two technical innovations which are relevant here. These are: the CIPFA/SIS clustering service and the establishment of an integrated database held in the CIPFA computer [B.12].

**17.1.6**
First, let us consider the *clustering service*. The desirability and now the necessity of inter-authority statistical comparison have been referred to above. That managers in local government have seriously concerned themselves with this kind of comparison as a management tool is borne out by the fact that an entire INLOGOV seminar, sponsored by SOLACE, was given over to the subject–see *Inter-Authority Comparisons* [B.24]. However, the problem to be solved in practice in making useful comparisons reduces to that of comparing like with like. The obvious technical solution is to derive clusters of "like" or "comparator" authorities. CIPFA/SIS having now done this, has set up a clustering service which individual authorities may use. Details are obtainable from CIPFA, but [B.12] contains a useful description couched in non-technical language.

**17.1.7**
Work on compiling the CIPFA/SIS database began in 1983. Background information and full details of the operation and perspectives of the base were set out in an article by Chris Griffin in *Statistical News*: see "CIPFA Database Services" [B.6]. Further details of progress, file content, nature of the service and changes are set out in a CIPFA leaflet "CIPFA–Database News". In addition to subjects not in this review, the database contains the most recent data published in almost all of the CIPFA/SIS statistics bulletins reviewed in this volume. It is understood that back data for all subjects is now being added to the database.

## 17.2 Shortcomings

### 17.2.1

Perhaps the major shortcoming of local authority published statistics, especially in the financial area, is a lack of explanation of terms and meaningful commentary. All too often publications, throughout the United Kingdom presume, a level of knowledge and expertise which the general user is unlikely to possess. For instance the following items are introduced without definition or expansion in *some* publications: incidence of risk; 80% and 100% grant earning activities; full time equivalents; section 94 consents. One noteable exception to this criticism and one which might serve as a model for imitation is the Welsh Office [QRL.92] which contains not only useful comments on the statistics themselves but also a very full glossary of terms and a good deal of background information. To be fair and accurate; the situation is improving but has a long way to go.

### 17.2.2

Linked to the "lack of explanation" criticism is a "lack of cohesion" weakness. Once again in the financial area there is evidence that in many instances the bodies publishing local government statistics do not appear to agree definition and term, with the result that the figures published are at worst misleading and at best confusing. This is disappointing as the institutional bases of cooperation *via* SIS, IDLG and SSLG are now firmly in place. The examples of lack of consistency we have in mind are:

a) a wide discrepancy between local authority "total borrowing" as estimated by central government and "total amount of debt owed" as estimated by CIPFA/SIS;

b) inconsistency as between the various parts of Great Britain in the method of estimating average domestic rate bill;

c) as discussed in paragraph 4.4.12, a wide disparity without explanation between the calculation of the local authority share of GDP made respectively by CSO and CIPFA/SIS;

d) as shown in section 14.1 a wide difference in definition between the CIPFA/SIS concept of "environmental health" and that of "environmental services" used by government departments.

Worse still from the viewpoint of the user of the statistics publications, these differences are not commented upon in the publication concerned. However we are encouraged to learn that within the ambit of IDLG, the whole subject of consistency of definition among others, is now being pursued. Also we understand that DOE have a Working Group on Local Government Financial Statistics which is looking at the subject specifically in the financial area.

### 17.2.3

With the increasing involvement of local authorities, in industrial and commercial promotion as a means of generating employment in their areas, the absence of published statistics on the subject (e.g number of jobs created; cost per job) might well seem puzzling to the layman. The fact is that technically, the subject bristles with difficulties. Two at least stand out. First, to make sense, the employment recorded must be *new* employment (all too often jobs in new premises are simply transfers within a given area); monitoring can be complicated. Second, the identification and allocation of costs of job creation to a local authority can be difficult as a good deal of the time and effort invested is administrative, partial, difficult to trace (e.g work by solicitors, agents, clerks). There is the added difficulty that in any given area, different levels of authority (counties and districts) may be involved, thus incurring costs in the same piece of industrial or commericial promotion. However, "difficult" does not mean "impossible". Once again within the local authority statistics area, the institutional means exist of solving the problems referred to. CIPFA/SIS, the IDLG, SSLG and LARIA all have the potential to grapple with, what seems to us, a matter of some importance.

**17.2.4**
The Central Statistical Office in its publication *Financial Statistics–Explanatory Handbook* [B.19] makes the point that in the area of local authority financial statistics, the best sources of information for its purposes are the statutory annual returns for fiscal years (i.e [QRL.41], [QRL.92] and [QRL.77]) but that one of the drawbacks is that they are returned well in arrears. We echo this criticism and appeal to the DOE/Welsh Office and Scottish Office to speed up publication. We are encouraged by the fact that time lags now seem to be shortening.

### 17.3 Suggestions

**17.3.1**
Given the wide span of subject matter and the mass of publications and data sources identified in this and companion volumes listed in Appendix 1, ready access to information on the part of the non-specialist is unthinkable. Indeed inaccessibility must often defeat the purpose of publication by discouraging the use of the statistics produced. This being so, it seems obvious that the production of, say, an annual abstract of United Kingdom local government statistics would greatly increase the utility of what is now provided. If it is argued that however desirable, the resources simply do not exist at present to produce such a publication, it has to be replied that by far the major part of the expense involved in provision is already incurred *viz* the collection, analysis and tabulation of data. The only new major resource required is that involved in the initial planning and design of the proposed abstract, a once-for-all resource which could surely be funded by the local authority associations or some such body. The on-going cost of collection in later years could possibly be met from a redeployment of some of the resources already devoted to producing current publications. At all events, the institutional means of planning and master-minding this suggestion already exist. We will be pleased to learn that is is being seriously considered.

**17.3.2**

A second suggestion, perhaps, plea is a better word, arises from the uneveness or disparity of local authority statistics coverage between the various parts of the United Kingdom. Chapters 12 to 15 of the review especially, contain many examples of service statistics sparsely covered in Scotland and Northern Ireland by comparison with England and Wales. To be specific, we would appeal to CIPFA/SIS to consider extending its "service" financial statistics publications to Scotland and where appropriate to Northern Ireland, precedents having already been set in the cases of *Capital Expenditure and Debt Financing* [QRL.8] and *Direct Labour Organisations* [QRL.13]. Also, in the case of Northern Ireland only, a much fuller publication of rate fund outturn financial statistics would be more helpful to users than the summary figures published in the Department of the Environment (Northern Ireland) *District Councils Summary of Statements of Accounts* [QRL.15]. Tabulations along the lines of *District Council Rate Statistics* [QRL.14] would at least permit quite basic comparison of estimate against outturn. That the necessary primary data sources already exist for this purpose is shown by the production of the Abstract of Accounts (see paragraph 4.5.2).

**17.3.3**

One final suggestion, the value of much of the hard won financial data published by CIPFA/SIS would be greatly enhanced by some fairly straight forward statistical analysis. As statistics is largely about variability or the distribution of "observations" about a mean, inference and illumination would derive from the publication of frequency distributions, in diagrammatic form. Users are likely to be interested not only in the data relating to a particular local authority but to the characteristics of the set of authorities (e.g metropolitan districts, London borough) to which it belongs. Admittedly this point is already met to a degree by the CIPFA/SIS comparisons service which enables an authority to obtain a computer print relating data for an authority to various group averages. But this service does not and cannot provide that invaluable quality–ready access for the general, non-specialist user of statistics.

# QUICK REFERENCE LIST TABLE OF CONTENTS

**General Subjects** 132
Expenditure 132
Income 136
Borrowing 141
Net Borrowing 142
Manpower 142
Complaints 143
Penny Rate Product 144
Rate Poundage 144
Rateable Values 145
Rates Collection 146
Rates 146
Rate Rebates 147
Rate Support Grant 147
Rateable Subjects 147
Rateable Valuation 147
Block Grant Indicators 147
Block Grant Multipliers 148
Receipts 148
Financial Accounts and Balancing Item 149
Government Grant Factors 149
Grant Related Poundage 149
Gross Domestic Product 149
Hereditaments 149
Payments 150
Interest Rates 150
Investments 150
Loan Debt 150
Capital Account 150
Current Account 151
Debt 151
Debt Management 152
Debt Outstanding 152
Elections 152
Service Indicators 152
Superannuation Funds 153

Public Sector Borrowing Requirement 153

RSG Needs Element Indicators 153

**Services** 153
Fire 153
Public Protection 157
Police 157
Probation 161
Administration of Justice 163
Magistrates Courts 163
District Courts 164
Coroners Courts 165

Library Service 165
Libraries 166
Public Libraries and Museums 167
Museums and Art Galleries 168
Libraries, Museums and Galleries 168
Water 169
Public Water 169
Water Supply 169
Sewerage 170
Sewers and Drains 170
Cleansing 170
Environmental Health 171
Environmental Services 173
Highways and Transportation 173
Roads and Transport 177
Transport 177
Airports 180
Trading Standards, Consumer Protection 180
Consumer Protection 181
Refuse Collection 182
Waste Collection 182
Refuse Disposal 183
Waste Disposal 183
Refuse Collection and Disposal 184

Civil Defence 184
Emergency Planning, Civil Defence 184
Cemeteries 185
Crematoria 185
Burial Grounds and Cremetoria 186
Smallholdings 186
School Crossing Patrols 187
Direct Labour Organisations 188
Registration of Electors 188
Direct Works 189

# QUICK REFERENCE LIST

| Type of data | Breakdown | Area | Frequency | QRL Publication | Text reference |
|---|---|---|---|---|---|
| **General Subjects** | | | | | |
| *Expenditure* | | | | | |
| Expenditure | By England, Wales, Scotland; by current, by capital; by outturn, provisional outturn | GB | Annual | [QRL.27] | 4.4.2 |
| Planned expenditure | By England, Wales, Scotland and Northern Ireland; by revenue, by capital | UK | Annual | [QRL.27] | 4.4.5 |
| Expenditure | By class of authority; by revenue, by capital | England and Wales | Annual | [QRL.43] | 4.2.4 |
| Expenditure | By service; authority combined; by current, by capital | Scotland | Annual | [QRL.55] | 4.4.4, 4.4.5 Status of figures (i.e outturn, provisional outturn, planned) described in introductory note |
| Expenditure | By economic category; authorities combined; by current, by capital | Scotland | Annual | [QRL.55] | 4.4.4, 4.4.5 Status of figures (i.e outturn, provisional outturn, planned) described in introductory note |
| Block grant related assessment | Authorities combined; by service | England | Annual | [QRL.28] | Appendix 12 paragraph 2 |
| Block grant related assessment | Authorities combined; by service | England | Annual | [QRL.84] | Appendix 12 paragraph 5 |

| | | | | | |
|---|---|---|---|---|---|
| Block grant related assessment indicator | By service | England | Annual | [QRL.84] | Appendix 12 paragraph 5 |
| Block grant related assessment | By service; by class of authority; by assessment component | England | Annual | [QRL.84] | Appendix 12 paragraph 5 |
| Block grant related assessment | By authority; by assessment component | England | Annual | [QRL.84] | Appendix 12 paragraph 5 |
| Block grant related assessment | By authority | England | Annual | [QRL.84] | Appendix 12 paragraph 5 |
| Block grant; ranking for target; estimated | By authority | England and Wales | Annual | [QRL.7] | Appendix 12 paragraph 10 |
| Current and capital less debt interest | By percentage GDP expenditure; by percentage public expenditure; authorities combined | UK | Annual | [QRL.43] | 4.4.2, 4.4.7, 4.4.13 |
| Grant related | By authority | England | Annual | [QRL.60] | Appendix 12 paragraph 5 |
| Grant related | By authority | Wales | Annual | [QRL.61] | Appendix 12 paragraph 5 |
| Grant related | By authority | Wales | Annual | [QRL.62] | 6.4.7 and Appendix 12 paragraph 5 |
| Grant related expenditure | By authority | England | Annual | [QRL.6] | Appendix 12 paragraph 5 |
| Loan charges | By authority; by service | Scotland | Annual | [QRL.65] | 4.2.2 |
| Local authority component of GDP | Authorities combined | UK | Monthly | [QRL.16] | 4.4.1, 4.4.8 |
| Non-relevant (in block grant related expenditure assessment) | Authorities combined; by service | England | Annual | [QRL.84] | Appendix 12 paragraph 5 |
| RSG guidelines | By authority | Scotland | Annual | [QRL.65] | Appendix 12 paragraph 17 |
| Ranking for block grant; estimated | By authority | England and Wales | Annual | [QRL.7] | Appendix 12 paragraph 8 |
| Rate fund | Authorities combined | Wales | Annual | [QRL.62] | 6.4.7 |
| Rate fund | Authorities combined; by revenue, by capital; by service; by economic category | Scotland | Annual | [QRL.77] | 4.2.4, 4.3.5 |
| Rate fund | By revenue, by capital; by region, islands area; by service | Scotland | Annual | [QRL.77] | 4.2.4, 4.3.5 |

| Type of data | Breakdown | Area | Frequency | QRL Publication | Text reference |
|---|---|---|---|---|---|
| Rate fund, actual net | By authority, by service | Scotland | Annual | [QRL.65] | 4.2.2 |
| Rate fund capital | By authority; by service; by payments, by consents | Scotland | Annual | [QRL.65] | 4.3.5, 4.3.7 |
| Rate fund, capital estimates | By authority; by service | Scotland | Annual | [QRL.64] | 4.3.5 |
| Rate fund, current | Authorities combined | Wales | Annual | [QRL.62] | 6.4.7 |
| Rate fund, estimated; increased costs and prices underlying estimates | By cost, by price component; authorities combined | Scotland | Annual | [QRL.71] | 4.2.8 |
| Rate fund, estimates | By authority; by service | England and Wales | Annual | [QRL.20] | 4.2.5 |
| Rate fund, estimates | By authority; by service | Scotland | Annual | [QRL.64] | 4.2.6 |
| Rate fund, net rate borne | By county; by service | England and Wales | Annual | [QRL.21] | 4.2.4 |
| Rate fund, and grant borne | By county; by service | England and Wales counties only | Annual | [QRL.21] | 4.2.4 |
| Rate fund, relevant | Authorities combined | Wales | Annual | [QRL.61] | 6.4.5 |
| Rate fund, relevant | Authorities combined | Wales | Annual | [QRL.62] | 6.4.7 |
| Rate fund, relevant expenditure | Authorities combined | England | Annual | [QRL.60] | 6.4.5 |
| Rate fund, relevant expenditure | Authorities combined | Scotland | Annual | [QRL.72] | 6.4.10 |
| Rate fund, net revenue trends | By authority; by service | Scotland | Annual | [QRL.65] | 4.2.3 |
| Rate fund revenue account | By service; by economic category; authorities combined | England and Wales | Annual | [QRL.43] | 4.2.4 |
| Relevant (in block grant related expenditure assessment) | Authorities combined; by service | England | Annual | [QRL.84] | Appendix 12 paragraph 5 |
| Revenue estimates | By authority; by service | Northern Ireland | Annual | [QRL.14] | 4.5.3 |

| | | | | | |
|---|---|---|---|---|---|
| Revenue total | By account; by service; authorities combined | Northern Ireland | Annual | [QRL.15] | 4.5.2 |
| Specific and supplementary grants (in block grant related expenditure assessment) | Authorities combined, by service | England | Annual | [QRL.84] | Appendix 12 paragraph 5 |
| Common good fund | By revenue, by capital; by economic category, authorities combined | Scotland | Annual | [QRL.77] | 4.2.10, 4.3.8 |
| Rate fund estimates | By service; by class of authority; by authority | Wales | Annual | [QRL.92] | 4.2.5 |
| Grant related | By authority | Wales | Annual | [QRL.92] | Appendix 12, paragraph 5 |
| Rate fund | By class of authority; by revenue, by capital; by county area; by class of authority; by service | Wales | Annual | [QRL.92] | 4.2.4, 4.3.3 |
| All funds combined | Authorities combined; by economic category; by revenue, by capital | Wales | Annual | [QRL.92] | Historical series; 4.2.2, 4.3.3 |
| Trading services | By revenue, by capital; by service; by England, by Wales | England and Wales | Annual | [QRL.41] | 3.3, 4.3.3 |
| Trading services | By revenue, by capital; authorities combined; by service; by economic category | England | Annual | [QRL.41] | 3.3, 4.2.2 |
| Trading services | By authorities combined; by revenue, by capital; by economic category; by service | Wales | Annual | [QRL.92] | 4.2.2, 4.3.3 |
| Trading services | Authorities combined; by revenue capital; by service; by economic category | Scotland | Annual | [QRL.77] | 4.2.10, 4.3.5 |
| Rate fund | Authorities combined; by revenue, by capital; by service; by economic category | Wales | Annual | [QRL.92] | 4.2.4, 4.3.3 |
| Rate fund | By RSG settlement, by outturn; by relevant, by non-relevant; by England, by England and Wales | England and Wales | Annual | [QRL.41] | 6.4.5 England figures relate to Budgets or RSG Settlement provisions |
| Rate fund | By revenue, by capital; by class of authority; by service; by England, by Wales | England and Wales | Annual | [QRL.41] | 3.3.3, 4.2.4, 4.3.3 |
| Special funds | Authorities combined; by fund, by account | England | Annual | [QRL.41] | 3.3, 4.2.2 |
| Special funds | Authorities combined; by fund, by account | Wales | Annual | [QRL.92] | 4.2.2 |
| Special funds | Authorities combined; by economic category; by fund, by account | Scotland | Annual | [QRL.77] | 4.2.10, 4.3.5 |

| Type of data | Breakdown | Area | Frequency | QRL Publication | Text reference |
|---|---|---|---|---|---|
| Superannuation funds | By heads of account; authorities combined | England | Annual | [QRL.41] | 4.2.10 |
| Superannuation funds | Authorities combined, by heads of account | Wales | Annual | [QRL.92] | 4.2.10 |
| Superannuation funds | Authorities combined; by heads of account | Scotland | Annual | [QRL.77] | 4.2.10 |
| Rate fund; financed by grants, rates and balances | Authorities combined; by service | England | Annual | [QRL.41] | Historical series, 4.2.4 |
| Relevant | Authorities combined; by service; by RSG settlement, by outturn | England | Annual | [QRL.41] | 6.4.5 |
| Rate fund | By revenue, by capital; by service; by economic category; authorities combined | England | Annual | [QRL.41] | 3.3.3, 4.2.4, 4.3.3 |
| Expenditure | By current, by capital; by percentage of all domestic expenditure; authorities combined | UK | Annual | [QRL.41] | 4.4.2, 4.4.7, 4.4.13 |
| Expenditure | By current, by capital; by service; by percentage of total public expenditure; authorities combined | UK | Annual | [QRL.41] | 4.4.2, 4.4.7, 4.4.13 |
| *Income* | | | | | |
| Rates | By individual authority | England and Wales | Annual | [QRL.63] | 5.2.1 |
| Rate fund | By county | England and Wales | Annual | [QRL.26] | 5.2.1 Income from rate referred to as "county precept" |
| Rate fund | By rating authority (England and Wales); by district, islands council (Scotland) | GB | Annual | [QRL.26] | 5.2.1 Income from rates referred to as "total product" |
| Rate fund, balances applied | By authority | Scotland | Annual | [QRL.65] | 8.2.1 |
| Rate fund, RSG provision | By service; by class of authority | Scotland | Annual | [QRL.65] | 6.4.12 |
| Rate fund, RSG resources element | By authority | Scotland | Annual | [QRL.65] | 6.4.12 |
| Rate fund, RSG needs element | By authority | Scotland | Annual | [QRL.65] | 6.4.12 |

| | | | | | |
|---|---|---|---|---|---|
| Rate fund, specific grants | By authority | Scotland | Annual | [QRL.65] | 6.2.5 |
| Block grant; penalty, abatement; estimated | By authority | England and Wales | Annual | [QRL.7] | Appendix 12, paragraph 12 |
| Rate fund; block grant; estimated | By authority; before penalty or abatement, after penalty or abatement | England and Wales | Annual | [QRL.16] | 6.4.6 |
| Revenue; government grants; general | Authorities combined | Northern Ireland | Annual | [QRL.15] | 6.5.3 |
| Revenue; government grants specific | By account; by service; authorities combined | Northern Ireland | Annual | [QRL.15] | 6.5.3 |
| Revenue; rates | Authorities combined | Northern Ireland | Annual | [QRL.15] | 5.6.3 |
| Revenue; total | By account; by income source; by service; authorities combined | Northern Ireland | Annual | [QRL.15] | 3.4.3 |
| Revenue; miscellaneous | By account; by service; authorities combined | Northern Ireland | Annual | [QRL.15] | 8.7.2 |
| Rate fund; government grants; distribution formula components (pre-1981/82) | All authorities | England and Wales (except London) | Annual | [QRL.45] | 6.4.16 No longer published; final issue 1980/81 |
| Rate fund; RSG needs element (pre-1981/82) | By authority | England and Wales (except London) | Annual | [QRL.45] | 6.4.16 No longer published; final issue 1980/81 |
| Rate fund; RSG | Authorities combined | Wales | Annual | [QRL.62] | 6.4.7 |
| Rate fund; block grant | Authorities combined | Wales | Annual | [QRL.62] | 6.4.7 |
| Rate fund; aggregate Exchequer grant | Authorities combined | Wales | Annual | [QRL.62] | 6.4.7 |
| Rate fund; domestic rate relief | Authorities combined | Wales | Annual | [QRL.62] | 6.4.7 |
| Revenue; government grants; general | By authority; by type of grant | Northern Ireland | Annual | [QRL.14] | 6.5.3 |
| Revenue; government grants; specific | By authority; by type of grant | Northern Ireland | Annual | [QRL.14] | 6.5.3 |
| Revenue; miscellaneous or other income; estimates | By authority; by service | Northern Ireland | Annual | [QRL.14] | 8.7.2 |
| Rates; estimates | By authority | Northern Ireland | Annual | [QRL.14] | 5.6.3 |

| Type of data | Breakdcwn | Area | Frequency | QRL Publication | Text reference |
|---|---|---|---|---|---|
| Rate fund; aggregate exchequer grant | Authorities combined | Scotland | Annual | [QRL.72] | 6.1.1 |
| Rate fund; RSG | Authorities combined | Scotland | Annual | [QRL.72] | 6.1.1, 6.4.10 |
| Rate fund; specific grant | Authorities combined; by type of grant | Scotland | Annual | [QRL.72] | 6.2.3 |
| Rates fund; RSG domestic element | Authorities combined | Scotland | Annual | [QRL.72] | 6.4.10 |
| Rate fund; RSG resources element | Authorities combined | Scotland | Annual | [QRL.72] | 6.4.10 |
| Rate fund; RSG needs element | By authority | Scotland | Annual | [QRL.72] | 6.4.10 |
| Rate fund; aggregate exchequer grant | Authorities combined | Wales | Annual | [QRL.61] | 6.1.1 |
| Rate fund; RSG | Authorities combined | Wales | Annual | [QRL.61] | 6.1.1, 6.4.2 |
| Rate fund; block grant | Authorities combined | Wales | Annual | [QRL.61] | 6.1.1, 6.4.2 |
| Rate fund; specific grants | Authorities combined; type of grant | Wales | Annual | [QRL.61] | 6.2.3 |
| Rate fund capital; transport supplementary grant | Authorities combined | Wales | Annual | [QRL.61] | 6.2.4 |
| Rate fund; national parks; supplementary grant | Authorities combined | Wales | Annual | [QRL.61] | 6.2.4 |
| Rate fund; domestic rate relief grant | Authorities combined | Wales | Annual | [QRL.61] | 6.1.1, 6.4.2 |
| Rate fund; London Equalisation grant | By authority | London | Annual | [QRL.60] | 6.4.4 |
| Rate fund; aggregate exchequer grant | Authorities combined | England | Annual | [QRL.60] | 6.1.1 |
| Rate fund; RSG | Authorities combined | England | Annual | [QRL.60] | 6.1.1, 6.4.2 |
| Rate fund; block grant | Authorities combined | England | Annual | [QRL.60] | 6.1.1, 6.4.2 |
| Rate fund; specific grants | Authorities combined; type of grant | England | Annual | [QRL.60] | 6.2.3 |
| Rate fund; transport supplementary grant | By county | England | Annual | [QRL.60] | 6.2.4 |

| | | | | | |
|---|---|---|---|---|---|
| Rate fund; national parks supplementary grants | By park authority | England | Annual | [QRL.60] | 6.2.4 |
| Rate fund; domestic rate relief grant | Authorities combined | England | Annual | [QRL.60] | 6.1.1, 6.4.2 |
| Rates | By county | England and Wales | Annual | [QRL.21] | 5.2.3 |
| Rate income | By rating authority | Scotland | Annual | [QRL.77] | 3.3, 5.2.3 |
| Block grant | By county | England and Wales | Annual | [QRL.21] | 6.4.7 |
| Rate fund | By source; authorities combined | England and Wales | Annual | [QRL.43] | 4.2.4 |
| Revenue; supplementary grants | By England, by Wales | England and Wales | Annual | [QRL.27] | 6.1.1 |
| Revenue; aggregate exchequer grant | By England; by Wale; by Scotland | GB | Annual | [QRL.27] | 6.1.1 |
| Revenue; specific government grants | By England; by Wales; by Scotland | GB | Annual | [QRL.27] | 6.1.1 |
| Revenue; block grant | By England; by Wales | England and Wales | Annual | [QRL.27] | 6.1.1 |
| Revenue; domestic rate relief grant | By England; by Wales | England and Wales | Annual | [QRL.27] | 6.1.1 |
| Revenue; RSG | By England; by Wales; by Scotland | GB | Annual | [QRL.27] | 6.1.1 |
| Revenue; RSG | Authorities combined; by domestic, resources, needs elements | Scotland | Annual | [QRL.27] | 6.1.1 |
| Rate fund; balances applied; estimates | By authority | Scotland | Annual | [QRL.64] | 8.2.2 |
| Rate fund; RSG resources element; estimates | By authority | Scotland | Annual | [QRL.64] | 6.4.11 |
| Rate fund; RSG needs element; estimates | By authority | Scotland | Annual | [QRL.64] | 6.4.11 |
| Rate fund; specific government grants; estimates | By authority | Scotland | Annual | [QRL.64] | 6.2.3 |
| Rate fund; rates; estimates | By authority | England and Wales | Annual | [QRL.20] | 5.2.1 |

| Type of data | Breakdown | Area | Frequency | QRL Publication | Text reference |
|---|---|---|---|---|---|
| Revenue; block grant claim | By authority | England and Wales | Annual | [QRL.20] | 6.4.6 |
| Revenue; anticipated grant claim, receipt | By authority | England and Wales | Annual | [QRL.20] | 6.4.6 |
| Revenue; specific and supplementary grants combined; estimates | By authority | England and Wales | Annual | [QRL.20] | 6.2.1, 6.2.2, 6.2.3 |
| Revenue; RSG domestic rate relief; estimates | By authority | England and Wales | Annual | [QRL.20] | 6.4.2 |
| Revenue; government grants; London equalisation; estimates | By authority | London | Annual | [QRL.20] | 6.4.4, 6.4.5 |
| Rate fund; estimates | By authority; by source | England and Wales | Annual | [QRL.20] | 3.3.2 |
| Rate fund | Authorities combined; by revenue; by capital; by service; by source | Scotland | Annual | [QRL.77] | 5.2, 6.2, 6.4, 8.2.1, 8.2.3 |
| Common good fund | By revenue, by capital; authorities combined; by source | Scotland | Annual | [QRL.77] | 8.5.3 |
| All funds combined | Authorities combined | Wales | Annual | [QRL.92] | 6.2, 6.3, 6.4, 7.2.2, 8.1.2 Historical series |
| Rate fund; capital | By class of authority; by source; by service | Wales | Annual | [QRL.92] | 8.2.3 |
| Specific grants | Authorities combined; by account; by revenue, by capital; by rate fund service | Wales | Annual | [QRL.92] | 6.2.5 |
| Rate support grant | By county area; by district; by block grant, by domestic rate relief grant | Wales | Annual | [QRL.92] | 6.4.2, 6.4.7 |
| National parks supplementary grant | By park authority | Wales | Annual | [QRL.92] | 6.2.5 |
| Transport supplementary grants | By authority; by grant per head | Wales | Annual | [QRL.92] | 6.2.5 |
| Rate fund | By class of authority; by economic category; by service | Wales | Annual | [QRL.92] | 5.2, 6.2, 6.4, 8.2.1 |

| | | | | | |
|---|---|---|---|---|---|
| Rate fund | Authorities combined; by source | England | Annual | [QRL.41] | 5.2, 5.5, 6.2, 6.3, 6.4, 8.2.1 Time series budgets or RSG settlement provision |
| Rate fund | Authorities combined; by source | England and Wales | Annual | [QRL.41] | 3.3, 5.2, 5.5, 6.2, 6.3, 6.4, 8.2.1 Time series |
| Superannuation funds | Authorities combined; by source of income | England | Annual | [QRL.41] | 7.3.5, 8.6.1 |
| Superannuation funds | Authorities combined; by source | Scotland | Annual | [QRL.77] | 3.3.5, 8.6.1 |
| Superannuation funds | Authorities combined; by source | Wales | Annual | [QRL.92] | 3.3.5, 8.6.1 |
| Special funds | Authorities combined; by type of fund; by source of income | England | Annual | [QRL.41] | 3.3.5, 8.5.1 |
| Special funds | Authorities combined; by source; by type of fund | Wales | Annual | [QRL.92] | 3.3.5, 8.5.1 |
| Special funds | Authorities combined; by fund; by account | Scotland | Annual | [QRL.77] | 3.3.5, 8.5.2 |
| Rate fund | By revenue, by capital; authorities combined; by service; by source | England | Annual | [QRL.41] | 5.2, 6.2, 6.4, 8.2.1, 8.2.3 |
| Trading services | Authorities combined; by revenue, by capital; by service; by source | England | Annual | [QRL.41] | 8.4.1, 8.4.3 |
| Trading services | Authorities combined; by revenue, by capital; by source; by service | Scotland | Annual | [QRL.77] | 8.4.1, 8.4.3 |
| Trading services | Authorities combined; by revenue; by capital; by source; by service | Wales | Annual | [QRL.92] | 8.4.1, 8.4.3 |
| Rate fund; rates | Authorities combined; by domestic, non-domestic shares | England and Wales | Annual | [QRL.41] | 5.2.3 |
| *Borrowing* | | | | | |
| Current; all accounts (including trading undertaking) | By authority | Northern Ireland | Annual | [QRL.36] | 7.5.1 |
| Loans raised in current year | By account; by service; authorities combined | Northern Ireland | Annual | [QRL.15] | 7.5.1 |
| Local authority sector | By type; by other sector source of finance; by year; by quarter; authorities combined | UK | Monthly | [QRL.22] | 7.2.2, 7.2.5 |
| Net; local authority sector | By source of finance; by term | England | Annual | [QRL.41] | 7.2.2 |

| Type of data | Breakdown | Area | Frequency | QRL Publication | Text reference |
|---|---|---|---|---|---|
| Local authority sector requirement | Temporary, PWLB, other long term borrowing; by percentage of public sector borrowing requirement | UK | Annual | [QRL.41] | 7.2.2 |
| Rate fund | Authorities combined; by service | Scotland | Annual | [QRL.77] | 7.2.1 |
| Rate fund; loans raised in current year | By service; authorities combined | England | Annual | [QRL.41] | 7.2.1 |
| Requirement; local authority sector | By source of finance; authorities combined | UK | Annual | [QRL.88] | 7.2.3 |
| Total rate | By term; by source of finance; authorities combined | Wales | Annual | [QRL.92] | 7.2.2 |
| Trading services | Authorities combined; by service | Scotland | Annual | [QRL.77] | 7.2.1 |
| Requirement; local authority sector | By year, by quarter; authorities combined | UK | Monthly | [QRL.16] | 7.2.3 |
| Requirement; local authority sector | By year, by quarter; authorities combined | UK | Monthly | [QRL.22] | 7.2.3 |
| *Net Borrowing* | | | | | |
| Local authority sector | By source of finance; by term of borrowing; authorities combined; by England, by Wales, by Scotland and Northern Ireland | UK | Monthly | [QRL.22] | 7.2.6 |
| *Manpower* | | | | | |
| Manpower | By authority; by manual, non-manual; by full-time, part-time | Scotland | Quarterly | [QRL.33] | 10.1.7 |
| Manpower | Authorities combined; by service; by full-time, part-time, full-time equivalent | Scotland | Quarterly | [QRL.33] | 10.1.7 |
| Manpower | By authority; by full-time, part-time | Wales | Quarterly | [QRL.37] | 10.1.5 |
| Manpower | Authorities combined; by service; by full-time, part-time, full-time equivalent | Wales | Quarterly | [QRL.37] | 10.1.5 |
| Manpower | By authority; by full-time, part-time | England | Quarterly | [QRL.32] | 10.1.5 |
| Manpower | Authorities combined; by service; by full-time, part-time, full-time equivalent | England | Quarterly | [QRL.32] | 10.1.5 |
| Manpower | By full-time, part-time, full-time equivalent; by England and Wales; by service | England and Wales | Quarterly | [QRL.42] | 10.1.3 |

| | | | | | |
|---|---|---|---|---|---|
| Manpower | By full-time, part-time, full-time equivalent; by England, Wales and Scotland; by service | GB | Quarterly | [QRL.18] | 10.1.6, 10.1.7 |
| Manpower | Authorities combined | Northern Ireland | Annual | [QRL.47] | 10.3.1 Since 1983 only |
| Manpower | Full-time, part-time, full-time equivalent; by source | Scotland | Annual | [QRL.77] | 10.1.4 |
| Manpower | By authority; by full-time, part-time | Wales | Annual | [QRL.92] | 10.1.6 |
| Manpower | Authorities combined; by manual, non-manual; by service | Wales | Annual | [QRL.92] | 10.1.6 |
| Manpower | By service; full-time, part-time, full-time equivalent | England | Annual | [QRL.41] | 10.1.3 |
| Registered disabled persons employed | By authority | GB | Annual | [QRL.18] | 10.1.13 |
| Earnings | Authorities combined | GB | Annual | [QRL.46] | 10.2.1 Analysis relates to local government sector as a whole |
| Earnings | Authorities combined; by collective agreement, wages board or council | GB | Annual | [QRL.46] | 10.2.1 |
| Hours worked | Authorities combined; by collective agreement, wages board or council | GB | Annual | [QRL.46] | 10.2.3 |
| *Complaints* | | | | | |
| Complaints | County, City of Belfast, City of Londonderry; by subject | Northern Ireland | Annual | [QRL.4] | 16.2.1 |
| Complaints | Authorities combined; by subject | Scotland | Annual | [QRL.66] | 16.1.2, 16.1.3 |
| Complaints | Type of authority | Scotland | Annual | [QRL.66] | 16.1.2, 16.1.3 |
| Complaints | By service; authorities combined; by progress stage | Wales | Annual | [QRL.3] | 16.1.2, 16.1.3 |
| Complaints | Authorities combined | England | Annual | [QRL.93] | 16.1.2, 16.1.3 |
| Complaints | Authorities combined | England | Annual | [QRL.93] | 16.1.2, 16.1.3 |
| Complaints | Authority complained against | Northern Ireland | Annual | [QRL.4] | 16.2.1 |
| Investigation | Authorities combined | Scotland | Annual | [QRL.66] | 16.1.2, 16.1.3 |
| Investigation terminated | London, Essex, Kent, Surrey; South, West, West Midlands, East Anglia (except Essex); North, East Midlands | England | Annual | [QRL.93] | 16.1.2, 16.1.3 |

| Type of data | Breakdown | Area | Frequency | QRL Publication | Text reference |
|---|---|---|---|---|---|
| Not accepted | By reason for non-acceptance; authorities combined | Wales | Annual | [QRL.3] | 16.1.2, 16.1.3 |
| Not accepted for investigation | Authorities combined | Scotland | Annual | [QRL.66] | 16.1.2, 16.1.3 |
| Report findings, maladministration | Authorities combined | England | Annual | [QRL.93] | 16.1.2, 16.1.3 |
| *Penny Rate Produce* | | | | | |
| Penny rate produce | By authority | Northern Ireland | Annual | [QRL.38] | 9.3.3 |
| Estimated | By individual authority | England and Wales | Annual | [QRL.63] | 9.2.2 |
| Grant exclusive | By authority | GB | Annual | [QRL.26] | 9.2.2 |
| Penny rate product | By authority | England and Wales | Annual | [QRL.44] | 9.2.2 |
| Penny rate product | By authority; excluding, including block grant | England and Wales | Annual | [QRL.20] | 9.2.2 |
| Penny rate product | By authority; by grant inclusive, exclusive | Wales | Annual | [QRL.92] | 9.2.2 |
| Estimated | By authority | Northern Ireland | Annual | [QRL.14] | 9.3.3 |
| Grant exclusive | By authority | England and Wales | Annual | [QRL.58] | 9.2.2 |
| Grant inclusive | By county | England and Wales | Annual | [QRL.21] | 9.2.2 |
| Net | By authority | Scotland | Annual | [QRL.64] | 9.2.4 |
| *Rate Poundage* | By region, islands area | Scotland | Annual | [QRL.65] | 5.1.3, 5.3.5, 14.6.10 Also includes 12 year time series |
| Rate poundage | By authority | Scotland | Annual | [QRL.65] | 5.3.4 |
| Rate poundage | By authority | GB | Annual | [QRL.26] | 5.3.2 |

| | | | | | |
|---|---|---|---|---|---|
| Rate poundage | By authority; by service; by borough, town district, county-at-large | Northern Ireland | Annual | [QRL.38] | 5.6.4 In the case of services, notional apportionment |
| Rate poundage | By authority | Northern Ireland | Annual | [QRL.14] | 5.6.4 |
| Rate poundage | By authority | England and Wales | Annual | [QRL.58] | 5.3.2 |
| Estimate | By authority; regional, district, domestic water | Scotland | Annual | [QRL.64] | 5.3.4 Actual rate poundage for previous two years also included |
| Estimate | By authority; general, special, Parish, County and Metropolitan Police precept | England and Wales | Annual | [QRL.20] | 5.3.2 |
| Average | By individual authority | England and Wales | Annual | [QRL.63] | 5.3.2 |
| Average | By rating authority; domestic, non-domestic, domestic water rate | Scotland | Annual | [QRL.77] | 5.3.5, 14.6.14 |
| Average | By authority, by general, domestic; by county, district, community council purpose | Wales | Annual | [QRL.92] | 5.3.2 |
| Average | Class of authority; by domestic, non-domestic | England | Annual | [QRL.41] | 5.3.2 |
| Average | Authorities combined; by domestic, non-domestic | Wales | Annual | [QRL.41] | 5.3.2 |
| Domestic water | | | | | |
| *Rateable Values* | | | | | |
| Rateable values | By individual authority; by type of hereditament | England and Wales | Annual | [QRL.63] | 9.1.4 |
| Summary | By class of hereditament; by class of authority | England and Wales | Annual | [QRL.31] | 9.1.3 |
| Rateable value | By authority | GB | Annual | [QRL.26] | 9.1.4 |
| Rateable value | By authority | UK | Annual | [QRL.44] | 9.1.4, 9.1.6, 9.3.2 |
| For block grant purposes | By authority | England and Wales | Annual | [QRL.7] | Appendix 12 paragraph 9 |
| Rateable value | By net annual valuation, rateable valuation; by authority | Northern Ireland | Annual | [QRL.14] | 9.3.2 |

| Type of data | Breakdown | Area | Frequency | QRL Publication | Text reference |
|---|---|---|---|---|---|
| For block grant purposes | By authority | England | Annual | [QRL.6] | Appendix 12 paragraph 9 |
| Rateable value | By authority; by category of hereditament | England and Wales | Annual | [QRL.58] | 9.1.4 |
| Total | By county | England and Wales | Annual | [QRL.21] | 9.1.4 |
| Rateable value | By authority; by band of rateable value; by domestic, commercial, other | Scotland | Annual | [QRL.64] | 9.1.6 |
| Rateable value | By authority, by domestic, other | England and Wales | Annual | [QRL.20] | 9.1.4 |
| Rateable value | By rating authority; by total, by domestic | Scotland | Annual | [QRL.77] | 9.1.6 |
| Rateable value | By rating authority; by total, by domestic; by size distribution of hereditament | Wales | Annual | [QRL.92] | 9.1.4 |
| Rateable value | By class of authority; domestic, non-domestic | England | Annual | [QRL.41] | 9.1.4 |
| Rateable value | Authorities combined; domestic, non-domestic | Wales | Annual | [QRL.41] | 9.1.4 |
| *Rates Collection* | | | | | |
| Total collectable | By authority | England and Wales | Annual | [QRL.58] | 5.4.1 |
| Constraints | By authority; by type of constraint | England and Wales | Annual | [QRL.58] | 5.4.1 |
| Arrears as proportion of rates collectable | By authority | England and Wales | Annual | [QRL.58] | 5.4.1 |
| Proportion collected per quarter | By authority | England and Wales | Annual | [QRL.58] | 5.4.1 |
| Rates collection | By authority; by method | England and Wales | Annual | [QRL.58] | 5.4.1 |
| Cost | By authority | England and Wales | Annual | [QRL.58] | 5.4.1 |
| Staff employed | By authority | England and Wales | Annual | [QRL.58] | 5.4.1 |
| *Rates* | | | | | |

| | | | | | |
|---|---|---|---|---|---|
| Average domestic rate bill | By authority | Scotland | Annual | [QRL.65] | 5.3.8 |
| Average domestic rate bill | By authority | Scotland | Annual | [QRL.64] | 5.3.8 |
| Average payment per domestic hereditament | By authority | England and Wales | Annual | [QRL.20] | 5.3.8 |
| Average domestic payment | By rating authority | Scotland | Annual | [QRL.77] | 5.3.8 |
| Average domestic rate bill | By authority | Wales | Annual | [QRL.92] | 5.3.8 |
| *Rate Rebates* | | | | | |
| Rate rebates | By district and London borough; by recipients, value of rebates; by percentage households receiving | England and Wales | Annual | [QRL.59] | 5.5.3 Discontinued, last issue relates to financial year 1982/83 |
| Rate rebates | By authority; by number granted; by average amount | England and Wales | Annual | [QRL.58] | 5.5.3 |
| Rate rebates | By authority; by total amount granted; by number of recipients; by average rebate | Scotland | Annual | [QRL.77] | 5.2.3 |
| Rate rebates | By rates authority; by amount granted; by number of recipients; by average rebate; by percentage domestic hereditaments receiving | Wales | Annual | [QRL.92] | 5.5.3 |
| *Rate Support Grant* | | | | | |
| Domestic element | By rating authority | Scotland | Annual | [QRL.77] | 6.4.9 |
| *Rateable Subjects* | | | | | |
| Rateable subjects | Domestic water; by authority | Scotland | Annual | [QRL.26] | 9.1.6 |
| Rateable subjects | By category; by authority | Scotland | Annual | [QRL.64] | 9.1.6 |
| Domestic | By rating authority; by value, by number | Scotland | Annual | [QRL.77] | 9.1.6 |
| *Rateable Valuation* | | | | | |
| Net annual value | By authority | Northern Ireland | Annual | [QRL.38] | 9.3.2 |
| *Block Grant Indicators* | | | | | |
| Block grant indicators | Authorities combined | Wales | Annual | [QRL.61] | Appendix 12 paragraph 2 |
| Block grant indicators | By authority; by indicator component | England | Annual | [QRL.6] | Appendix 12 paragraph 3 |

| Type of data | Breakdown | Area | Frequency | QRL Publication | Text reference |
|---|---|---|---|---|---|
| Block grant indicators | By class of authority | England | Annual | [QRL.60] | Appendix 12 paragraph 2 |
| *Block Grant Multipliers* | | | | | |
| Block grant multipliers | By authority | England | Annual | [QRL.60] | Appendix 12 paragraph 9 and 11 |
| Block grant multipliers | By authority | Wales | Annual | [QRL.62] | 6.4.7 |
| Block grant multipliers | By authority | Wales | Annual | [QRL.61] | Appendix 12 paragraph 9 and 11 |
| *Receipts* | | | | | |
| Revenue; miscellaneous; estimated | By authority; by service | Northern Ireland | Annual | [QRL.38] | 8.7.3 |
| Government grants; specific; estimated | By authority; by service | Northern Ireland | Annual | [QRL.38] | 6.5.4 |
| General Exchequer contribution; estimated | By authority; by service | Northern Ireland | Annual | [QRL.38] | 6.5.4 In the case of services, notional apportionment only |
| Rates; estimated | By authority; by service | Northern Ireland | Annual | [QRL.38] | 5.6.3 |
| Capital (excluding trading undertakings) | By authority; by service | Northern Ireland | Annual | [QRL.36] | 6.5.4, 7.5.1, 8.7.3 |
| General Revenue Account | By authority; by service | Northern Ireland | Annual | [QRL.36] | 6.5.4, 8.7.3 |
| Revenue; all accounts (including trading undertakings) | Authorities combined | Northern Ireland | Annual | [QRL.36] | 5.6.3, 6.5.4, 8.7.3 |
| Capital; government grants | Authorities combined | Northern Ireland | Annual | [QRL.15] | 6.5.3 |
| Capital | By account; by source; by service; authorities combined | Northern Ireland | Annual | [QRL.15] | 3.4.3 |

| | | | | | |
|---|---|---|---|---|---|
| Capital; other | By account; by services; authorities combined | Northern Ireland | Annual | [QRL.15] | 8.7.2 including "decreases in balances" |
| *Financial Account and Balancing Item* | | | | | |
| Local authority sector | By type of transaction in financial liabilities (net) and assets (net) | UK | Monthly | [QRL.22] | 7.4.1 |
| Local authority sector | Authorities combined | UK | Annual | [QRL.88] | 7.4.1 |
| *Government Grant Factors* | | | | | |
| Pre-1981/82 | By authority | England and Wales except London | Annual | [QRL.45] | 6.1.16 No longer published, final issue 1980/81 |
| *Grant Related Poundage* | | | | | |
| Grant related Poundage | By authority; before hold back, actual | England and Wales | Annual | [QRL.7] | Appendix 12 paragraph 7 |
| Grant related poundage | By authority | Wales | Annual | [QRL.62] | 6.4.7 and Appendix 12 paragraph 2 |
| Grant related poundage | By class of authority | Wales | Annual | [QRL.61] | Appendix 12 paragraph 6 |
| Grant related poundage | By class of authority | England | Annual | [QRL.60] | Appendix 12 paragraph 6 |
| *Gross Domestic Product* | | | | | |
| Gross domestic fixed capital formation by local authority sector | Financial year, calendar year, quarter; authorities combined | UK | Monthly | [QRL.22] | 4.4.1, 4.4.8, 4.4.12 |
| Final consumption by local authority sector | Financial year, calendar year, quarter; authorities combined | UK | Monthly | [QRL.22] | 4.4.1, 4.4.8 |
| *Hereditaments* | | | | | |
| Domestic | By value range; by authority | England and Wales | Annual | [QRL.63] | 9.1.4 |
| Hereditaments | By class of hereditament, by class of authority | England and Wales | Annual | [QRL.31] | 9.1.3 |
| Hereditaments | Domestic, other; by authority | England and Wales | Annual | [QRL.26] | 9.1.4 |

| Type of data | Breakdown | Area | Frequency | QRL Publication | Text reference |
|---|---|---|---|---|---|
| Hereditaments | By category; by authority | England and Wales | Annual | [QRL.58] | 9.1.4 |
| Hereditaments | By type; by county; by rateable value; by rating authority | Wales | Annual | [QRL.92] | 9.1.4 |
| Domestic | By class of authority | England | Annual | [QRL.41] | 9.1.4 |
| Domestic | Authorities combined | Wales | Annual | [QRL.41] | 9.1.4 |
| *Payments* | | | | | |
| Revenue; estimated | By authority; by service | Northern Ireland | Annual | [QRL.38] | 4.5.5 |
| General revenue account | By authority; by service | Northern Ireland | Annual | [QRL.36] | 4.5.5 |
| Capital (excluding trading undertakings) | By authority; by service | Northern Ireland | Annual | [QRL.36] | 4.5.5 |
| Revenue; all accounts (including trading undertakings) | Authorities combined | Northern Ireland | Annual | [QRL.36] | 4.5.5 |
| Capital | By account; by service; authorities combined | Northern Ireland | Annual | [QRL.15] | 4.5.2 |
| Rate fund; capital | By authority; by service | UK | Annual | [QRL.8] | 4.3.3, 4.3.5, 4.5.2 |
| Capital | Authorities combined; by service; by economic category | Wales | Annual | [QRL.92] | 4.3.3 |
| *Interest Rates* | | | | | |
| Paid by local authorities | PWLB by term, other by term and source of finance; authorities combined | UK | Monthly | [QRL.22] | 7.2.7 |
| *Investments* | | | | | |
| Local authority sector | By term, by type of asset or investment; authorities combined | UK | Monthly | [QRL.22] | 7.3.6 |
| *Loan Debt* | | | | | |
| All accounts (including trading undertakings) | By authority | Northern Ireland | Annual | [QRL.36] | 7.5.2 |
| *Capital Account* | | | | | |

| | | | | | |
|---|---|---|---|---|---|
| Local authority sector | Authorities combined; by receipts, expenditure | UK | Annual | [QRL.88] | 4.4.7 and 4.4.17 Expenditure figures show separately a *gross domestic fixed capital formation* (by service) and by *transfer payments* (i.e capital grants); see reference 4.4.6 |
| Local authority sector; capital grants to personal sector and public corporations | Financial year, calendar year, quarter; authorities combined | UK | Monthly | [QRL.22] | 4.4.12 |
| *Current Account* | | | | | |
| Estimated, local authority sector | Authorities combined; by receipts, expenditure | UK | Annual | [QRL.88] | 4.4.7, 4.4.10 Expenditure figures show separately *final consumption* (by service) and *transfer payments*; see reference 4.4.6 |
| *Debt* | | | | | |
| All funds; outstanding | By authority; by service; by source of finance; by maturity; by administering body | UK | Annual | [QRL.8] | 7.3.1, 7.5.2 |
| Average interest rates by local authority | By authority | UK | Annual | [QRL.8] | 7.2.7, 7.5.2 |
| Capital; estimated | By rate fund service; by authority | Scotland | Annual | [QRL.64] | 7.3.1 |
| Loans outstanding | By account; by service; authorities combined | Northern Ireland | Annual | [QRL.15] | 7.5.1 |
| Net; estimated | By authority | England and Wales | Annual | [QRL.20] | 7.3.1 |

| Type of data | Breakdown | Area | Frequency | QRL Publication | Text reference |
|---|---|---|---|---|---|
| Outstanding | By county area; by fund; by rate fund service; by class of authority | Wales | Annual | [QRL.92] | 7.3.1 |
| Outstanding; rate fund | By service; by region, islands areas | Scotland | Annual | [QRL.77] | 7.3.1 |
| Outstanding; trading services | By region, islands area | Scotland | Annual | [QRL.77] | 7.3.1 |
| Rate fund; outstanding loan debt | By service; authorities combined | England | Annual | [QRL.41] | 7.3.1 |
| Trading services | Authorities combined; by service | England | Annual | [QRL.41] | 7.3.1 |
| *Debt Management* | | | | | |
| Expenses | By authority | UK | Annual | [QRL.8] | 7.3.5, 7.5.2 |
| *Debt Outstanding* | | | | | |
| Local authority sector | By borrowing source, by term; authorities combined | UK | Monthly | [QRL.22] | 7.3.3 |
| *Elections* | | | | | |
| Election analysis | Councillors returned unopposed; councillors returned at contested elections; electorate in contested electoral divisions or wards; ballot papers in ballot boxes; percentage of electorate voting; region, island council, district | Scotland | Election years | [QRL.5] | 11.3.2 |
| Electors | Total electors, attainers; by local authority area | GB | Annual | [QRL.48] | 11.2.1 |
| Electors | Total electors, attainers, service voters; England, county, districts, London borough; Wales, county, district | England and Wales | Annual | [QRL.17] | 11.2.1 |
| Electors | By district | Northern Ireland | Annual | [QRL.2] | 11.4.2 |
| Electors | By authority | Scotland | Annual | [QRL.49] | 11.2.1 |
| Political composition of councils | Party, non-party representation; England (county, London borough, district); Wales (county, district); Scotland (region, islands council, district); Northern Ireland (district) | UK | Annual | [QRL.44] | 11.3.1, 11.4.3 |
| *Service Indicators* | | | | | |

| | | | | | |
|---|---|---|---|---|---|
| Service indicators | Financial, manpower, output; by individual authority | England and Wales | Annual | [QRL.40] | 17.1.3 All relevant primary statistics and their sources have been identified elsewhere in this review. Data standardised (e.g cost per unit; manpower per 1000 population) to permit inter-authority comparison |
| *Superannuation Funds* | | | | | |
| Superannuation funds | Fund size, turnover, activity, class and size of investment; by selected authority | GB | Annual | [QRL.39] | 8.6.4 |
| *Public Sector Borrowing Requirement* | | | | | |
| Local authority contribution | By year, by quarter; authority combined | UK | Monthly | [QRL.22] | 7.2.3 |
| *RSG Needs Element Indicators* | | | | | |
| RSG needs element indicators | Authorities combined | Scotland | Annual | [QRL.72] | Appendix 12 paragraph 16 |
| **Services** | | | | | |
| *Fire* | | | | | |
| Expenditure | By revenue, by capital; authorities combined; by economic category | Scotland | Annual | [QRL.77] | 12.2.9 |
| Expenditure | By class of authority; by economic category; by county area; by revenue, by capital | Wales | Annual | [QRL.92] | 12.2.4 |
| Expenditure | By revenue, by capital; authorities combined | England | Annual | [QRL.41] | 12.2.4 |
| Expenditure | By revenue, by capital; by class of authority | England and Wales | Annual | [QRL.41] | 12.2.4 |
| Expenditure, current | Authorities combined | England and Wales | Annual | [QRL.69] | 12.2.4 |

| Type of data | Breakdown | Area | Frequency | QRL Publication | Text reference |
|---|---|---|---|---|---|
| Expenditure, estimated | By authority; by net revenue, by capital | Scotland | Annual | [QRL.64] | 12.2.9 |
| Expenditure, estimated | By authority | England and Wales | Annual | [QRL.20] | 12.2.4 |
| Expenditure, estimated per 1000 population | By authority | England and Wales | Annual | [QRL.24] | 12.2.4 |
| Expenditure, estimated | By class of authority | England and Wales | Annual | [QRL.24] | 12.2.4 |
| Expenditure met from rates and RSG | By authority | England and Wales | Annual | [QRL.23] | 12.2.4 |
| Expenditure, met from rates and RSG, estimated | By authority | England and Wales | Annual | [QRL.24] | 12.2.4 |
| Expenditure, net revenue | By authority | Scotland | Annual | [QRL.65] | 12.2.9 Includes trends and expenditure per head |
| Expenditure per head | By county | England and Wales | Annual | [QRL.21] | 12.2.4 |
| Expenditure, planned | Authorities combined; by current, by capital | Scotland | Annual | [QRL.55] | 12.2.9 |
| Expenditure, RSG provisions | Authorities combined | Scotland | Annual | [QRL.65] | 12.2.9 |
| Expenditure, total | By class of authority | England and Wales | Annual | [QRL.23] | 12.2.4 |
| Income | By revenue, by capital; by source; authorities combined | Scotland | Annual | [QRL.77] | 12.2.9 |
| Income | By revenue, by capital; by source; by class of authority | Wales | Annual | [QRL.92] | 12.2.4 |
| Income | By revenue, by capital; authorities combined | England | Annual | [QRL.41] | 12.2.4 |
| Income | By class of authority | England and Wales | Annual | [QRL.23] [QRL.69] | 12.2.4 |
| Income, estimated | By class of authority; by source | England and Wales | Annual | [QRL.24] | 12.2.4 |
| Income, estimated per 1000 population | By authority; by source | England and Wales | Annual | [QRL.24] | 12.2.4 |

| | | | | | |
|---|---|---|---|---|---|
| Income, per 1000 population | By authority; by source | England and Wales | Annual | [QRL.23] | 12.2.4 |
| Loan charges, estimates | By authority | Scotland | Annual | [QRL.64] | 12.2.9 |
| Capital debt, estimated | By authority | Scotland | Annual | [QRL.64] | 12.2.9 |
| Debt outstanding per head | By county area; by class of authority | Wales | Annual | [QRL.92] | 12.2.4 |
| Loan debt outstanding | Authorities combined | England | Annual | [QRL.41] | 12.2.4 |
| Manpower | Authorities combined; by regular, others; by full-time, part-time, full time equivalent | England | Annual | [QRL.41] | 12.2.4 |
| Manpower | By county | England and Wales | Annual | [QRL.21] | 12.2.4 |
| Manpower | Authorities combined; by regular firemen, others; by full-time, part-time, full-time equivalent | Scotland | Annual | [QRL.77] | 12.2.8 |
| Manpower | Authorities combined; by manual, non-manual; by regular, other | Wales | Annual | [QRL.92] | 12.2.4 |
| Manpower | By authority | Scotland | Annual | [QRL.70] | 12.2.8 |
| Manpower | Authorities combined (England); authorities combined (Wales); authorities combined (Scotland); by full-time, part-time, full-time equivalent | GB | Quarterly | [QRL.18] | 12.2.4, 12.2.8 |
| Manpower | Authorities combined (England); authorities combined (Wales); by regular, others; by full-time, part-time, full-time equivalent | England and Wales | Quarterly | [QRL.42] | 12.2.4 |
| Manpower | Authorities combined; by regulars, others; by full-time, part-time, full-time equivalent | England | Quarterly | [QRL.32] | 12.2.4 |
| Manpower | Authorities combined; by regular, other; by full-time, part-time, full time equivalent | Wales | Quarterly | [QRL.37] | 12.2.4 |
| Manpower | Authorities combined; by regular firemen, other staff; by full-time, part time, full-time equivalent | Scotland | Quarterly | [QRL.33] | 12.2.8 |
| Manpower | By authority; authorised and serving establishment by type of staff | England and Wales | Annual | [QRL.24] | 12.2.4 |
| Manpower | By authority; by authorised and serving establishment, by type; full-time by rank | England and Wales | Annual | [QRL.23] | 12.2.4 |
| Loan charges | By authority | Scotland | Annual | [QRL.65] | 12.2.9 |

| Type of data | Breakdown | Area | Frequency | QRL Publication | Text reference |
|---|---|---|---|---|---|
| Hours (operational ranks below station officer) | Average weekly by normal basic, by overtime; authorities combined | GB | Annual | [QRL.46] | 12.2.5, 12.2.9 |
| Earnings (operational ranks below station officer) | Average gross weekly, hourly; authorities combined | GB | Annual | [QRL.46] | 12.2.5, 12.2.9 |
| Casualties | By England and Wales, Scotland; by brigade area | GB | Annual | [QRL.25] | 12.2.5, 12.2.9 |
| Calls answered | By brigade area | GB | Annual | [QRL.25] | 12.2.3, 12.2.8 |
| Fire prevention inspections | By authority; by type of inspection | England and Wales | Annual | [QRL.23] | 12.2.3 |
| Fire calls answered | By authority; by type of call | England and Wales | Annual | [QRL.23] | 12.2.3 |
| Brigade areas classified by risk categories | By authority | England and Wales | Annual | [QRL.23] | 12.2.5 |
| Resources, fire appliances | By authority; by type of appliance | England and Wales | Annual | [QRL.23] | 12.2.4 |
| Resources, vehicles | By authority | England and Wales | Annual | [QRL.23] | 12.2.4 |
| Resources, stations | By authority; by risk category; by manning | England and Wales | Annual | [QRL.23] | 12.2.4 |
| Resources, fire appliances | By authority; by type of appliance | England and Wales | Annual | [QRL.24] | 12.2.4 |
| Resources, stations | By authority | England and Wales | Annual | [QRL.24] | 12.2.4 |
| | By authority | England and Wales | Annual | [QRL.69] | 12.2.4 |
| Resources, fire appliances | Authorities combined | England and Wales | Annual | [QRL.69] | 12.2.4 |
| Fire calls answered | Authorities combined; by fires, false alarms | England and Wales | Annual | [QRL.69] | 12.2.3 |
| Fire protection inspections | Authorities combined; by type of inspection | England and Wales | Annual | [QRL.69] | 12.2.3 |

| | | | | | |
|---|---|---|---|---|---|
| Casualties | Authorities combined | England and Wales | Annual | [QRL.69] | 12.2.5 Fatalities by location for previous 3 years but not for current year |
| Resources, fire appliances | Authorities combined | Scotland | Annual | [QRL.70] | 12.2.8 |
| Inspections | Authorities combined | Scotland | Annual | [QRL.70] | 12.2.8 |
| Calls answered | By authority | Scotland | Annual | [QRL.70] | 12.2.8 |
| Casualties (fatalities only) | By authority | Scotland | Annual | [QRL.70] | 12.2.10 |
| *Public Protection* | | | | | |
| *Police* | | | | | |
| Expenditure | Class of authority; by revenue, by capital | England and Wales | Annual | [QRL.41] | 12.1.5 |
| Expenditure | By force | Scotland | Annual | [QRL.68] | 12.1.11 |
| Expenditure | By net revenue, capital, capital consents; by authority | Scotland | Annual | [QRL.65] | 12.1.11 |
| Expenditure | By revenue, by capital; by economic category; authorities combined | Scotland | Annual | [QRL.77] | 12.1.11 |
| Expenditure | By revenue, by capital; by economic category; by class of authority; by county area | Wales | Annual | [QRL.92] | 12.1.5 |
| Expenditure, revenue | By force and regional crime squad | England and Wales | Annual | [QRL.50] | 12.1.5 |
| Expenditure, capital estimated | By force | England and Wales | Annual | [QRL.51] | 12.1.5 Relates to expenditure on vehicles only |
| Expenditure, revenue estimated | By force and regional crime squad | England and Wales | Annual | [QRL.51] | 12.1.5 |
| Expenditure per head | By county | England and Wales | Annual | [QRL.21] | 12.1.5 |
| Expenditure, estimated | By net revenue, by capital; by authority | Scotland | Annual | [QRL.64] | 12.1.11 |
| Expenditure, estimated | By authority | England and Wales | Annual | [QRL.20] | 12.1.5 |
| Expenditure, planned | Authorities combined; by current, by capital | Scotland | Annual | [QRL.55] | 12.1.11 |
| Expenditure, RSG provisions | Authorities combined | Scotland | Annual | [QRL.65] | 12.1.11 |
| Income | By force and regional crime squad | England and Wales | Annual | [QRL.50] | 12.1.5 |

| Type of data | Breakdown | Area | Frequency | QRL Publication | Text reference |
|---|---|---|---|---|---|
| Income | By revenue, by capital; authorities combined | England | Annual | [QRL.41] | 12.1.5 |
| Income | By force, regional crime squad; by source | England and Wales | Annual | [QRL.51] | 12.1.5 |
| Income | By revenue, by capital; by source; by class of authority | Wales | Annual | [QRL.92] | 12.1.5 |
| Income | By force | Scotland | Annual | [QRL.68] | 12.1.11 |
| Income | By revenue, by capital; by source; authorities combined | Scotland | Annual | [QRL.77] | 12.1.11 |
| Offences notifiable | By offence group; forces combined | England and Wales | Annual | [QRL.10] | 12.1.3 |
| Deaths in police custody | By case; by police force | England and Wales | Annual | [QRL.67] | 12.1.6 |
| Compliants against police | Forces combined (excluding metropolitan police) | England and Wales | Annual | [QRL.67] | 12.1.6 |
| Resources, animals | By forces combined | England and Wales | Annual | [QRL.67] | 12.1.4 |
| Resources, motor vehicles and motor cycles | By forces combined | England and Wales | Annual | [QRL.67] | 12.1.4 |
| Manpower | By authority (full-time, part-time); authorities combined; by all ranks, cadets, traffic wardens, civilians; by full-time, part-time, full-time equivalent | England | Quarterly | [QRL.32] | 12.1.4 |
| Manpower | By authority (full-time, part-time); authorities combined; by all ranks, cadets, traffic wardens, civilians; by full-time, part-time, full-time equivalent | Wales | Quarterly | [QRL.37] | 12.1.4 |
| Manpower | Authorities combined; by officers, cadets; by full-time, part-time, full-time equivalent | Scotland | Quarterly | [QRL.33] | 12.1.10 |
| Manpower | By force | England and Wales | Annual | [QRL.67] | 12.1.4 |
| Manpower | Authorities combined; by all ranks, cadets, traffic wardens, civilians; by full-time, part-time, full-time equivalent | England | Annual | [QRL.41] | 12.1.4 |

| | | | | | |
|---|---|---|---|---|---|
| Manpower | By regional crime squad; by rank, secondments, actual strength | England and Wales | Annual | [QRL.51] | 12.1.4 |
| Manpower | By force | England and Wales | Annual | [QRL.51] | 12.1.4 |
| Manpower | By force | Scotland | Annual | [QRL.68] | 12.1.10 |
| Manpower, authorised establishment | By regional crime squad; by rank, secondments, average strength | England and Wales | Annual | [QRL.50] | 12.1.4 |
| Manpower | By force | England and Wales | Annual | [QRL.50] | 12.1.4 |
| Manpower | By county | England and Wales | Annual | [QRL.21] | 12.1.4 |
| Manpower | Authorities combined; by officers, cadets, traffic wardens, other; full time, part-time, full-time equivalent | Scotland | Annual | [QRL.77] | 12.1.10 |
| Manpower | Authorities combined (England); authorities combined (Wales); authorities combined (Scotland); by full-time, part-time, full-time equivalent | GB | Quarterly | [QRL.18] | 12.1.4, 12.1.10 |
| Manpower | Authorities combined (England); authorities combined (Wales); by all ranks, cadets, traffic wardens, civilians; by full-time, part-time, full-time equivalent | England and Wales | Quarterly | [QRL.42] | 12.1.4 |
| Manpower | Authorities combined; by manual, non-manual; by all ranks, others | Wales | Annual | [QRL.92] | 12.1.4 |
| Resources, vehicles | By vehicle type, type of duty; by force | Scotland | Annual | [QRL.68] | 12.1.10 |
| Crimes recorded | By force | Scotland | Annual | [QRL.68] | 12.1.9 |
| Offences recorded | By force; by offence group | Scotland | Annual | [QRL.68] | 12.1.9 |
| Complaints against police | By force; by number, results | Scotland | Annual | [QRL.68] | 12.1.12 |
| Deaths in police custody | By case; by force | Scotland | Annual | [QRL.68] | 12.1.12 |
| Terrorism, persons detained | Forces combined | Scotland | Annual | [QRL.68] | 12.1.9 |
| Crimes recorded | Forces combined; by category of crime | Scotland | Annual | [QRL.12] | 12.1.9 Publication now discontinued |
| Offences | Forces combined; by offence category; by recorded, cleared up | Scotland | Annual | [QRL.12] | 12.1.9 |

| Type of data | Breakdown | Area | Frequency | QRL Publication | Text reference |
|---|---|---|---|---|---|
| Crimes reported | By force | England and Wales | Annual | [QRL.51] | 12.1.3 |
| Training establishments | By force | England and Wales | Annual | [QRL.51] | 12.1.4 |
| Offences reported | By force | England and Wales | Annual | [QRL.51] | 12.1.3 |
| Breath tests | By force | England and Wales | Annual | [QRL.51] | 12.1.3 |
| Resources, animals | By force; by type of animal | England and Wales | Annual | [QRL.51] | 12.1.4 |
| Resources, vehicles | By force | England and Wales | Annual | [QRL.51] | 12.1.4 |
| Loan charges, estimated | By authority | Scotland | Annual | [QRL.64] | 12.1.11 |
| Capital debt, estimated | By authority | Scotland | Annual | [QRL.64] | 12.1.11 |
| Debt outstanding per head | By county area; by class of authority | Wales | Annual | [QRL.92] | 12.1.5 |
| Loan debt outstanding | Authorities combined | England | Annual | [QRL.41] | 12.1.5 |
| Misuse of drugs, narcotic drug addicts | Authorities combined | UK | Annual | [QRL.81] | 12.1.3, 12.1.9 |
| Misuse of drugs, persons found guilty of or cautioned for drugs offences | Authorities combined | UK | Annual | [QRL.81] | 12.1.3, 12.1.9 |
| Misuse of drugs, addicts known to the Home Office | Authorities combined | UK | Annual | [QRL.81] | 12.1.3, 12.1.9 |
| Misuse of drugs, seizure of controlled drugs | Authorities combined | UK | Annual | [QRL.81] | 12.1.3, 12.1.9 |
| Roadside screening breath tests | By force area | England and Wales | Annual | [QRL.81] | 12.1.3 |
| Prevention of terrorism, persons detained | By force area | GB | Quarterly | [QRL.81] | 12.1.3, 12.1.9 |

| | | | | | |
|---|---|---|---|---|---|
| Notifiable offences recorded by police | By force area, by offence group | England and Wales | Quarterly | [QRL.81] | 12.1.3 |
| Persons arrested (Sect 2 of Criminal Law Act 1977) | By force area | England and Wales | Annual | [QRL.81] | 12.1.6 |
| Deaths in custody or otherwise with police | By force area | England and Wales | Annual | [QRL.81] | 12.1.6 |
| Loan charges | By authority | Scotland | Annual | [QRL.65] | 12.1.11 |
| Hours (for ranks below superintendent only) | Average weekly by normal basic , by overtime; authorities combined | GB | Annual | [QRL.46] | 12.1.6, 12.1.12 |
| Earnings (for ranks below superintendent only) | Average gross weekly, hourly; authorities combined | GB | Annual | [QRL.46] | 12.1.6, 12.1.12 |
| Notifiable offences recorded | By force area | England and Wales | Annual | [QRL.11] | 12.1.3 |
| Notifiable offence recorded, firearms reported used | By force area | England and Wales | Annual | [QRL.11] | 12.1.3 |
| Persons proceeded against at Magistrates Courts | By force area; by age band | England and Wales | Annual | [QRL.11] | 12.1.3 |
| Persons found guilty of all offences at Magistrates Courts | By force area; by age band | England and Wales | Annual | [QRL.11] | 12.1.3 |
| Persons found guilty of all offences at Crown Court | By force area; by age band | England and Wales | Annual | [QRL.11] | 12.1.3 |
| Persons cautioned | By force area | England and Wales | Annual | [QRL.11] | 12.1.3 |
| Persons tried and/or sentenced at Crown Court | By force area; by age band | England and Wales | Annual | [QRL.11] | 12.1.3 |
| *Probation* | | | | | |
| Expenditure | By class of authority | England and Wales | Annual | [QRL.52] | 12.3.4 |
| Expenditure | By administering authority (except Inner London) | England and Wales | Annual | [QRL.52] | 12.3.1, 12.3.4 |

| Type of data | Breakdown | Area | Frequency | QRL Publication | Text reference |
|---|---|---|---|---|---|
| Expenditure | Summary (central and local government) | England and Wales | Annual | [QRL.53] | 12.3.1, 12.3.4 |
| Expenditure | By revenue, by capital; by economic category; by class of authority; by county area | Wales | Annual | [QRL.92] | 12.3.4 |
| Expenditure, estimated | By class of authority | England and Wales | Annual | [QRL.54] | 12.3.4 |
| Expenditure, estimated | By administering authority (except Inner London) | England and Wales | Annual | [QRL.54] | 12.3.1, 12.3.4 |
| Income | By class of authority; by source | England and Wales | Annual | [QRL.52] | 12.3.4 |
| Income | By administering authority (except Inner London) | England and Wales | Annual | [QRL.52] | 12.3.1, 12.3.4 |
| Income | By revenue, by capital; by source; by class of authority | Wales | Annual | [QRL.92] | 12.3.4 |
| Income, estimated | By class of authority; by source | England and Wales | Annual | [QRL.54] | 12.3.4 |
| Income, estimated | By administering authority (except Inner London) | England and Wales | Annual | [QRL.54] | 12.3.1, 12.3.4 |
| Population at risk | By administering authority | England and Wales | Annual | [QRL.52] | 12.3.4 |
| Performance indicators | By administering authority (except Inner London) | England and Wales | Annual | [QRL.52] | 12.3.1, 12.3.4 |
| Manpower | By administering authority | England and Wales | Annual | [QRL.52] [QRL.54] | 12.3.4 |
| Manpower | By probation officers, other staff; summary (central and local government) | England and Wales | Annual | [QRL.53] | 12.3.1, 12.3.4 |
| Manpower | Authorities combined; officers, others; full-time, part-time, full-time equivalent | England | Annual | [QRL.41] | 12.3.4 |
| Manpower | Authorities combined; by officers, other; by full-time, part-time, full time equivalent; by authority (full-time, part-time) | England | Quarterly | [QRL.32] | 12.3.4 |

| | | | | | |
|---|---|---|---|---|---|
| Manpower | Authorities combined; by officers, others; by full-time, part-time, full time equivalent; by authority (full-time, part-time) | Wales | Quarterly | [QRL.37] | 12.3.4 |
| Manpower | Authorities combined (England); authorities combined (Wales); by officers, others; by full-time, part-time, full-time equivalent | England and Wales | Quarterly | [QRL.42] | 12.3.4 |
| Institutions for which service responsible | By adminstering authority | England and Wales | Annual | [QRL.54] | 12.3.4 |
| Persons proceeded against in Magistrates Courts | By administering authority | England and Wales | Annual | [QRL.54] | 12.3.4 |
| Population "at risk" | By administering authority | England and Wales | Annual | [QRL.54] | 12.3.4 |
| Performance indicators | By indicator; by administering authority | England and Wales | Annual | [QRL.54] | 12.3.4 |
| Cost per case/report | Summary (central and local government) | England and Wales | Annual | [QRL.53] | 12.3.1, 12.3.4 |
| Case load, reports per officer | By probation area | England and Wales | Annual | [QRL.53] | 12.3.4 |
| Population "at risk" | Summary (central and local government) | England and Wales | Annual | [QRL.53] | 12.3.1, 12.3.4 |
| Persons proceeded against in Magistrates Courts | Summary (central and local government) | England and Wales | Annual | [QRL.53] | 12.3.1, 12.3.4 |
| Offenders | By penal order; by region | Scotland | Annual | [QRL.80] | 12.3.5 |
| *Administration of Justice* | | | | | |
| Expenditure per head | By county | England and Wales | Annual | [QRL.21] | 12.4.7 |
| Expenditure, estimated | By authority | England and Wales | Annual | [QRL.20] | 12.4.7 |
| Expenditure | By revenue, by capital; by economic category; authorities combined | Scotland | Annual | [QRL.77] | 12.4.9, 12.4.10 |
| Expenditure, planned | Authorities combined; by current, by capital | Scotland | Annual | [QRL.55] | 12.4.10 |
| Income | By revenue, by capital; by source; authorities combined | Scotland | Annual | [QRL.77] | 12.4.9, 12.4.10 |

*Magistrates Courts*

| Type of data | Breakdown | Area | Frequency | QRL Publication | Text reference |
|---|---|---|---|---|---|
| Income | By revenue, by capital; by service; by class of authority | Wales | Annual | [QRL.92] | 12.4.3 |
| Expenditure | Authorities combined; by revenue, by capital | England | Annual | [QRL.41] | 12.4.3 |
| Expenditure | By revenue, by capital; by economic category; by county area; by class of authority | Wales | Annual | [QRL.92] | 12.4.3 |
| Expenditure, estimated | By per 1000 defendants, per 1000 population; by administering authority | England and Wales | Annual | [QRL.1] | 12.4.3 |
| Debt outstanding per head | By county area; by class of authority | Wales | Annual | [QRL.92] | 12.4.3 |
| Loan debt outstanding | Authorities combined | England | Annual | [QRL.41] | 12.4.3 |
| Manpower | Authorities combined; by full-time, part-time, full-time equivalent | England | Annual | [QRL.41] | 12.4.3 |
| Manpower | By authority (full-time, part-time); authorities combined; by full-time, part-time, full-time equivalent | England | Quarterly | [QRL.32] | 12.4.3 |
| Manpower | By authority (full-time, part-time); authorities combined; full-time, part time, full-time equivalent | Wales | Qaurterly | [QRL.37] | 12.4.3 |
| Manpower | Authorities combined; by full-time, part-time, full-time equivalent | England | Annual | [QRL.42] | 12.4.3 |
| Manpower | Authorities combined; by full-time, part-time, full-time equivalent | Wales | Annual | [QRL.42] | 12.4.3 |
| Income, estimated, per 1000 defendants proceeded against | By administering authority | England and Wales | Annual | [QRL.1] | 12.4.3 |
| Expenditure met from rates and RSG, estimated | By administering authority | England and Wales | Annual | [QRL.1] | 12.4.3 |
| Defendants proceeded against | By administering authority | England and Wales | Annual | [QRL.1] | 12.4.3 |
| *District Courts* | | | | | |
| Expenditure, net revenue | Authorities combined | Scotland | Annual | [QRL.65] | 12.4.10 |
| Expenditure, net revenue, estimated | Authorities combined | Scotland | Annual | [QRL.64] | 12.4.10 |

| | | | | | |
|---|---|---|---|---|---|
| Manpower | Authorities combined; by full-time, part-time, full-time equivalent | Scotland | Annual | [QRL.77] | 12.4.10 |
| Manpower | Authorities combined; by full-time, part-time, full-time equivalent | Scotland | Quarterly | [QRL.33] | 12.4.10 |
| Manpower | Authorities combined; by full-time, part-time, full-time equivalents | Scotland | Quarterly | [QRL.18] | 12.4.10 |
| *Coroners Courts* | | | | | |
| Deaths reported | Authorities combined | England and Wales | Annual | [QRL.82] | 12.4.3 |
| Inquests | Authorities combined | England and Wales | Annual | [QRL.82] | 12.4.3 |
| Post mortems | Authorities combined | England and Wales | Annual | [QRL.82] | 12.4.3 |
| Verdicts | Authorities combined | England and Wales | Annual | [QRL.82] | 12.4.3 |
| Expenditure met from rates and RSG, estimated | By administering authority | England and Wales | Annual | [QRL.1] | 12.4.3 |
| Expenditure, estimated | By administering authority; by per 1000 population, per 1000 elderly population living alone | England and Wales | Annual | [QRL.1] | 12.4.3 |
| Income per 1000 population, estimated | By administering authority | England and Wales | Annual | [QRL.1] | 12.4.3 |
| Deaths reported on which inquests held | By authority | England and Wales | Annual | [QRL.1] | 12.4.3 |
| Elderly population living alone | By administering authority | England and Wales | Annual | [QRL.1] | 12.4.3 |
| *Library Service* | | | | | |
| Expenditure per head | By county | England and Wales | Annual | [QRL.21] | 15.1.3 |
| Expenditure | By revenue, by capital; authorities combined | England | Annual | [QRL.41] | 15.1.3 |
| Expenditure | By revenue, by capital; class of authority | England and Wales | Annual | [QRL.41] | 15.1.3 |
| Expenditure | By revenue, by capital; by economic category; by county area; by class of authority | Wales | Annual | [QRL.92] | |
| Income | By revenue, by capital; authorities combined | England | Annual | [QRL.41] | 15.1.3 |
| Income | Class of authority; by revenue, by capital; by source | Wales | Annual | [QRL.92] | 15.1.3 |

| Type of data | Breakdown | Area | Frequency | QRL Publication | Text reference |
|---|---|---|---|---|---|
| Staff per 100,000 population | By county | England and Wales | Annual | [QRL.21] | 15.1.3 |
| Book stock per head | By county | England and Wales | Annual | [QRL.21] | 15.1.3 |
| | | | | | 15.1.3 |
| Debt outstanding per head | By county area; by class of authority | Wales | Annual | [QRL.92] | 15.1.3 |
| Loan debt outstanding | Authorities combined | England | Annual | [QRL.41] | 15.1.3 |
| *Libraries* | | | | | |
| Expenditure, planned | Authorites combined; by current, by capital | Scotland | Annual | [QRL.55] | 15.1.8 |
| Expenditure, RSG provision | By class of authority | Scotland | Annual | [QRL.65] | 15.1.8 |
| Expenditure | By class of authority; by England, Wales and Scotland | GB | Annual | [QRL.56] | 15.1.4, 15.1.8 |
| Expenditure, net rate borne | By library authority or education and library board (Northern Ireland) | UK | Annual | [QRL.56] | 15.1.4, 15.1.8, 15.5.2 Excluding services to educational establishments |
| Expenditure, net rate borne, estimated | By authority | England and Wales | Annual | [QRL.57] | 15.1.4 Excluding services to educational establishments in Great Britain |
| Income | By class of authority; by England, Wales and Scotland; by source | GB | Annual | [QRL.56] | 15.1.4, 15.1.8 |
| Income per 1000 population | By library authority or education and library board (Northern Ireland) | UK | Annual | [QRL.56] | 15.1.4, 15.1.8, 15.5.2 |
| Income per 1000 population, estimated | By authority, by source | England and Wales | Annual | [QRL.57] | 15.1.4 |
| Capital payments | By library authority or education and library board (Northern Ireland) | UK | Annual | [QRL.56] | 15.1.4, 15.1.8, 15.5.2 |

| | | | | | |
|---|---|---|---|---|---|
| Manpower (in post) | By library authority or education and library board (Northern Ireland) | UK | Annual | [QRL.56] | 15.1.4, 15.1.8, 15.5.2 |
| Staffing establishments | By library authority or education and library board (Northern Ireland) | UK | Annual | [QRL.56] | 15.1.4, 15.1.8, 15.5.2 |
| Staffing establishments | By authority; by type of post | England and Wales | Annual | [QRL.57] | 15.1.4 |
| Books and other materials available | By class of authority; by England, Wales and Scotland | GB | Annual | [QRL.56] | 15.1.4, 15.1.8 |
| Books and other material available | By library authority or education and library board (Northern Ireland) | UK | Annual | [QRL.56] | 15.1.4, 15.1.8, 15.5.2 |
| Issues from lending libraries | By authority | GB | Annual | [QRL.44] | 15.1.5, 15.1.8 |
| Service delivery points | By library authority or education and library board (Northern Ireland) | UK | Annual | [QRL.56] | 15.1.4, 15.1.8, 15.5.2 |
| Agency services provided, expenditure | By library authority or education and library board (Northern Ireland) | UK | Annual | [QRL.56] | 15.1.4, 15.1.8, 15.5.2 |
| Agency services provided, staff | By library authority or education and library board (Northern Ireland) | UK | Annual | [QRL.56] | 15.1.4, 15.1.8, 15.5.2 |
| Service delivery points | By authority; by mobile library, libraries in institutions, agency; by main libraries opening hours | England and Wales | Annual | [QRL.57] | 15.1.4 |
| Book stocks and other materials | By authority | England and Wales | Annual | [QRL.57] | 15.1.4 Book stocks estimated only |
| *Public Libraries and Museums* | | | | | |
| Manpower | Authorities combined; full-time, part-time, full-time equivalent | Wales | Quarterly | [QRL.37] | 15.1.3 |
| Manpower | By full-time, part-time, full-time equivalent; authorities combined (England); authorities combined (Wales) | England and Wales | Quarterly | [QRL.42] | 15.1.3 |
| Manpower | Authorities combined; full-time, part-time, full-time equivalent | England | Quarterly | [QRL.32] | 15.1.3 |
| Manpower | By full-time, part-time, full-time equivalent; authorities combined (England); authorities combined (Wales); authorities combined (Scotland) | GB | Quarterly | [QRL.18] | 15.1.3, 15.1.7 |

| Type of data | Breakdown | Area | Frequency | QRL Publication | Text reference |
|---|---|---|---|---|---|
| Expenditure, estimated | By net revenue, by capital; by authority | Scotland | Annual | [QRL.64] | 15.1.8 |
| Capital debt, estimated | By authority | Scotland | Annual | [QRL.64] | 15.1.8 |
| Loan charges, estimated | By authority | Scotland | Annual | [QRL.64] | 15.1.8 |
| Expenditure | By net revenue, capital consents; by authority | Scotland | Annual | [QRL.65] | 15.1.8 |
| Loan charges | By authority | Scotland | Annual | [QRL.65] | 15.1.8 |
| *Museums* | | | | | |
| Expenditure per head | By county | England and Wales | Annual | [QRL.21] | 15.1.3 |
| *Museums and Art Galleries* | | | | | |
| Expenditure | By revenue, by capital | England | Annual | [QRL.41] | 15.1.3 |
| Expenditure | By revenue, by capital; by economic category; by county area; by class of authority | Wales | Annual | [QRL.92] | 15.1.3 |
| Expenditure | By revenue, by capital; by class of authority | England and Wales | Annual | [QRL.41] | 15.1.3 |
| Expenditure, estimated | By authority; by loan charges, other | Northern Ireland | Annual | [QRL.14] | 15.5.1 |
| Expenditure, RSG provision | By class of authority | Scotland | Annual | [QRL.65] | 15.1.8 |
| Expenditure, planned | Authorities combined; by current, by capital | Scotland | Annual | [QRL.55] | 15.1.8 |
| Income | By revenue, by capital; authorities combined | England | Annual | [QRL.41] | 15.1.3 |
| Income | By revenue, by capital; by source; by class of authority | Wales | Annual | [QRL.92] | 15.1.3 |
| Income, estimated | By authority; by miscellaneous, specific government grants | Northern Ireland | Annual | [QRL.14] | 15.5.1 |
| Net cost, estimated | By authority; by amount, per head, rate in £ | Northern Ireland | Annual | [QRL.14] | 15.5.1 |
| Outstanding debt per head | By county area; by class of authority | Wales | Annual | [QRL.92] | 15.1.3 |
| Loan debt outstanding | Authorities combined | England | Annual | [QRL.41] | 15.1.3 |
| *Libraries, Museums and Galleries* | | | | | |

| | | | | | |
|---|---|---|---|---|---|
| Expenditure | By revenue, by capital; by economic category; by region, islands area | Scotland | Annual | [QRL.77] | 15.1.8 |
| Expenditure | By authority | England and Wales | Annual | [QRL.20] | 15.1.3 |
| Income | By revenue, by capital; by source; authorities combined | Scotland | Annual | [QRL.77] | 15.1.8 |
| Manpower | By full-time, part-time, full-time equivalent; authorities combined | Scotland | Annual | [QRL.77] | 15.1.8 |
| Manpower | Full-time, part-time, full-time equivalent; authorities combined | Scotland | Quarterly | [QRL.33] | 15.1.8 |
| Outstanding debt per head | By region, islands area | Scotland | Annual | [QRL.77] | 15.1.8 |
| *Water* | | | | | |
| Capital debt, estimated | By authority | Scotland | Annual | [QRL.64] | 14.6.12 |
| Expenditure, capital, estimated | By authority | Scotland | Annual | [QRL.64] | 14.6.12 |
| Loan charges, estimated | By authority | Scotland | Annual | [QRL.64] | 14.6.12 |
| Developed resources | By reservoirs and lochs; by river intakes; by underground sources; by authority | Scotland | Annual | [QRL.76] | 14.6.16 |
| Consumption | By potable metered, unmetered; by non-potable; by authority | Scotland | Annual | [QRL.76] | 14.6.16 |
| Manpower | Authorities combined; by full-time, part-time, full-time equivalent | Scotland | Annual | [QRL.77] | 14.6.15 |
| Manpower | Authorities combined; by full-time, part-time, full-time equivalent | Scotland | Quarterly | [QRL.33] | 14.6.15 |
| *Public Water* | | | | | |
| Expenditure, revenue | By economic category; by regional, islands area | Scotland | Annual | [QRL.77] | 14.6.11 |
| Expenditure, net revenue, estimated | By authority | Scotland | Annual | [QRL.64] | 14.6.12 |
| Expenditure, net revenue | By authority | Scotland | Annual | [QRL.65] | 14.6.12 |
| Income, revenue | Authorities combined; by source | Scotland | Annual | [QRL.77] | 14.6.11 |
| *Water Supply* | | | | | |
| Expenditure | Authorities combined; by revenue, by capital; by economic category | Scotland | Annual | [QRL.77] | 14.6.11 Relate to "Water Supply" as Trading Service |

| Type of data | Breakdown | Area | Frequency | QRL Publication | Text reference |
|---|---|---|---|---|---|
| Income | Authorities combined; by revenue, by capital; by source | Scotland | Annual | [QRL.77] | 14.6.11 Relate to "Water Supply" as Trading Service |
| *Sewerage* | | | | | |
| Manpower | By full-time, part-time, full-time equivalent; authorities combined | Scotland | Annual | [QRL.77] | 14.6.8 |
| Manpower | Authorities combined; by full-time, part-time, full-time equivalent | Scotland | Quarterly | [QRL.33] | 14.6.8 |
| *Sewers and Drains* | | | | | |
| Expenditure, net revenue | By authority | Scotland | Annual | [QRL.65] | 14.6.8 |
| Expenditure, RSG provision | Authorities combined | Scotland | Annual | [QRL.65] | 14.6.8 |
| Expenditure | By revenue, by capital; by economic category; authorities combined | Scotland | Annual | [QRL.77] | 14.6.8 |
| Expenditure, estimated | By net revenue, by capital; by authority | Scotland | Annual | [QRL.64] | 14.6.8 |
| Income | By revenue, by capital; by source; authorities combined | Scotland | Annual | [QRL.77] | 14.6.8 |
| Capital debt, estimated | By authority | Scotland | Annual | [QRL.64] | 14.6.8 |
| Loan charges | By authority | Scotland | Annual | [QRL.65] | 14.6.8 |
| Loan charges, estimated | By authority | Scotland | Annual | [QRL.64] | 14.6.8 |
| *Cleansing* | | | | | |
| Expenditure, net revenue | By service component | Scotland | Annual | [QRL.65] | 14.6.2 |
| Expenditure | By authority; by net revenue, by capital consents | Scotland | Annual | [QRL.65] | 14.6.2 |
| Expenditure, RSG provision | Authorities combined | Scotland | Annual | [QRL.65] | 14.6.2 |
| Expenditure, estimated | By net revenue, by capital; by authority | Scotland | Annual | [QRL.64] | 14.6.2 |
| Loan charges | By authority | Scotland | Annual | [QRL.65] | 14.6.2 |
| Loan charges, estimated | By authority | Scotland | Annual | [QRL.64] | 14.6.2 |
| Capital debt, estimated | By authority | Scotland | Annual | [QRL.64] | 14.6.2 |
| Manpower | By full-time, part-time, full-time equivalent; authorities combined | Scotland | Annual | [QRL.77] | 14.6.3 |

| | | | | | |
|---|---|---|---|---|---|
| Manpower | Authorities combined; by full-time, part-time, full-time equivalent | Scotland | Quarterly | [QRL.33] | 14.6.3 |
| Manpower | By full-time, part-time, full-time equivalent; authorities combined | Scotland | Quarterly | [QRL.18] | 14.6.3 |
| *Environmental Health* | | | | | |
| Expenditure, net revenue, estimated | By authority | Scotland | Annual | [QRL.64] | 14.6.5, 14.6.6 |
| Expenditure, capital, estimated | Authorities combined | Scotland | Annual | [QRL.64] | 14.6.5, 14.6.6 |
| Expenditure, estimated | By authority | England and Wales | Annual | [QRL.20] | 14.2.2 |
| Expenditure | By revenue, by capital; by economic category; by county area; by class of authority | Wales | Annual | [QRL.92] | 14.2.2 |
| Expenditure | Authorities combined; by revenue, by capital | England | Annual | [QRL.41] | 14.2.2 |
| Expenditure | Class of authority; by revenue, by capital | England and Wales | Annual | [QRL.41] | 14.2.2 |
| Expenditure, RSG provision | Authorities combined | Scotland | Annual | [QRL.65] | 14.6.5, 14.6.6 |
| Expenditure, net revenue | By authority | Scotland | Annual | [QRL.65] | 14.6.5, 14.6.6 |
| Income | Authorities combined; by revenue, by capital | England | Annual | [QRL.41] | 14.2.2 |
| Income | By revenue, by capital; by source; by class of authority | Wales | Annual | [QRL.92] | 14.2.2 |
| Capital, debt, estimated | Authorities combined | Scotland | Annual | [QRL.64] | 14.6.5, 14.6.6 |
| Loan, charges, estimated | Authorities combined | Scotland | Annual | [QRL.64] | 14.6.5, 14.6.6 |
| Clean air, expenditure | By revenue, by capital; by economic category; authorities combined | Scotland | Annual | [QRL.77] | 14.6.5, 14.6.6 |
| Clean air, income | By revenue, by capital; by source; authorities combined | Scotland | Annual | [QRL.77] | 14.6.5, 14.6.6 |
| Public conveniences, baths and washouses, expenditure | By revenue, by capital; by economic category; authorities combined | Scotland | Annual | [QRL.77] | 14.6.5, 14.6.6 |
| Public conveniences, baths and washouses, income | By revenue, by capital; by source; authorities combined | Scotland | Annual | [QRL.77] | 14.6.5, 14.6.6 |
| Other public health, expenditure | By revenue, by capital; by economic category; authorities combined | Scotland | Annual | [QRL.77] | 14.6.5, 14.6.6 |

| Type of data | Breakdown | Area | Frequency | QRL Publication | Text reference |
|---|---|---|---|---|---|
| Other public health, income | By revenue, by capital; by source; authorities combined | Scotland | Annual | [QRL.77] | 14.6.5, 14.6.6 |
| Debt outstanding per head | By county area; by class of authority | Wales | Annual | [QRL.92] | 14.2.2 |
| Loan debt outstanding | Authorities combined | England | Annual | [QRL.41] | 14.2.2 |
| Manpower | By full-time, part-time, full-time equivalent | England | Annual | [QRL.41] | 14.2.4 |
| Manpower | Authorities combined; by full-time, part-time, full-time equivalent | England | Quarterly | [QRL.32] | 14.2.4 |
| Manpower | Authorities combined; by manual, non-manual | Wales | Annual | [QRL.92] | 14.2.4 |
| Manpower | Full-time, part-time, full-time equivalent; authorities combined | Scotland | Quarterly | [QRL.33] | 14.6.5, 14.6.6 |
| Manpower | By full-time, part-time, full-time equivalent; authorities combined | Scotland | Annual | [QRL.77] | 14.6.5, 14.6.6 |
| Manpower | Authorities combined; by full-time, part-time, full-time equivalent | Wales | Quarterly | [QRL.37] | 14.2.4 |
| Manpower | By full-time, part-time, full-time equivalent; authorities combined (England); authorities combined (Wales) | England and Wales | Quarterly | [QRL.42] | 14.2.4 |
| Loan charges | By authority | Scotland | Annual | [QRL.65] | 14.6.5, 14.6.6 |
| Core functions, staffing | By authority; by function; by staff activity | England and Wales | Annual | [QRL.19] | 14.2.6, 14.2.7 |
| Core functions, hereditaments | By authority; by function; by class of hereditament | England and Wales | Annual | [QRL.19] | 14.2.6, 14.2.7 |
| Core functions, expenditure | By authority | England and Wales | Annual | [QRL.19] | 14.2.6, 14.2.7 |
| Core functions, income | By authority; by source | England and Wales | Annual | [QRL.19] | 14.2.6, 14.2.7 |
| Core functions, enforcement actions | By authority; by function; by statutory notices served, prosecutions | Englnd and Wales | Annual | [QRL.19] | 14.2.6, 14.2.7 |
| Meat inspections | By authority; by type of animal | England and Wales | Annual | [QRL.19] | 14.2.6, 14.2.7 |

| | | | | | |
|---|---|---|---|---|---|
| Housing improvement grants | By authority | England and Wales | Annual | [QRL.19] | 14.2.6, 14.2.7 |
| Pest control | By authority | England and Wales | Annual | [QRL.19] | 14.2.6, 14.2.7 |
| *Environmental Services* | | | | | |
| Expenditure, payments | By revenue, by capital; by component service; authorities combined | Northern Ireland | Annual | [QRL.15] | 14.7.1 |
| Expenditure, estimated | By loan charges, other; by component service; by authority | Northern Ireland | Annual | [QRL.14] | 14.7.1 |
| Income, estimated | By miscellaneous, specific government grants; by component service; by authority | Northern Ireland | Annual | [QRL.14] | 14.7.1 |
| Income, receipts | By revenue, by capital; by component service; authorities combined | Northern Ireland | Annual | [QRL.15] | 14.7.1 |
| Loans outstanding | By component service; authorities combined | Northern Ireland | Annual | [QRL.15] | 14.7.1 |
| Payments | By revenue, by capital; by authority; by component service | Northern Ireland | Annual | [QRL.36] | 14.7.2 |
| Receipts | By revenue, by capital; by component service; by authority | Northern Ireland | Annual | [QRL.36] | 14.7.2 |
| Payments, revenue, estimated | By component service; by authority | Northern Ireland | Annual | [QRL.38] | 14.7.2 |
| Receipts, estimated | By revenue, rates, government grant; by component service; by authority | Northern Ireland | Annual | [QRL.38] | 14.7.2 |
| Net cost, estimated | By amount, per head, rate in £; by component service; by authority | Northern Ireland | Annual | [QRL.14] | 14.7.1 |
| *Highways and Transportation* | | | | | |
| Expenditure | By county; by current, by capital | England and Wales | Annual | [QRL.29] | 13.1.4 |
| Expenditure, estimated | By class of authority; by county; by current, by capital, by component service | England and Wales | Annual | [QRL.30] | 13.1.4 |
| Expenditure, loan charges, estimated | By district | England and Wales | Annual | [QRL.30] | 13.1.4 |
| Expenditure, capital financing charges, estimated | By county | England and Wales | Annual | [QRL.30] | 13.1.4 |

| Type of data | Breakdown | Area | Frequency | QRL Publication | Text reference |
|---|---|---|---|---|---|
| Expenditure on roads by local authorities | By county | England | Annual | [QRL.87] | 13.1.4 Includes expenditure by Commission for New Towns |
| Expenditure on motorways and trunk roads | By region, islands area | Scotland | Annual | [QRL.78] | 13.2.4 |
| Income, TSG, estimated | By county | England and Wales | Annual | [QRL.30] | 13.1.4 Also includes data on "submitted" and "eligible" grant expenditure |
| Income, transport Supplementary Grant | By county | England and Wales | Annual | [QRL.87] | 13.1.4 |
| Road accidents | By police force area | England and Wales | Annual | [QRL.51] | 12.1.3 |
| Road accidents | By county | England | Annual | [QRL.74] | 13.1.5 |
| Road accidents per 100km of road | By county | England | Annual | [QRL.74] | 13.1.5 |
| Vehicles and casualties per road accident | By county | England | Annual | [QRL.74] | 13.1.5 |
| Road accidents on main roads | By county | England | Annual | [QRL.74] | 13.1.5 |
| Reasons for requiring breath tests | By county | England and Wales | Annual | [QRL.74] | 13.1.5 |
| Breath tests, drivers involved in accidents | By county | Wales | Annual | [QRL.75] | 13.1.5 |
| Road accidents | By county | Wales | Annual | [QRL.75] | 13.1.5 |
| Road lengths | By police force area | England and Wales | Annual | [QRL.51] | 12.1.3 |
| Road lengths | By county; by class of road | England and Wales | Annual | [QRL.30] | 13.1.5 |

| | | | | | |
|---|---|---|---|---|---|
| Road lengths | By class of road; by county | Wales | Occasional | [QRL.83] | 13.1.5 First published 1985 |
| Motorways and trunk road maintenance, current expenditure, estimated | By county | England and Wales | Annual | [QRL.30] | 13.1.4 |
| Professional and technical services, gross expenditure and recharges, estimated | By county; by district | England and Wales | Annual | [QRL.30] | 13.1.4 |
| Bus and underground, rail revenue support, total expenditure, estimated | By district; by London Borough | England and Wales | Annual | [QRL.30] | 13.1.4 |
| Provision of parking places; net surplus, deficit; estimated | By authority | London | Annual | [QRL.30] | 13.1.4 |
| Provision of parking places, expenditure | By London Borough; by net capital, gross revenue, loan charges | London | Annual | [QRL.30] | 13.1.4 |
| Car parks, expenditure, estimated | By district | England and Wales | Annual | [QRL.30] | 13.1.4 |
| Net expenditure, estimated | By district | England and Wales | Annual | [QRL.30] | 13.1.4 |
| Other transport services, expenditure, estimated | By district; by London Borough; by current, by capital | England and Wales | Annual | [QRL.30] | 13.1.4 For London Borough, loan charges separately identified |
| Provision of parking places, income, estimated | By authority | London | Annual | [QRL.30] | 13.1.4 |
| Motorway and trunk road construction, expenditure, estimated | By county | England and Wales | Annual | [QRL.30] | 13.1.4 |

| Type of data | Breakdown | Area | Frequency | QRL Publication | Text reference |
|---|---|---|---|---|---|
| Principal roads and other county roads maintenance, gross expenditure, estimated | By county | England and Wales | Annual | [QRL.30] | 13.1.4 |
| Concessionary fares, expenditure, estimated | By county; by district; by London Borough | England and Wales | Annual | [QRL.30] | 13.1.4 |
| Motorways and trunk road construction, gross expenditure | By county | England and Wales | Annual | [QRL.29] | 13.1.4 |
| Motorways and trunk road maintenance, expenditure | By county | England and Wales | Annual | [QRL.29] | 13.1.4 |
| Road lengths | By class of road | England and Wales | Annual | [QRL.29] | 13.1.5 |
| Street lighting maintenance, expenditure | By county | England and Wales | Annual | [QRL.29] | 13.1.4 |
| Professional and technical services, expenditure | By county | England and Wales | Annual | [QRL.29] | 13.1.4 |
| Professional and technical services, income, recharges | By county | England and Wales | Annual | [QRL.29] | 13.1.4 |
| Bus, underground, rail revenue support, expenditure | By county | England and Wales | Annual | [QRL.29] | 13.1.4 |
| Concessionary fares, total expenditure | By county | England and Wales | Annual | [QRL.29] | 13.1.4 |
| Eligible expenditure for TSG | By county | England and Wales | Annual | [QRL.87] | 13.1.4 |
| Motor vehicles currently licensed | By county (England and Wales); by region, islands area (Scotland) | GB | Annual | [QRL.87] | 13.1.5 |
| Road lengths | By county (England and Wales); by region, islands area (Scotland); by class of road | GB | Annual | [QRL.87] | 13.1.5 |

| | | | | | |
|---|---|---|---|---|---|
| Road casualties | By county (England and Wales); by region (Scotland) | GB | Annual | [QRL.73] | 13.1.5, 13.2.5 |
| Road accidents | By county (England and Wales); by region (Scotland) | GB | Annual | [QRL.73] | 13.1.5, 13.2.5 |
| Harbours, government grants | By construction, improvement; by recipient authority | Scotland | Annual | [QRL.78] | 13.2.4 |
| Road accidents | By police force area; by number, severity; by casualties, severity | Scotland | Annual | [QRL.78] | 13.2.5 |
| Motor vehicles licensed | By region, island area; by vehicle type, vehicle weight | Scotland | Annual | [QRL.78] | 13.2.5 |
| Road lengths | By region, islands area; by class of road | Scotland | Annual | [QRL.78] | 13.2.5 |
| Car ownership | By region, islands area | Scotland | Annual | [QRL.78] | 13.2.5 |
| Bus, ferry services subsidies, expenditure | Authorities combined | Scotland | Annual | [QRL.78] | 13.2.4 |
| Concessionary fares, expenditure | Authorities combined | Scotland | Annual | [QRL.78] | 13.2.4 |
| Local transport, expenditure by central government | Scotland summary; by current, by capital | Scotland | Annual | [QRL.78] | 13.2.4 |
| Roads and transport, net revenue expenditure | By region, islands council | Scotland | Annual | [QRL.78] | 13.2.4 |
| *Roads and Transport* | | | | | |
| Expenditure | By revenue, by capital; by service component; authorities combined | Scotland | Annual | [QRL.77] | 13.2.3 |
| Expenditure | By revenue, by capital; by economic category; by region, islands area | Scotland | Annual | [QRL.77] | 13.2.3 |
| Income | By revenue, by capital; by service component; by source; authorities combined | Scotland | Annual | [QRL.77] | 13.2.3 |
| Outstanding debt per head | By region, islands area | Scotland | Annual | [QRL.77] | 13.2.3 |
| *Transport* | | | | | |
| Public passenger, expenditure | By county area; by class of authority; by revenue, by capital; by economic cetegory | Wales | Annual | [QRL.92] | 13.1.4 |
| Public passenger, income | By revenue, by capital; by source; by class of authority | Wales | Annual | [QRL.92] | 13.1.4 |
| Public passenger, debt outstanding per head | By county area; by class of authority | Wales | Annual | [QRL.92] | 13.1.4 |

| Type of data | Breakdown | Area | Frequency | QRL Publication | Text reference |
|---|---|---|---|---|---|
| Highways and other transport, expenditure | By county area; by class of authority; by revenue, by capital; by economic category | Wales | Annual | [QRL.92] | 13.1.4 |
| Highways and other transport, income | By revenue, by capital; by source; by class of authority | Wales | Annual | [QRL.92] | 13.1.4 |
| Highways and other transport; debt outstanding per head | By county area; by class of authority | Wales | Annual | [QRL.92] | 13.1.4 |
| Concessionary fares, expenditure, revenue | By county area; by class of authority; by economic category | Wales | Annual | [QRL.92] | 13.1.4 |
| Concessionary fares, income, revenue | By source; by class of authority | Wales | Annual | [QRL.92] | 13.1.4 |
| Highways and other transport, expenditure | Authorities combined; by revenue, by capital | England | Annual | [QRL.41] | 13.1.4 |
| Highways and other transport, income | By revenue, by capital; authorities combined | England | Annual | [QRL.41] | 13.1.4 |
| Highways and other transport, expenditure | By class of authority; by revenue, by capital | England and Wales | Annual | [QRL.41] | 13.1.4 |
| Highways and other transport, loan debt outstanding | Authorities combined | England | Annual | [QRL.41] | 13.1.4 |
| Concessionary fares, expenditure, revenue | Authorities combined | England | Annual | [QRL.41] | 13.1.4 |
| Concessionary fares, income, revenue | Authorities combined | England | Annual | [QRL.41] | 13.1.4 |
| Concessionary fares, expenditure, revenue | By class of authority | England and Wales | Annual | [QRL.41] | 13.1.4 |
| Public passenger expenditure | Authorities combined; by revenue, by capital | England | Annual | [QRL.41] | 13.1.4 |
| Public passenger, income | Authorities combined; by revenue, by capital | England | Annual | [QRL.41] | 13.1.4 |
| Public passenger expenditure | Class of authority; by revenue, by capital | England and Wales | Annual | [QRL.41] | 13.1.4 |
| Public passenger, loan debt outstanding | Authorities combined | England | Annual | [QRL.41] | 13.1.4 |

| | | | | | |
|---|---|---|---|---|---|
| Expenditure, planned | Authorities combined; by current, by capital | Scotland | Annual | [QRL.55] | 13.2.3 |
| Roads and transport, expenditure, RSG provision | By class of authority | Scotland | Annual | [QRL.65] | 13.2.3 |
| Roads and transport, expenditure, capital | By authority | Scotland | Annual | [QRL.65] | 13.2.3 Includes also capital consents |
| Roads and lighting, loan charges | By authority | Scotland | Annual | [QRL.65] | 13.2.3 |
| Concessionary fares, expenditure, net revenue | By authority | Scotland | Annual | [QRL.65] | 13.2.3 |
| Subsidies, expenditure, net revenue | By authority | Scotland | Annual | [QRL.65] | 13.2.3 |
| Roads and lighting, expenditure, net revenue | By authority | Scotland | Annual | [QRL.65] | 13.2.3 |
| Concessionary fares, expenditure, estimated | By authority | England and Wales | Annual | [QRL.20] | 13.1.4 |
| Passenger, expenditure, estimated | By authority | England | Annual | [QRL.20] | 13.1.4 |
| Highways and local transport, expenditure, estimated | By authority | England and Wales | Annual | [QRL.20] | 13.1.4 |
| Subsidies; expenditure; net revenue; estimated | By authority | Scotland | Annual | [QRL.64] | 13.2.3 |
| Roads and lighting, expenditure | By net revenue, by capital; by authority | Scotland | Annual | [QRL.64] | 13.2.3 |
| Roads and lighting; loan charges; estimated | By authority | Scotland | Annual | [QRL.64] | 13.2.3 |
| Roads and lighting; capital debt; estimated | By authority | Scotland | Annual | [QRL.64] | 13.2.3 |
| Concessionary fares; expenditure; net revenue; estimated | By authority | Scotland | Annual | [QRL.64] | 13.2.3 |
| Highways, expenditure, per head | By county | England and Wales | Annual | [QRL.21] | 13.1.4 |
| Concessionary fares, expenditure per head | By county | England and Wales | Annual | [QRL.21] | 13.1.4 |

| Type of data | Breakdown | Area | Frequency | QRL Publication | Text reference |
|---|---|---|---|---|---|
| Passenger expenditure per head | By county | England and Wales | Annual | [QRL.21] | 13.1.4 |
| Harbours; expenditure, estimated | By loan charges, other; by authority | Northern Ireland | Annual | [QRL.14] | 13.5.1 |
| Harbours; income, estimated | By authority; by source | Northern Ireland | Annual | [QRL.14] | 13.5.1 |
| Harbours; net cost, estimated | By authority; by current, per head, rate in | Northern Ireland | Annual | [QRL.14] | 13.5.1 |
| *Airports* | | | | | |
| Expenditure, estimated | By local authority airport; by revenue, by capital | GB | Annual | [QRL.35] | 13.4.3 |
| Expenditure | By local authority airport; by revenue, by capital | GB | Annual | [QRL.34] | 13.4.3 |
| Income, revenue | By source; by local authority airport | GB | Annual | [QRL.34] | 13.4.3 |
| Income; revenue account, estimated | By local authority airport | GB | Annual | [QRL.35] | 13.4.3 |
| Passengers, estimated | By local authority airport | GB | Annual | [QRL.35] | 13.4.3 |
| Air cargo, estimated | By local authority airport | GB | Annual | [QRL.35] | 13.4.3 |
| Aircraft movements, estimated | By local authority airport | GB | Annual | [QRL.35] | 13.4.3 |
| Work load units, estimated | By local authority airport | GB | Annual | [QRL.35] | 13.4.3 |
| Air transport movements, estimated | By local authority airport | GB | Annual | [QRL.35] | 13.4.3 |
| Passengers | By local authority airport | GB | Annual | [QRL.34] | 13.4.3 |
| Cargo handled | By local authority airport | GB | Annual | [QRL.34] | 13.4.3 |
| Aircraft movements | By local authority airport | GB | Annual | [QRL.34] | 13.4.3 |
| Work load units | By local authority airport | GB | Annual | [QRL.34] | 13.4.3 |
| Scheduled services | By local authority airport | GB | Annual | [QRL.34] | 13.4.3 |
| Employees | By local authority airport | GB | Annual | [QRL.34] | 13.4.3 |
| *Trading Standards, Consumer Protection* | | | | | |

| | | | | | |
|---|---|---|---|---|---|
| Expenditure | By authority or consortium | England and Wales | Annual | [QRL.85] | 12.5.3 |
| Expenditure | By authority or consortium; by function | England and Wales | Annual | [QRL.85] | 12.5.3 |
| Expenditure, estimated | By authority or consortium | England and Wales | Annual | [QRL.86] | 12.5.3 |
| Income | By testing fees, other; by authority or consortium | England and Wales | Annual | [QRL.86] | 12.5.3 |
| Manpower | By authority or consortium; by qualified, other; by function | England and Wales | Annual | [QRL.86] | 12.5.3 |
| Inspections | By authority or consortium | England and Wales | Annual | [QRL.86] | 12.5.3 |
| Tests made | By authority or consortium; by type of test | England and Wales | Annual | [QRL.86] | 12.5.3 |
| Prosecutions, informations | By authority or consortium; by type of court | England and Wales | Annual | [QRL.86] | 12.5.3 |
| Complaints and infringements dealt with | By authority or consortium | England and Wales | Annual | [QRL.86] | 12.5.3 |
| *Consumer Protection* | | | | | |
| Expenditure, net revenue | By authority | Scotland | Annual | [QRL.65] | 12.5.4 |
| Expenditure, RSG provision | Authorities combined | Scotland | Annual | [QRL.65] | 12.5.4 |
| Expenditure, revenue | Authorities combined; by economic category | Scotland | Annual | [QRL.77] | 12.5.4 |
| Expenditure | authorities combined; by revenue, by capital | England | Annual | [QRL.41] | 12.5.3 |
| Expenditure per head | By county | England and Wales | Annual | [QRL.21] | 12.5.3 |
| Expenditure | By class of authority; by revenue, by capital | England and Wales | Annual | [QRL.41] | 12.5.3 |
| Expenditure; net revenue, estimated | Authorities combined | Scotland | Annual | [QRL.64] | 12.5.4 |
| Expenditure | By economic category; by county area; by class of authority; by revenue, by capital | Wales | Annual | [QRL.92] | 12.5.3 |
| Income | By revenue, by capital; by source; by class of authority | Wales | Annual | [QRL.92] | 12.5.3 |
| Income | Authorities combined; by revenue, by capital | England | Annual | [QRL.41] | 12.5.3 |
| Income, revenue | By source; authorities combined | Scotland | Annual | [QRL.77] | 12.5.4 |

| Type of data | Breakdown | Area | Frequency | QRL Publication | Text reference |
|---|---|---|---|---|---|
| Manpower | Authorities combined; by full-time, part-time, full-time equivalent | Scotland | Quarterly | [QRL.33] | 12.5.4 |
| Loan debt outstanding | Authorities combined | England | Annual | [QRL.41] | 12.5.3 |
| Debt outstanding per head | By county area; by class of authority | Wales | Annual | [QRL.92] | 12.5.3 |
| *Refuse Collection* | | | | | |
| Expenditure, estimated | By authority | England and Wales | Annual | [QRL.20] | 14.3.2 |
| Expenditure | By county area; by revenue, by capital; by economic category; by class of authority | Wales | Annual | [QRL.92] | 14.3.2 |
| Income | By revenue, by capital; by source; by class of authority | Wales | Annual | [QRL.92] | 14.3.2 |
| Income | Authorities combined; by revenue, by capital | England | Annual | [QRL.41] | 14.3.2 |
| Debt outstanding per head | By county area; by class of authority | Wales | Annual | [QRL.92] | 14.3.2 |
| Expenditure | By revenue, by capital; authorities combined | England | Annual | [QRL.41] | 14.3.2 |
| Expenditure | Class of authority; by revenue, by capital | England and Wales | Annual | [QRL.41] | 14.3.2 |
| Expenditure | By authority; by revenue, by capital | England and Wales | Annual | [QRL.89] | 14.3.2 |
| Loan debt outstanding | Authorities combined | England | Annual | [QRL.41] | 14.3.2 |
| *Waste Collection* | | | | | |
| Current income | By authority; by charges, sales, contribution from other authorities | England and Wales | Annual | [QRL.89] | 14.3.5 |
| Costs | By authority; by gross, by net | England and Wales | Annual | [QRL.89] | 14.3.5 |
| Resources | By authority; by vehicles, manpower | England and Wales | Annual | [QRL.89] | 14.3.5 |
| Tonnage collected | By authority | England and Wales | Annual | [QRL.89] | 14.3.5 |
| Waste reclaimed | By authority | England and Wales | Annual | [QRL.89] | 14.3.5 |

| | | | | | |
|---|---|---|---|---|---|
| Collection charges | By authority | England and Wales | Annual | [QRL.89] | 14.3.5 No longer included, record of questionnaire responses can be purchased directly from CIPFA/SIS |
| *Refuse Disposal* | | | | | |
| Expenditure, estimated | By authority | England and Wales | Annual | [QRL.20] | 14.4.2 |
| Expenditure | By revenue, by capital; by economic category; by county area; by class of authority | Wales | Annual | [QRL.92] | 14.4.2 |
| Expenditure | By revenue, by capital; by class of authority | England and Wales | Annual | [QRL.41] | 14.4.2 |
| Expenditure | Authorities combined; by revenue, by capital | England | Annual | [QRL.41] | 14.4.2 |
| Income | By revenue, by capital; by source; by class of authority | Wales | Annual | [QRL.92] | 14.4.2 |
| Debt outstanding per head | By county area; by class of authority | Wales | Annual | [QRL.92] | 14.4.2 |
| Income | Authorities combined; by revenue, by capital | England | Annual | [QRL.41] | 14.4.2 |
| Loan debt outstanding | Authorities combined | England | Annual | [QRL.41] | 14.4.2 |
| *Waste Disposal* | | | | | |
| Expenditure | By authority, by revenue, by capital | England and Wales | Annual | [QRL.90] | 14.4.3 |
| Expenditure, estimated | By authority; by revenue, by capital | England and Wales | Annual | [QRL.91] | 14.4.3 |
| Income | By authority | England and Wales | Annual | [QRL.90] | 14.4.3 |
| Income, estimated | By authority; by charges, other | England and Wales | Annual | [QRL.91] | 14.4.3 |
| Manpower | By authority; by Waste Disposal Authority, agent authority, contractor | England and Wales | Annual | [QRL.90] | 14.4.3 |
| Treatment costs | By gross, net; by method of treatment; by weight; by authority | England and Wales | Annual | [QRL.90] | 14.4.3 |
| Waste | By authority; by total arising, disposed of, treated | England and Wales | Annual | [QRL.90] | 14.4.3 |

| Type of data | Breakdown | Area | Frequency | QRL Publication | Text reference |
|---|---|---|---|---|---|
| Waste treatment method | By authority; by tonnage, cost | England and Wales | Annual | [QRL.90] | 14.4.3 |
| Waste treatment method | By tonnage, cost; by authority | England and Wales | Annual | [QRL.91] | 14.4.3 |
| Waste treatment, disposal, costs | By authority | England and Wales | Annual | [QRL.91] | 14.4.3 |
| Waste arising | By disposed of, treated, total arising; by authority | England and Wales | Annual | [QRL.91] | 14.4.3 |
| Manpower | By authority | England and Wales | Annual | [QRL.91] | 14.4.3 |
| *Refuse Collection and Disposal* | | | | | |
| Expenditure | By revenue, by capital; by economic category; authorities combined | Scotland | Annual | [QRL.77] | 14.6.2 |
| Income | By revenue, by capital; by source; authorities combined | Scotland | Annual | [QRL.77] | 14.6.2 |
| *Civil Defence* | | | | | |
| Expenditure | By class of authority; by revenue, by capital | England | Annual | [QRL.41] | 12.4.4 |
| Expenditure | Authorities combined; by revenue, by capital | Wales | Annual | [QRL.41] | 12.4.4 |
| Expenditure, revenue | Authorities combined; by economic category | Scotland | Annual | [QRL.77] | 12.4.11 |
| Expenditure | By revenue, by capital; by economic category; by county area; by class of authority | Wales | Annual | [QRL.92] | 12.4.4 |
| Expenditure, planned | Authorities combined; by current, by capital | Scotland | Annual | [QRL.55] | 12.4.11 |
| Income, revenue | Authorities combined; by source | Scotland | Annual | [QRL.77] | 12.4.11 |
| Income | By revenue, by capital; by source; by class of authority | Wales | Annual | [QRL.92] | 12.4.4 |
| Income | Authorities combined; by revenue, by capital | England | Annual | [QRL.41] | 12.4.4 |
| Debt outstanding per head | By county area; by class of authority | Wales | Annual | [QRL.92] | 12.4.4 |
| Loan debt outstanding | Authorities combined | England | Annual | [QRL.41] | 12.4.4 |
| *Emergency Planning, Civil Defence* | | | | | |

| | | | | | |
|---|---|---|---|---|---|
| Expenditure; net revenue, estimated | Authorities combined | Scotland | Annual | [QRL.64] | 12.4.11 |
| *Cemeteries* | | | | | |
| Expenditure | By authority | England and Wales | Annual | [QRL.9] | 14.5.4 |
| Income | By authority; by fees, charges, interest | England and Wales | Annual | [QRL.9] | 14.5.4 |
| Charges imposed | By authority | England and Wales | Annual | [QRL.9] | 14.5.4 |
| Deaths | By authority area | England and Wales | Annual | [QRL.9] | 14.5.4 |
| Grave spaces maintained | By "inperpetuity", war graves, annual account, authority; by authority | England and Wales | Annual | [QRL.9] | 14.5.4 |
| New graves excavated | By authority; by manually, mechanically | England and Wales | Annual | [QRL.9] | 14.5.4 |
| Interments | By authority | England and Wales | Annual | [QRL.9] | 14.5.4 |
| Area | By authority | England and Wales | Annual | [QRL.9] | 14.5.4 |
| Manpower | By authority | England and Wales | Annual | [QRL.9] | 14.5.4 |
| Cemeteries | By authority | England and Wales | Annual | [QRL.9] | 14.5.4 |
| *Crematoria* | | | | | |
| Expenditure per cremation | By crematorium | GB | Annual | [QRL.9] | 14.5.4 |
| Income, per cremation | By crematorium | GB | Annual | [QRL.9] | 14.5.4 |
| Cremations | By crematorium | GB | Annual | [QRL.9] | 14.5.4 |
| Cremators | By crematorium | GB | Annual | [QRL.9] | 14.5.4 |
| Means of financing any deficit | By crematorium | GB | Annual | [QRL.9] | 14.5.4 |
| Financial surplus, deficit | By crematorium; by method of recovering deficit | GB | Annual | [QRL.9] | 14.5.4 |
| Fees, charges | By crematorium; by range, purpose for which imposed | GB | Annual | [QRL.9] | 14.5.4 |
| *Burial Grounds and Crematoria* | | | | | |

| Type of data | Breakdown | Area | Frequency | QRL Publication | Text reference |
|---|---|---|---|---|---|
| Expenditure, net revenue | By authority | Scotland | Annual | [QRL.65] | 14.6.17 |
| Loan charges | By authority | Scotland | Annual | [QRL.65] | 14.6.17 |
| Expenditure; net revenue, estimated | By authority | Scotland | Annual | [QRL.64] | 14.6.17 |
| Expenditure; capital, estimated | Authorities combined | Scotland | Annual | [QRL.64] | 14.6.17 |
| Expenditure | By revenue, by capital; by economic category; authorities combined | Scotland | Annual | [QRL.77] | 14.6.17 |
| Income | By revenue, by capital; by source; authorities combined | Scotland | Annual | [QRL.77] | 14.6.17 |
| Capital debt, estimated | Authorities combined | Scotland | Annual | [QRL.64] | 14.6.17 |
| Loan charges, estimated | Authorities combined | Scotland | Annual | [QRL.64] | 14.6.17 |
| *Cemeteries and Crematoria* | | | | | |
| Expenditure | By revenue, by capital; by class of authority | England and Wales | Annual | [QRL.41] | 14.5.2 |
| Expenditure | Authorities combined; by revenue, by capital | England | Annual | [QRL.41] | 14.5.2 |
| Expenditure, estimated | By authority | England and Wales | Annual | [QRL.20] | 14.5.2 |
| Expenditure | By revenue, by capital; by economic category; by county area; by class of authority | Wales | Annual | [QRL.92] | 14.5.2 |
| Income | Authorities combined; by revenue, by capital | England | Annual | [QRL.41] | 14.5.2 |
| Income | By revenue, by capital; by source; by class of authority | Wales | Annual | [QRL.92] | 14.5.2 |
| Debt outstanding per head | By county area; by class of authority | Wales | Annual | [QRL.92] | 14.5.2 |
| Loan debt outstanding | Authorities combined | England | Annual | [QRL.41] | 14.5.2 |
| *Smallholdings* | | | | | |
| Expenditure | By revenue, by capital; authorities combined | England | Annual | [QRL.41] | 15.3.4 |
| Expenditure | By county; by revenue, by capital | England and Wales | Annual | [QRL.79] | 15.3.2 |
| Expenditure per head | By county | England and Wales | Annual | [QRL.21] | 15.3.4 |

| | | | | | |
|---|---|---|---|---|---|
| Expenditure | By revenue, by capital; by economic category; by county area; by class of authority | Wales | Annual | [QRL.92] | 15.3.4 |
| Income | By revenue, by capital; by source; by class of authority | Wales | Annual | [QRL.92] | 15.3.4 |
| Income | By county; by operational, non-operational | England and Wales | Annual | [QRL.79] | 15.3.2 |
| Area per head | By county | England and Wales | Annual | [QRL.21] | 15.3.4 |
| Administration costs | By county | England and Wales | Annual | [QRL.79] | 15.3.2 |
| Operational surplus, deficit | By county; by gross, net | England and Wales | Annual | [QRL.79] | 15.3.2 |
| Net surplus, deficit | By county | England and Wales | Annual | [QRL.79] | 15.3.2 |
| Debt charges | By county | England and Wales | Annual | [QRL.79] | 15.3.2 Includes title redemption annuities |
| Rents | By county; by equipped holdings, bare land holdings, other land and dwellings | England and Wales | Annual | [QRL.79] | 15.3.2 |
| Staff | By county; by number, cost; by administrative and manual | England and Wales | Annual | [QRL.79] | 15.3.2 |
| Holdings equipped | By county; by fully, partially | England and Wales | Annual | [QRL.79] | 15.3.2 |
| Tenants | By county | England and Wales | Annual | [QRL.79] | 15.3.2 |
| Holdings | By county; by type, commercial status | England and Wales | Annual | [QRL.79] | 15.3.2 |
| Size of holding | By county; by total, average size, commercial status | England and Wales | Annual | [QRL.79] | 15.3.2 |
| *School Crossing Patrols* | | | | | |
| Expenditure; net revenue, estimated | Authorities combined | Scotland | Annual | [QRL.64] | 12.4.12 |
| Expenditure, revenue | Authorities combined; by economic category | Scotland | Annual | [QRL.77] | 12.4.12 |
| Expenditure, revenue | By economic category; by class of authority | Wales | Annual | [QRL.92] | 12.4.6 |
| Expenditure, revenue | By class of authority | England and Wales | Annual | [QRL.41] | 12.4.6 |
| Expenditure, revenue | By authority | Scotland | Annual | [QRL.65] | 12.4.12 |

| Type of data | Breakdown | Area | Frequency | QRL Publication | Text reference |
|---|---|---|---|---|---|
| Income, revenue | By source; by class of authority | Wales | Annual | [QRL.92] | 12.4.6 |
| Income, revenue | By class of authority | England and Wales | Annual | [QRL.41] | 12.4.6 |
| *Direct Labour Organisations* | | | | | |
| Appropriation account | Balance at start, contribution to or from General Rate Fund, or DLO Reserve Fund, balance at end; by authority | GB | Annual | [QRL.13] | 15.2.2 |
| Manpower | Direct Employment, Indirect Employment; by authority; category of work | GB | Annual | [QRL.13] | 15.2.2 |
| Percentage of total income by method of charging for work | By type of contract; by authority | GB | Annual | [QRL.13] | 15.2.2 |
| Historical cost account | Expenditure, income, surplus or deficit; by authority | GB | Annual | [QRL.13] | 15.2.2 |
| Current cost account | General highways work, major works of new construction, minor works of new construction, general maintenance work; by authority | GB | Annual | [QRL.13] | 15.2.2 Scotland only; additional breakdown category, "water and sewerage works" |
| Division of work between DLO and outside contractors | General highways work, major works of new construction, minor works of new construction, general maintenance work; by authority | GB | Annual | [QRL.13] | 15.2.2 Scotland only; additional breakdown category, "water and sewerage category" |
| *Registration of Electors* | | | | | |
| Expenditure, revenue | By class of authority | England and Wales | Annual | [QRL.41] | 12.4.5 |
| Expenditure, revenue | By county area; by class of authority; by economic category | Wales | Annual | [QRL.92] | 12.4.5 |
| Income, revenue | By source; by class of authority | Wales | Annual | [QRL.92] | 12.4.5 |

| | | | | | |
|---|---|---|---|---|---|
| Income, revenue | Authorities combined | England | Annual | [QRL.41] | 12.4.5 |
| *Direct Works* | | | | | |
| Manpower | By full-time, part-time, full-time equivalent; authorities combined | Scotland | Quarterly | [QRL.33] | 15.2.4 |

# QUICK REFERENCE LIST KEY TO PUBLICATIONS

| Reference | Author or Organisation | Title | Publisher | Frequency or Date | Remarks |
|---|---|---|---|---|---|
| [QRL.1] | Chartered Institute of Public Finance and Accountancy – Statistical Information Service (CIPFA/SIS) | *Administration of Justice (Estimates)* | CIPFA | Annual | |
| [QRL.2] | Chief Electoral Officer for Northern Ireland | *Analyses of Register of Electors* | Chief Electoral Officer for Northern Ireland | Annual | |
| [QRL.3] | Commissioner for Local Administration in Wales | *Annual Report of Commissioner for Local Administration in Wales* | Commissioner for Local Administration in Wales | Annual | |
| [QRL.4] | House of Commons Paper | *Annual Report of the Northern Ireland Commissioner for Complaints* | HMSO | Annual | |
| [QRL.5] | Registrar General for Scotland | *Annual Report of Registrar General for Scotland* | HMSO | Annual | |
| [QRL.6] | Society of County Treasurers in association with Society of Metropolitan Treasurers and Association of District Council Treasurers | *Block Grant Indicators* | Society of County Treasurers | Annual | First publication September 1981 |
| [QRL.7] | CIPFA/SIS | *Block Grant Statistics* | CIPFA | Annual | Originally published by Cambs County Council |
| [QRL.8] | CIPFA/SIS | *Capital Expenditure of Debt Financing Statistics* | CIPFA | Annual | |
| [QRL.9] | CIPFA/SIS | *Cemeteries and Crematoria Statistics (Actuals)* | CIPFA | Annual | |
| [QRL.10] | Home Office | *Criminal Statistics –England and Wales* | HMSO | Annual | |

| | | | | | |
|---|---|---|---|---|---|
| [QRL.11] | Home Office | *Criminal Statistics –England and Wales. Supplementary Tables –Volume 3* | Home Office | Annual | |
| [QRL.12] | House of Commons Command Paper | *Criminal Statistics (Scotland)* | HMSO | Annual | Now replaced by *Statistical Bulletin* obtainable from SHHD, St. Andrews House, Edinburgh |
| [QRL.13] | CIPFA/SIS | *Direct Labour Organisations Statistics (Actuals)* | CIPFA | Annual | First publication 1981/82 |
| [QRL.14] | Department of Environment –Northern Ireland | *District Council Rate Statistics* | DOE –Northern Ireland | Annual | |
| [QRL.15] | Department of Environment –Northern Ireland | *District Councils Summary of Statements of Accounts* | HMSO, Belfast | Annual | |
| [QRL.16] | Central Statistical Office (CSO) | *Economic Trends* | HMSO | Monthly | |
| [QRL.17] | Office of Population Censuses and Surveys (OPCS) | *Electoral Statistics* | HMSO | Annual | For years up to 1973, see *Registrar General's Statistical Review of England and Wales Part II* |
| [QRL.18] | Department of Employment | *Employment Gazette* | Department of Employment | Monthly | Formerly *Ministry of Labour Gazette* |

| Reference | Author or Organisation | Title | Publisher | Frequency or Date | Remarks |
|---|---|---|---|---|---|
| [QRL.19] | CIPFA/SIS | *Environmental Health Statistics –Actuals* | CIPFA | Annual | New publication, first issue 1982/83 |
| [QRL.20] | CIPFA/SIS | *Finance and General Statistics* | CIPFA | Annual | |
| [QRL.21] | Society of County Treasurers (SCT) | *Financial and General Statistics of County Councils* | SCT | Annual | |
| [QRL.22] | CSO | *Financial Statistics* | HMSO | Monthly | |
| [QRL.23] | CIPFA/SIS | *Fire Statistics (Actuals)* | CIPFA | Annual | |
| [QRL.24] | CIPFA/SIS | *Fire Statistics (Estimates)* | CIPFA | Annual | |
| [QRL.25] | Home Office | *Fire Statistics –United Kingdom* | HMSO | Annual | Now includes Fire Brigades Supplement |
| [QRL.26] | The Rating and Valuation Association | *General Rate Poundages and Products (England and Wales)* | The Rating and Valuation Association | Annual | |
| [QRL.27] | House of Commons Paper | *The Government's Expenditure Plans (White Paper)* | HMSO | Annual | |
| [QRL.28] | Department of the Environment –Local Government Finance Directorate | *Grant Related Expenditure –How the Expenditure Heads of Local Authorities are Assessed for Block Grant* | DOE | Annual | |
| [QRL.29] | CIPFA/SIS | *Highways and Transport Statistics (Actuals)* | CIPFA | Annual | |
| [QRL.30] | CIPFA/SIS | *Highways and Transportation Statistics (Estimates)* | CIPFA | Annual | |
| [QRL.31] | Board of Inland Revenue | *Inland Revenue Statistics* | HMSO | Annual | |
| [QRL.32] | Joint Manpower Watch Group | *Joint Manpower Watch Returns – England* | Department of the Environment | Quarterly | Press release available on request from DOE |

| | | | | | |
|---|---|---|---|---|---|
| [QRL.33] | Scottish Office and COSLA | *Joint Manpower Watch Survey Results* | Scottish Office | Quarterly | Press release available on request from the Scottish Office |
| [QRL.34] | CIPFA/SIS | *Local Authority Airports –Accounts and Statistics* | CIPFA | Annual | Formerly *Financial Costs and Statistics of Local Authority Airports* |
| [QRL.35] | CIPFA/SIS | *Local Authority Airports –Accounts and Statistics (Estimates)* | CIPFA | Annual | Now ceased |
| [QRL.36] | Government of Northern Ireland –Ministry of Development | *Local Authority Financial Returns* | HMSO, Belfast | Annual | No longer published – 1971/72 last edition |
| [QRL.37] | Joint Manpower Watch Group | *Local Authority Manpower in Wales* | Welsh Office | Quarterly | Press release available on request from the Welsh Office |
| [QRL.38] | Government of Northern Ireland –Ministry of Development | *Local Authority Rate Statistics* | HMSO, Belfast | Annual | No longer published – 1972/73 last edition |
| [QRL.39] | CIPFA and SCT | *Local Authority Superannuation Fund – Investment Statistics (years)* | CIPFA/SCT | Annual | First published 1982 |
| [QRL.40] | CIPFA/SIS | *Local Government Comparative Statistics* | CIPFA | Annual | First published 1982 |
| [QRL.41] | Department of the Environment and Watch Office | *Local Government Financial Statistics – England and Wales* | HMSO | Annual | |

| Reference | Author or Organisation | Title | Publisher | Frequency or Date | Remarks |
|---|---|---|---|---|---|
| [QRL.42] | Local Authorities Conditions of Service Advisory Board (LACSAB) | *Local Government Manpower* | LACSAB | Alternate months beginning February | |
| [QRL.43] | CIPFA | *Local Government Trends* | CIPFA | Annual | |
| [QRL.44] | Municipal Publications Limited | *The Municipal Year Book* | Municipal Publications Limited | Annual | |
| [QRL.45] | SCT | *Needs Element Statistics 1980/81* | SCT | | In 1979/80 entitled *Rate Support Grant Statistics*. These were the only issues –now replaced by [QRL.6] and [QRL.62] |
| [QRL.46] | Department of Employment | *New Earnings Survey (Parts A and B)* | HMSO | Annual | |
| [QRL.47] | Policy Planning and Research Unit, Department of Finance and Personnel, Stormont | *Northern Ireland Annual Abstract of Statistics* | HMSO, Belfast | Annual | Formerly *Northern Ireland Digest of Statistics* |
| [QRL.48] | OPCS | *OPCS Monitor –Reference EL* | OPCS | Annual | |
| [QRL.49] | Registrar General for Scotland | *Parliamentary and Local Government Electors* | Registrar General for Scotland | Annual | Available in June each year on application to Registrar General for Scotland |

| | | | | | |
|---|---|---|---|---|---|
| [QRL.50] | CIPFA/SIS | *Police Statistics (Actuals)* | CIPFA | Annual | |
| [QRL.51] | CIPFA/SIS | *Police Statistics (Estimates)* | CIPFA | Annual | |
| [QRL.52] | CIPFA/SIS | *Probation Statistics (Actuals)* | CIPFA | Annual | |
| [QRL.53] | Home Office | *Probation Statistics –England and Wales* | HMSO | Annual | |
| [QRL.54] | CIPFA/SIS | *Probation Statistics (Estimates)* | CIPFA | Annual | |
| [QRL.55] | Scottish Office | *Public Expenditure to 1987/88 (A Commentary on the Scotland Programme)* | Scottish Office | Annual | |
| [QRL.56] | CIPFA/SIS | *Public Libraries Statistics (Actuals)* | CIPFA | Annual | |
| [QRL.57] | CIPFA/SIS | *Public Libraries Statistics (Estimates)* | CIPFA | Annual | |
| [QRL.58] | CIPFA/SIS Information Service (CIPFA/SIS) | *Rate Collection Statistics (Actuals)* | CIPFA | Annual | |
| [QRL.59] | Department of the Environment –Welsh Office | *Rate Rebates* | Department of the Environment | Annual | Now ceased –last issue 1982/83 |
| [QRL.60] | House of Commons Paper | *Rate Support Grant Report (England)* | HMSO | Annual | Reference relates to both Main and Supp. Reports |
| [QRL.61] | House of Commons Paper | *Rate Support Grant Report (Wales)* | HMSO | Annual | Reference relates to both Main and Supp. Reports |
| [QRL.62] | Association of County Councils; Welsh Counties Committee; Association of District Councils; Council for the Principality | *Rate Support Grant (Wales)* | Association of County Councils; Welsh Counties Committee; Association of District Councils; Council for the Principality | Annual | |
| [QRL.63] | Department of the Environment and the Welsh Office | *Rating and Rateable Values in England and Wales* | HMSO | Annual | Last published for 1975/76 |

| Reference | Author or Organisation | Title | Publisher | Frequency or Date | Remarks |
|---|---|---|---|---|---|
| [QRL.64] | CIPFA Scottish Branch Branch | *Rating Review (Part I)* | CIPFA –Scottish Branch | Annual | |
| [QRL.65] | CIPFA –Scottish Branch | *Rating Review (Part 2)* | CIPFA –Scottish Branch | Annual | First Issue – May 1984 |
| [QRL.66] | Commissioner for Local Administration in Scotland | *Report of Commissioner for Local Administration in Scotland* | Commissioner for Local Administration in Scotland | Annual | |
| [QRL.67] | House of Commons Paper | *Report of HM Chief Inspector of Constabulary* | HMSO | Annual | |
| [QRL.68] | House of Commons Command Paper | *Report of HM Chief Inspector of Constabulary for Scotland* | HMSO | Annual | |
| [QRL.69] | House of Commons Paper | *Report of HM Chief Inspector of Fire Services* | HMSO | Annual | |
| [QRL.70] | House of Commons Command Paper | *Report of HM Chief Inspector of Fire Services (Scotland)* | HMSO | Annual | |
| [QRL.71] | Convention of Scottish Local Authorities (COSLA) | *Report on Rate Support Grant Order* | COSLA | Annual | Available on application to COSLA, 16 Moray Place, Edinburgh |
| [QRL.72] | House of Commons Paper | *Report on Rate Support Grant (Scotland) Order* | HMSO | Annual | Reference relates to both Main and Supp Reports |
| [QRL.73] | Department of Transport; Scottish Development Department; Welsh Office | *Road Accidents –Great Britain* | HMSO | Annual | |
| [QRL.74] | Department of Transport | *Road Accidents Statistics –English Regions* | Department of Transport | Annual | First edition 1981 |
| [QRL.75] | Welsh Office | *Road Accidents –Wales* | Welsh Office | Annual | |

| | | | | | |
|---|---|---|---|---|---|
| [QRL.76] | The Scottish Office | *Scottish Abstract of Statistics* | HMSO | Annual | |
| [QRL.77] | Scottish Office | *Scottish Local Government Financial Statistics* | HMSO | Annual | |
| [QRL.78] | The Scottish Office | *Scottish Transport Statistics* | The Scottish Office | Annual | |
| [QRL.79] | SCT | *Smallholdings Statistics* | Society of County Treasurers | Annual | |
| [QRL.80] | Scottish Education Department –Social Work Services Group | *Social Work Case Statistics –Statistical Bulletin* | Scottish Education Department | Annual | First published for year 1976/77 – now ceased. Statistics are now available from: Social Work Services Group, Statistics Branch, Room 423, 43 Jeffery Street, Edinburgh EH1 1DN |
| [QRL.81] | Home Office | *Statistical Bulletins* | Home Office | Various | Copies obtainable from Home Office, Statistics Dept., Tolworth Tower, Surbiton, Surrey |

| Reference | Author or Organisation | Title | Publisher | Frequency or Date | Remarks |
|---|---|---|---|---|---|
| [QRL.82] | Home Office | *Statistics on Deaths Reported to Coroners: England and Wales* | Home Office | Annual | This is a Home Office Statistical Bulletin |
| [QRL.83] | Welsh Office | *Statistics on Road Lengths in Wales* | Welsh Office | Annual | |
| [QRL.84] | Department of the Environment –Local Government Policy Directorate | *The Technical Handbook of Grant Related Expenditure* | DOE | Annual | |
| [QRL.85] | SCT | *Trading Standards Statistics (Actuals)* | Society of County Treasurers | Annual | |
| [QRL.86] | SCT | *Trading Standards Statistics (Estimates)* | Society of County Treasurers | Annual | |
| [QRL.87] | Department of Transport; Scottish Development Department; Welsh Office | *Transport Statistics –Great Britain* | HMSO | Annual | |
| [QRL.88] | CSO | *United Kingdom National Accounts (The CSO Blue Book)* | HMSO | Annual | Formerly: *National Income and Expend.* |
| [QRL.89] | CIPFA/SIS | *Waste Collection Statistics (Actuals)* | CIPFA | Annual | Formerly *Refuse Collection Statistics (Actuals)*. Title changed 1976/77 issue |
| [QRL.90] | CIPFA/SIS | *Waste Disposal Statistics (Actuals)* | CIPFA | Annual | |
| [QRL.91] | CIPFA/SIS | *Waste Disposal Statistics (Estimates)* | CIPFA | Annual | |
| [QRL.92] | Welsh Office | *Welsh Local Government Financial Statistics* | Welsh Office | Annual | |

| | | | | |
|---|---|---|---|---|
| [QRL.93] | Commission for Local Administration in England | *Your Local Ombudsman –Report for the Year* | Commission for Local Administration in England | Annual |

# BIBLIOGRAPHY


[B.1] Barker W. *Local Government Statistics*, Institute of Municipal Treasurers and Accountants (now CIPFA), 1965

[B.2] Copeman H. *The National Accounts–A Short Guide*, HMSO, 1981

[B.3] Esslemont G.B. *The Common Good in Scotland*, Local Government Finance, Vol. L111, No.10, October 1949

[B.4] Gilder P. *A General Guide to the Block Grant System in England 1983/84*, CIPFA, May 1983

[B.5] Griffin C. "Developments in Local Authority Comparative Statistics", *Statistical News*, August 1981

[B.6] Griffin C. "CIPFA's Database Services", *Statistical News*, November 1983

[B.7] Heald D. *Public Expenditure*, Martin Robertson, 1983

[B.8] Ilersic A.R. *Rate Equalisation in London*, Institute of Municipal Treasurers and Accountants, 1967

[B.9] Jenkins J. and Rose M. *Rate Support Grant: the System and Its Application to London*, GLC Research Memorandum RM 479, 1976

[B.10] Peretz J. and Cooke R. "The National and Local Government Statistical Liaison Committee", *Statistical News*, May 1975

[B.11] Prentice G.R. *The Statistical Information Service*, Chartered Institute of Public Finance and Accountancy (CIPFA), June 1976

[B.12] Sellwood R. and Griffin C. "New Developments in Statistics at CIPFA", *Statistical News*, November 1982

[B.13] Department of Employment, *British Labour Statistics: Historical Abstracts 1886/1968*, HMSO, 1982

[B.14] Department of Employment, *British Labour Statistics Year Book*, HMSO, annual 1969 to 1976

[B.15] Maud Committee, *Committee on the Management of Local Government, Vol.1*, HMSO, 1967

[B.16] Audit Commission for Local Authorities in England and Wales, *Cost Trees for (Authority) 1984/85*, Audit Commission for Local Authorities in England and Wales, undated

[B.17] Department of the Environment, *Digest of Environmental Protection and Water Statistics*, HMSO, annual

[B.18] Office of Population Censuses and Surveys, Social Survey Division, *Electoral Registration in 1981*, HMSO, 1982

[B.19] Office of Population Censuses and Surveys, Social Survey Division, *The Electoral Registration Process in the United Kingdom*, HMSO, 1982

[B.20] Chartered Institute of Public Finance and Accounatncy, Statistical Information Service, *Environmental Health Core Function Statistics–Report of Pilot Survey*, CIPFA/SIS, 1982

[B.21] Chartered Institute of Public Finance and Accountancy, *Financial Information Service*, 25 volumes, CIPFA, various

[B.22] Central Statistical Office, *Financial Statistics, Explanatory Handbook*, HMSO, annual

[B.23] Chartered Institute of Public Finance and Accountancy, *Guide to Local Authority Finance*, CIPFA, June 1984

[B.24] Institute of Local Government Studies (University of Birmingham), *Inter Authority–Comparisons (Report on Seminar sponsored by SOLACE in April/May 1979)*, Institute of Local Government Studies, December 1979

[B.25] Chartered Institute of Public Finance and Accountancy, *Issues in Public Sector Accounting No.3*, CIPFA, January 1984

[B.26] Department of the Environment, Welsh Office, *Local Authority Annual Reports*, HMSO, 1981

[B.27] Central Office of Information, *Local Government in Britain–Reference Pamphlet 1*, HMSO, February 1980

[B.28] Scottish Information Office, *Local Government in Scotland, Fact Sheet No.28*, Scottish Information Office, October 1984

[B.29] Audit Commission for Local Authorities in England and Wales, *Profile of (Authority) 1983/84*, Audit Commission for Local Authorities in England and Wales, October 1983

[B.30] H.M. Treasury, *Public Expenditure: Planning in cash (Economic Progress Report No.139)*, H.M. Treasury, November 1981

[B.31] Scottish Office, *Publication of Financial and Other Information by Local Authorities*, HMSO, 1982

[B.32] ACC/AMA/ADC/LBA/GLC, *Rate Support Grant (England) 1983/84*, ACC/AMA/ADC/LBA/GLC, April 1983

[B.33] Association of Metropolitan Authorities, *The RSG Settlement 1983/84–A Guide*, Association of Metropolitan Authorities, December 1982

[B.34] Chartered Institute of Public Finance and Accountancy, Statistical Information Service, *Statistical Information Service*, CIPFA/SIS, annual (March)

[B.35] Northern Ireland Information Service, *Ulster Year Book*, Northern Ireland Information Service, annual

[B.36] Central Statistical Office, *United Kingdom National Accounts, Sources and Methods*, HMSO, 1985

# LIST OF APPENDICES


| | |
|---|---|
| 1 Companion volumes dealing with local government | 205 |
| 2 Excerpt from *Local Government Statistics* [B.1] | 206 |
| 3 Organisation of the Scottish Office | 210 |
| 4 Revenue and Capital Outturn Forms–Lists | 211 |
| 5 Compilation of CIPFA/SIS Volumes | 216 |
| 6 Revenue and Capital Outturn Forms–Specimens | 218 |
| 7 Budget Estimates | 231 |
| 8 Rate Fund Services | 243 |
| 9 Quarterly Financial Returns to Central Government | 250 |
| 10 Capital Expenditure and Debt Financing | 256 |
| 11 Local Financial Return (LFR 11) | 264 |
| 12 Rate Support Grants | 267 |
| 13 Special Funds | 272 |
| 14 Rateable Values | 273 |
| 15 Manpower Statistics | 285 |
| 16 Public Protection: Fire | 310 |
| 17 Electorate and Election Results | 312 |
| 18 Environmental Health Statistics | 316 |
| 19 Direct Labour Organisation Statistics | 326 |

# APPENDIX 1

The statistics of various aspects of the services provided by local authorities in the UK are dealt with in the following companion volumes:

| Volume | Title |
|---|---|
| Volume I | Personal Social Services |
| | Voluntary Organisations in the Personal Social Services |
| Volume III | Housing in Great Britain |
| | Housing in Northern Ireland |
| Volume IV | Leisure |
| | Tourism |
| Volume VII | Road Passenger Transport |
| Volume VIII | Land Use |
| Volume X | Ports and Inland Waterways |
| | Civil Aviation |
| Volume XV | Crime |

# APPENDIX 2

*Local Government Statistics* (W. Barker)

(A guide to statistics on local government finance and services in the United Kingdom at August, 1964)

This Appendix which is referred to in paragraph 1.2.1 consists of the introduction to the above named study and contains a catalogue of previous guides to local government statistics.

## INTRODUCTION

Several guides to local government statistics have from time to time been published. In 1951, the then Institute of Public Administration published *Local Government Financial Statistics in England and Wales* written by Mr. V. J. Oxley, Deputy Borough Treasurer of Ipswich. This was followed in 1953 by the publication by H.M.S.O. of *Guides to Official Sources No. 3—Local Government Statistics*, the work of the Interdepartmental Committee on Social and Economic Research, of which the Chairman of the Sub-Committee on Local Government Statistics was Mr. D. N. Chester. This work concentrated mainly on general and financial statistics collected by Government departments and by the Institute of Municipal Treasurers and Accountants and the Society of County Treasurers. The Guide contains a very useful section on historical local government statistics, beginning with the Poor Rate Return of 1839—which itself contains reference to figures for earlier years—up to reports of Departmental Committees and of Royal Commissions of the present century. Information is also given on historical statistics of Scottish local government.

In 1955, the Journal of the Royal Statistical Society published a paper by Mr. B. J. Brocks, Research Officer of the Institute of Municipal Treasurers and Accountants, on "Statistics available on the Local Government Services". It was one of a series of thirty-nine articles concerned with sources of statistics and later published in two volumes for the Society:

*Sources and Nature of Statistics of the United Kingdom*
edited by Maurice G. Kendall, published by Oliver and Boyd, Edinburgh, Volume 1 (1952), Volume 2 (1957).

Articles of local government interest include:

Volume I "Housing Statistics"—Marian Bowley
"Rail and Road"—Frederick A. A. Menzler

II "Population and Vital Statistics"—Eugene Grebenik

"Medicine and Medical Services in the U.K."—Bernard Benjamin
"Local Government Statistics"—Bernard J. Brocks
"Statistics of Education"—Doris M. Lee
"Regional Economic Statistics"—Kenneth S. Lomax
"Criminal Statistics"—Tom S. Lodge

In preparing the present publication acknowledgement must therefore be made of the work already done in this field, which makes the task of tracing sources published up to 1955 so much easier.

Since that time, however, there has been a very great increase in the volume of published material available, and the decision to include sources of operational statistics which might be of use to research workers has widened the scope of the present survey, so that it has become much more comprehensive than any that has gone before in this field.

The plan of this guide has been to list current sources of statistics and to give a brief factual description of each to indicate its scope. For reasons of space much has had to be passed over in the narrative, and research workers will often find in the publication itself more information than could be mentioned here. Sometimes, too, with annual publications, particularly annual reports of departments, the material dealt with changes from year to year.

A warning must be given that many of the statistics referred to should be used with caution. Obviously, local conditions have an effect on the costs of particular services, and the figures themselves can rarely give an indication of the quality or standard of the service they represent.

Sources of statistics have been grouped in sections according to the principal headings of information they contain, and have then been numbered within the sections. Cross-referencing has been necessary when a publication provides statistics on more than one service. Finally an index is provided.

Not all the sources contain original material: mention is made of some like the *Annual Abstract of Statistics* and several year books and other works of reference, because they may be more handy than the original source.

Most are annual publications, but a number of White Papers, committee reports and research studies are referred to because they are useful as a source of information as at a certain date. For reasons of space, with few exceptions, only those recently published have been considered. A number of Government publications concerned in some way with local government, but omitted from this survey because they are now only of historical interest, or because they are statements of policy rather than of fact, will be found in the *Annual Catalogue of Government Publications* (3s) and the *Sectional Lists* (free) obtainable from H.M.S.O. The lists are revised from time to time and gather together the publications for which individual departments are responsible.

Other sources, such as articles in the technical Press and Ministers' answers to questions in the House, have only occasionally been covered, but these are often fruitful sources of statistics. The *Municipal Review*, published by the Association of Municipal Corporations, the *County Councils Gazette*, the *Municipal Journal*, *Local Government Chronicle*, and other local government journals, contain regular digests of items of local government interest in *Hansard*. A number of articles referred to in this survey have appeared in *Local Government Finance*, which is published monthly by the Institute.

The fact that a source of statistics in a particular field of local government activity is not mentioned here does not necessarily mean that no information is available. The societies listed at the end of this book, Government departments and individual local authorities, frequently collect information of an *ad hoc* nature on many subjects, which is privately circulated. Research workers seeking such information should approach the appropriate body.

Some of the services for which few statistics relating to individual local authorities are at present available will shortly be covered by the Institute. Returns on sewage purification and disposal costs, baths and laundries, airports, art galleries, and housing advances and housing maintenance, are in draft, and others are planned to follow.

There is, indeed, a great and rapidly growing amount of information available if the seeker after knowledge knows where to look, or knows the most likely source of such information even if it is not published. This guide has been produced in the knowledge that it cannot be complete, but it is hoped that it may help all who are interested in local government statistics to find their way around.

*September 1964*

# APPENDIX 3

**Organisation of the Scottish Office**

The Scottish Office comprises the following sub-divisions or "departments" relevant to this review:

a) Scottish Development Department (SDD)
b) Scottish Education Department (SED)
c) Scottish Home and Health Department (SHHD)
d) Finance Division

Frequently in statistics publications or data source documents reference is made to the sub-division or "department" rather than to the Scottish Office itself.

# APPENDIX 4

## Revenue and Capital Outturn Forms –Lists

As explained in section 3.2, every local authority makes annual statutory financial returns to central government. These relate to all the accounts identified in section 3.3.

In England and Wales, revenue and capital returns are made on separate forms – see separate Revenue Outturn and Capital Outturn lists in this Appendix. By contrast the Scottish Local Financial Returns (LFR) relate to capital as well as revenue transactions. Selected specimens of the forms themselves are contained in Appendix 6.

The forms listed herein constitute the major basic sources of the publication, by central government, of local authority outturn financial statistics in Great Britain. (See also RS1B in Appendix 8.)

**ANNEX A**

**REVENUE OUTTURN 1981-82 – LIST OF FORMS (Note 1)/CHECK LIST**

| Form number | Service | Date returned to DOE/Welsh Office |
|---|---|---|
| RO1 | Education | |
| RO2 | Libraries, museums and art galleries | |
| RO3A | Personal social services<br>: residential care (children in care) | |
| RO3B | : residential care (other) | |
| RO3C | : support services (community care) | |
| RO3D | : support services (other) | |
| RO3E | : field work; research and development<br>and Port Health Administration | |
| RO4 | Housing other than HRA<br>Advances under the Housing and SDA Acts | |
| RO5 | Careers service administration<br>Sheltered employment and workshops | |
| RO7 | Police service | |
| RO8 | School crossing patrols<br>Registration of electors | |
| RO9 | Fire service | |
| RO10 | Magistrates and other courts<br>Civil defence | |
| RO11 | Probation and after care | |
| RO12 | Agriculture and fisheries<br>Consumer protection | |
| RO13A | General administration<br>Cost of rate collection | } |
| RO13B | Analysis of recharged administration | |
| RO14 | Refuse collection and disposal<br>Environmental health<br>Cemeteries and crematoria | |
| RO15 | Recreation, parks and swimming pools<br>Town and country planning | |
| RO17 | Miscellaneous Services | |
| RO18 | Highways<br>Public lighting | |
| RO19 | Parking of vehicles<br>Road Safety Activities<br>Administration (local transport)<br>Public passenger transport (non-trading)<br>Concessionary fares | |
| | Housing Revenue Account (Note 2) | |
| TSR1 | Trading services | |
| TSR2 | Passenger transport (trading undertakings) | |
| RS1A | Rate fund services summary of RO forms | |
| RS1B | Rate fund services summary revenue expenditure and income not entered on RO/CO forms | |
| RS2 | Supplementary information not on RS1B | |
| RS3 | Claim form for exemption from holdback | |
| **Other Returns** | | |
| SF3 | Local Government Superannuation Funds and Police and Firemens' Pension scheme | |

Note 1: RO6, RO16 and (RO20 to RO22) are not used.

2: Housing Revenue Account data is collected by means of Housing Subsidy Claims form FH82.02. Form HRA1 has been discontinued.

Department of the Environment/Welsh Office

## NOTES FOR GUIDANCE IN COMPLETING 1981-82 REVENUE OUTTURN FORMS

1. These notes are in 4 parts, viz

| | |
|---|---|
| (a) General Notes and list of forms | Ref: DOE 12515A |
| (b) Revenue outturn forms (except RO1) | DOE 12514 |
| (c) Form RO1 (Education) | DOE 12512 |
| (d) Additional Notes: Forms RO3A/B/C/D/E (Personal Social Services) | DOE 12511 |

Changes and additions since last year are sidelined in the margin.

### GENERAL NOTES AND LIST OF FORMS/CHECK LIST (ANNEX A OVERLEAF)

2. The notes below provide general instructions on the completion of all forms in RO series, except from RO1 (Education) for which separate notes exist (see above). Additional notes are printed on many of the RO forms, and there are also separate additional notes for guidance in completing forms RO3A/B/C/D/E (see above). Notes for guidance in respect of all other outturn forms (RS1A/B, RS2/3, TSR1/2 and SF3) are printed thereon.

### STATUS OF FIGURES

3. To avoid undue delay in submitting returns where figures based on final accounts are not available in time to meet the deadline, authorities are asked to give their best estimates of the outturn with an appropriate footnote saying so. It is important to emphasise that the return of the forms should not be delayed until final audited figures are available if this would mean that figures could not be submitted in time for their use in the rate support grant negotiations.

4. In order to avoid duplicating requests for information, authorities' outturn forms are used by agreement with the local authority associations and CIPFA, as thebasis for CIPFA 'Actuals' statistical publications covering expenditure on various services by individual authorities.

### FORM RS1A URBAN PROGRAMME EXPENDITURE

5. Collection of data on Urban Programme expenditure is unchanged from 1980-81. Revenue outturn expenditure will be collected at summary level of detail only on form RS1A.

### SHADING AND "DUMMY" COLUMNS

6. Shading on the forms has been discontinued in respect of the sub-total or total lines, but has been retained in those boxes where entries are not expected.

In order to achieve compatibility within computer programs, forms TSR1, TSR2, and SF3 contain "dummy" columns. These columns have been pre-printed to show that a zero entry is appropriate.

### DEFINITIONS

7. The term LOCAL AUTHORITY means an authority that is a County Council, a Metropolitan County Council, a District Council, the Greater London Council, a London Borough Council, the Common Council of the City of London or the Isles of Scilly Council. OTHER BODIES means all local bodies, other than companies trading for profit, and authorities who have power to levy rates, taxes, tolls or dues or to require them to be levied, or joint authorities incurring expenditure on behalf of constituent authorities.

### RETURNING INSTRUCTIONS

8. Authorities should return the number of copies of each form that is shown on the front of the form. All the forms for a particular service are circulated to all authorities capable of performing that service. Authorities are asked to co-operate by sending back "nil returns" for any form where this is appropriate.

### INQUIRIES

9. Any inquiries on the Revenue Notes for Guidance should be made to the Department of the Environment telephone 01-212 4475 or where appropriate to the Department whose telephone number is shown on the front of the form concerned.

## CAPITAL OUTTURN 1981/82 – LIST OF FORMS

**Capital Outturn**

| Form number | Service | Date Returned to DOE |
|---|---|---|
| CO1 | Education | |
| CO2 | Libraries, museums and art galleries | |
| CO3 | DHSS service | |
| CO4 | Housing services | |
| CO5 | Careers service & sheltered workshops | |
| CO7 | Home Office Services | |
| CO12 | Agriculture, Fisheries & Consumer Protection | |
| CO13 | General Administration & Refuse Collection & Disposal | |
| CO14 | Environmental Health & Cemeteries & Crematoria | |
| CO15 | Recreation, parks and swimming pools | |
| CO16 | Town and country planning | |
| CO17 | Miscellaneous | |
| CO18 | Public Lighting & Tolled Facilities | |
| CO19 | Parking of Vehicles & Road Safety | |
| CO20 | Public Passenger Transport | |
| CO21 | Trading services (I) | |
| CO22 | (II) | |
| CS1 | Capital Summary | |
| **Other Returns** | | |
| SF1 | Special Funds (excluding capital funds and superannuation funds) | |
| SF2 | Capital & Capital Reserve Funds | |

**Forms Received by Local Authorities**

| CLASS | | FORMS |
|---|---|---|
| SHIRE COUNTIES | – | All Forms |
| MET COUNTIES | – | All Forms except CO1, CO5 |
| MET DISTRICTS | – | All Forms |
| ISLES OF SCILLY | – | All Forms |
| GLC | – | All Forms |
| CITY OF LONDON | – | All Forms except CO1 |
| INNER LONDON | – | All Forms except CO1 |
| OUTER LONDON | – | All Forms |
| SHIRE DISTRICTS | – | All Forms except CO1, CO5, CO7 |
| MISCELLANEOUS AUTHORITIES | – | Various Forms |

Department of the Environment/Welsh Office

## GENERAL NOTES AND LIST OF CAPITAL OUTTURN FORMS

1. The notes below provide general instructions on the completion of forms in the CO series. Additional notes are printed on some of the CO forms.

### STATUS OF FIGURES

2. To avoid undue delay in submitting returns where figures based on final accounts are not available in time to meet the deadline, authorities are asked to give their best estimates of the outturn with an appropriate footnote saying so. It is important to emphasise that the return of the forms should not be delayed until final audited figures are available if this would mean that figures could not be submitted in time for their use in the rate support grant negotiations.

3. In order to avoid duplicating requests for information, authorities' outturn forms are used by agreement with the local authority associations and CIPFA, as the basis for CIPFA 'Actuals' statistical publications covering expenditure on various services by individual authorities.

### FORM CS1 URBAN PROGRAMME EXPENDITURE

4. Collection of data on Urban Programme expenditure is unchanged from 1980/81. Capital outturn expenditure will be collected at summary level of detail only on form CS1.

### SHADING AND "DUMMY" COLUMNS

5. Shading on the forms has been discontinued in respect of the sub-total or total lines, but has been retained in those boxes where entries are not expected.

In order to achieve compatibility within computer programs, forms SF1, SF2 contain "dummy" columns. These columns have been pre-printed to show that a zero entry is appropriate.

### DEFINITIONS

6. The term LOCAL AUTHORITY means an authority that is a County Council, a Metropolitan County Council, a District Council, the Greater London Council, a London Borough Council, the Common Council of the City of London or the Isles of Scilly Council. OTHER BODIES means all local bodies, other than companies trading for profit, and authorities who have power to levy rates, taxes, tolls or dues or to require them to be levied, or joint authorities incurring expenditure on behalf of constituent authorities.

### RETURNING INSTRUCTIONS

7. Authorities should return the number of copies of each form that is shown on the front of the form. All the forms for a particular service are circulated to all authorities capable of performing that service. Authorities are asked to co-operate by sending back "nil returns" for any form where this is appropriate. In the list of forms (see overleaf) there is a column to record when you make returns and to show which forms are not yet completed.

### INQUIRIES

8. Any inquiries on the Notes for Guidance should be made to the Department of the Environment (telephone 01-212 8583 or 4613 or where appropriate to the Department whose telephone number is shown on the front of the form concerned.

DOE 12515B

# APPENDIX 5

In paragraph 3.2.2 there is a brief description of the CIPFA/SIS method of administering and producing the wide range of statistics which it publishes. This Appendix contains the titles of the volumes mentioned in the review (as at summer 1983) with matching lists of (a) the Working Parties responsible for their content and (b) the local authorities or other bodies compiling the volumes. Within the compiling authorities, the responsible officer in each case is the financial officer or treasurer.

**Compilation of CIPFA/SIS Volumes**

*Summer 1983*

| Volume | Working Party | Compiling Authority; Other Body |
|---|---|---|
| Finance and General | Administration | Cambridgeshire County Council |
| Rate Collection | | SIS |
| Capital Expenditure | | SIS/DOE |
| Local Government Comparative Statistics | | Cambridgeshire County Council |
| Direct Labour Organisation | | SIS |
| Local Government Trends | | West Midlands County Council |
| Cemeteries | Environmental Health | London Borough of Richmond |
| Crematoria | | Calderdale Metropolitan District |
| Waste Collection | | SIS |
| Waste Disposal (Actuals) | | SIS |
| Waste Disposal (Estimates) | | SIS |
| Environmental Health | | SIS (Scrutiny by Arun District Council) |
| Fire (Actuals) | Public Protection | Leicestershire County Council |
| Fire (Estimates) | | Leicestershire County Council |
| Police (Actuals) | | Lancashire County Council |

| | | |
|---|---|---|
| Police (Estimates) | | Lancashire County Council |
| Probation (Actuals) | Public Protection | East Sussex County Council |
| Probation (Estimates) | | East Sussex County Council |
| Administration of Justice (Actuals) | | East Sussex County Council |
| Highways and Transportation (Actuals) | Transportation | Notts County Council |
| Highways and Transportation (Estimates) | | Notts County Council; London Borough Harrow; Eastbourne District Council |

# APPENDIX 6

**Revenue and Capital Outturn Forms–Specimens**

As the outturn statistics forms listed in Appendix 4 are too numerous to reproduce in this review, a specimen of each of the following is shown in this Appendix:

a) RO 18 Local Transport, Highways and Public Lighting;

b) CO 2 Libraries, Museums and Art Galleries;

c) LFR 5 Roads and Transport (excluding Trunk Roads).

In general layout these are typical of the series of forms listed in Appendix 4.

Please note that in England and Wales, in addition to the notes for guidance contained in the RO and CO forms, a further set of guidance notes is issued by the DOE/Welsh Office. These bear a five digit reference number and contain definitions which may be of value to the user of local government statistics.

Department of the Environment/Welsh Office

RO 18 LOCAL TRANSPORT SERVICES: HIGHWAYS AND PUBLIC LIGHTING EXCLUDING MOTORWAYS AND TRUNK ROAD

# Rate fund services revenue account 1981-1982

Reference should be made to the general notes for guidance in completing revenue out-turn forms and to the additional notes on page 3 of this form.

Authorities in England

This form should be completed using the latest available figures and 2 copies returned to Department of the Environment FLAS Division, Room N4/11, 2 Marsham Street London SW1P 3EB, by 31 August following the end of the financial year.

Telephone enquiries concerning the completion of this form may be made to Department of Transport, 01-212 8432.

Authorities in Wales

This form should be completed using the last available figures and 3 copies returned to the Welsh Office, Economic and Statistical Services Division (ESS8), Room 2004, New Crown Building, Cathays Park, Cardiff CF1 3NQ, by 31 August following the end of the financial year.

Telephone enquiries concerning the completion of this form may be made to the Welsh Office, Cardiff 02222-825067 or 823963.

Name, telephone and extension of council officer to whom any enquiries may be made:

..............................................................................

..............................................................................

Telephone (STD Code) ..................(No) ..................(Ext) ....

Name of Local Authority

..............................................................................

Name of Local Authority

## RO 18 LOCAL TRANSPORT – HIGHWAYS AND PUBLIC LIGHTING 1981/82

**Entries should be in whole pounds only**

| Service | RØ18 Service Code No | EXPENDITURE: Employees (1) | Running expenses (including recharges) (2) | Debt charges (see general notes) (3) | Revenue contributions to capital outlay (to agree with entry on CO form) * (4) | TOTAL EXPENDITURE * (5) | INCOME: Specific government grants (exclude TSG - see note)* (6) | Sales (7) | Fees and charges (excluding recharges) (8) | Interest receipts (see general notes) £ (9) |
|---|---|---|---|---|---|---|---|---|---|---|
| 1. HIGHWAYS – MAINTENANCE | 410 053 095 | | | | | | | | | |
| 2. Depots, Vehicles, Plant and equipment | 410 053 595 | | | | | | | | | |
| 3. Other | 410 054 095 | | | | | | | | | |
| 4. TOTAL HIGHWAYS* | | | | | | | | | | |
| PUBLIC LIGHTING – ROADS<br>5. – Installation | 410 100 595 | | | | | | | | | |
| 6. – Maintenance & Operation | 410 101 095 | | | | | | | | | |
| 7. TOTAL PUBLIC LIGHTING* | | | | | | | | | | |
| 8. TOTAL FORM RO18 | 410 999 990 | | | | | | | | | |

| Other income (including income from recharges) £ (10) | Total income* £ (11) |
|---|---|
| | |
| | |
| | |
| | |
| | |
| | |
| | |
| | |

## ADDITIONAL NOTES FOR GUIDANCE

### 1 Exclusions

Expenditure and income relating to:–

(a) a private street or other private works;

(b) roads not chargeable to the highways account, eg housing estate roads charged to the housing accounts or work for statutory undertakers that is not part of a highways improvement scheme charged to revenue;

(c) street cleansing attributable to public health etc services, which should be entered on form RO 14.

(d) trunk roads (if recoverable), including the cost of participation in Road Construction Units.

### 2 Non-reimbursable Trunk Road Expenditure

Any expenditure on trunk roads (other than administration) which is not reimbursable should be entered under "Other" ie line 3.
Any expenditure on trunk road administration which is not recoverable should be included in local transport administration on form RO19.

### 3 District Councils as Agent of the County Council

Note 4c of General Notes for Guidance in completing revenue out-turn forms (except RO1) should be strictly followed.

### 4 Treatment of Transport Supplementary Grant

County Councils and the GLC should report the entirety of TSG receivable in the manner specified below: other authorities should not therefore differentiate on the returns between TSG and other contributions from the county authority.

Transport supplementary grant taken to the revenue account should be entered on the Revenue Summary Forms RS1B and RS1C but not in columns 6 – 10 of form RO 18. If any transport supplementary grant is directed to the revenue account in the first instance for expenditure of a capital nature which is then capitalised, the entry in column 4 'Revenue contributions to capital outlay' (and column 5 of form CO 18) should include this part of transport supplementary grant. TSG taken direct to the capital account or to a suspense account should be recorded on form RS1C, and as grant income on CO 18 where appropriate.

### 5 Depots, Vehicles, Plant, and Equipment

The gross expenditure and income (including income from recharges, which should be shown in Col 10) of general highway holding account for these items should be included in line 2, the amounts charged out being included in the other lines as appropriate.

Where separate public lighting holding accounts are maintained for any or all of these items they should be entered under lines 5 or 6 as appropriate.

### 6. Debt Charges on Transferred Assets

Outstanding debt charges on assets transferred through local government re-organisation should be entered by the transferee authority, even though the loan fund is not similarly transferred. See note 13 of general notes for guidance. Agent authorities managing such debt on behalf of a county (transferee authority) may enter debt charges provided that no entry is made on the form RO 18 of the county.

### 7 Traffic management

Expenditure relating to aids to movements such as the maintaining of traffic signals, signs, pedestrian crossings and other road markings should be included in maintenance, unless directly related to a scheme involving new construction or improvement of a highway.

Department of the Environment/Welsh Office

CO2 LIBRARIES, MUSEUMS AND ART GALLERIES

# Rate fund services capital account 1981 - 1982

Reference should be made to the general notes for guidance in completing capital out-turn forms

**Authorities in England**

This form should be completed using the latest available figures and 2 copies returned to the Department of the Environment, FLAS Division, Room N4/11, 2 Marsham Street London SW1P 3EB by 31 August following the end of the financial year.

Telephone enquiries about completion of this form may be made to the Department of Education and Science, Office of Arts & Libraries 01-928 9222 Ext 3440.

**Authorities in Wales**

This form should be completed using the latest available figures and 3 copies returned to the Welsh Office, Economic and Statistical Services Division (ESS8), Room 2004, New Crown Building, Cathays Park, Cardiff CF1 3NQ, by 31 August following the end of the financial year.

Telephone enquiries about completion of this form may be made to the Welsh Office, Cardiff 0222–825067 or 823963

Name, telephone number and extension of council officer to whom enquiries may be made

.......................................................................

.......................................................................

Telephone (STD)..................(No) ..................(Ext) ..................

Name of local authority

.......................................................................

Name of local authority

## CO2A/CO2B: LIBRARIES, MUSEUMS AND ART GALLERIES 1981-82

**EXPENDITURE**

| Service | CØ 2A Service Code No | Acquisition of land and existing buildings and works £ (1) | New Construction and Works: Dwellings £ (2) | New Construction and Works: Other £ (3) | Vehicles (including ships and aircraft) £ (4) | Plant machinery and equipment £ (5) | Payments to local authorities acting as agents £ (6) | Other capital expenditure £ (7) | Amounts charged direct to revenue or other funds and not capitalised (these items should all be included in Cols 1 to 7 £ (8) |
|---|---|---|---|---|---|---|---|---|---|
| 1. Library Services | 019 019 993 | | | | | | | | |
| 2. Museums and Art Galleries | 019 039 993 | | | | | | | | |
| 3. TOTAL FORM CO2A (Lines 1 and 2)* | 019 999 990 | | | | | | | | |

**INCOME**

| Service | CO2B Service Code No | Sale of fixed assets, eg land, buildings, vehicles, plant etc originally charged to capital: To other local authorities £ (1) | Sale of fixed assets: To other sources £ (2) | Loans (including advances from CLF or mortgage pool) £ (3) | Government grants and reimbursements £ (4) | Revenue contributions to capital outlay (agreeing with relevant entry on form RO2) £ (5) | Transfers from special funds (including advances from capital funds and renewals funds) £ (6) | Receipts from other local authorities: Under agency arrangements £ (7) | Receipts from other local authorities: Other £ (8) |
|---|---|---|---|---|---|---|---|---|---|
| 1. Library Services | 019 019 993 | | | | | | | | |
| 2. Museums and Art Galleries | 019 039 993 | | | | | | | | |
| 3. TOTAL FORM CO2B (Lines 1 and 2)* | 019 999 990 | | | | | | | | |

**Entries should be in whole pounds only**

| Total of Cols 1 to 7 minus Col 8<br>£<br>(9) | Amounts repaid to lenders or transferred to sinking funds out of capital monies<br>£<br>(10) | Capital receipts transferred to other services and special funds<br>£<br>(11) | Total of Cols 9 + 10 + 11*<br>£<br>(12) |
|---|---|---|---|
| | | | |
| | | | |
| | | | |

| Capital Receipts | | Total of Cols 1 to 10* | Outstanding loan debt at end of year (include outstanding advances from capital funds repayable with interest)* |
|---|---|---|---|
| Transferred from other services<br>£<br>(9) | Other<br>£<br>(10) | £<br>(11) | £<br>(12) |
| | | | |
| | | | |
| | | | |

**Transfer to Capital Summary form CS1*

LFR 5 (1982)

# LOCAL FINANCIAL RETURNS 1981-82
# ROADS AND TRANSPORT
# (excluding TRUNK ROADS)

______________________________ Regional/Islands/Council

**Please read the GENERAL NOTES below before completing the form. The REVENUE ACCOUNT and CAPITAL ACCOUNT notes should be consulted as the form is being completed.**

GENERAL NOTES

1. RETURN OF FORM: The information requested in this form will be used in the compilation of the publication "Scottish Local Government Financial Statistics". It is also used to establish the level of actual expenditure by local authorities for the Government's Public Expenditure planning purposes as well as in connection with the determination of rate support grant and other central government grants. It would be helpful to the Department and in the general interests of local authorities if the information could be made available without undue delay. The returns of out-turn information should accord with that prepared by each authority for its annual abstract of account, although if in some case accounts have not been finalised, best estimates of actual expenditure should be given. Two copies of the form should be completed and returned to the Scottish Office Finance Division, Room 7/23, New St. Andrew's House, Edinburgh EH1 3TB as soon as possible and in any event not later than 31 August 1982.

2. VALUE ADDED TAX: The Commissioners of Customs and Excise have made arrangements for the refund of Value Added Tax falling on purchases of goods and services for Local Authorities' non-business activities, except Value Added Tax on certain motor vehicles which is not refundable. Tax which is refundable should be excluded from the data entered on LFR forms. Non-refundable tax chargeable to any item which is chargeable as reckonable expenditure should be included in the data entered on the appropriate LFR form.

3. INTER-AUTHORITY TRANSACTIONS: Where a service is provided for one Department by another or for one authority by another, the transaction should be included in the returns for both services or authorities.

4. STREET CLEANSING: Street cleansing undertaken by the authority should be included under ROAD MAINTENANCE.

5. ROAD LIGHTING: Road lighting only should be returned on this form. Footway and stair lighting should be returned on LFR 9.

6. QUARRIES: Where quarries are part of the Highways Account and not treated separately as a revenue-producing undertaking, they should be included in this form.

7. ROAD SAFETY: Expenditure entries against this heading should be restricted to schemes or activities designed specifically for road safety purposes, eg. the provision of crash barriers, road safety campaigns which include publicity, propaganda, contributions to ROSPA etc. Expenditure on roadworks, road signs, signals etc. should be shown under ROAD MAINTENANCE.

NOTES continued on page 6

LOCAL FINANCIAL RETURNS 1981-82

ROADS AND TRANSPORT (excluding TRUNK ROADS)

| REVENUE ACCOUNT OUT-TURN (Note 6) | ITEM | ADMINISTRATION | ROAD MAINTENANCE (excluding Lighting) (Note 4) | |
|---|---|---|---|---|
| | | | GRITTING AND SNOW CLEARING | OTHER |
| EXPENDITURE | | 1 | 2 | 3 |
| Staff costs: pay (Note 10) | 0049 | | | |
| : other | 0054 | | | |
| Total | 0060 | | | |
| Running expenses (Note 11) | 0437 | | | |
| RF subsidies to transport | 0336 | | | |
| Revenue contributions to capital | 0360 | | | |
| Loan charges (Note 13) | 0380 | | | |
| TOTAL EXPENDITURE | 0440 | | | |
| INCOME | | | | |
| Grants (Other Than RSG) | 0441 | | | |
| :Specific Grants (Note 14) | 0444 | | | |
| :Other Grants (Note 14) | 0448 | | | |
| TOTAL | 0450 | | | |
| Fees and Charges | 0453 | | | |
| Sales | 0466 | | | |
| Other accounts and authorities | 0468 | | | |
| Interest (Note 15) | 0469 | | | |
| Rents and Other income | | | | |
| TOTAL INCOME | 0470 | | | |
| NET EXPENDITURE (0440 minus 0470) | 0482 | | | |

ADDITIONAL INFORMATION (1)

PAYMENTS TO PERSONS AND BODIES:
(included in RUNNING EXPENSES above) (Note 12)

£000

| | Subsidies | Grants to Private Persons | Grants to Private Bodies |
|---|---|---|---|
| Item | 0307 | 0321 | 0322 |
| Administration | | | |
| Gritting and Snow clearing | | | |
| Other maintenance | | | |
| Road lighting | | | |
| Car and lorry parking | | | |
| Road safety: traffic management, surveys | | | |
| Concessionary fares | | | |
| TOTALS | | | |

ADDITIONAL INFORMATION (2)

SUPPORT FOR TRANSPORT:
NON-LA UNDERTAKINGS (Col 9 above)

| | £000 |
|---|---|
| Buses | |
| Rail | |
| Ferries | |
| Harbours and Piers | |
| Air | |
| Pte | |
| TOTAL | |

RE 88458/12 TEL

LFR 5 (1982)

£000

| ROAD LIGHTING (Note 5) | CAR AND LORRY PARKING | ROAD SAFETY: TRAFFIC MANAGEMENT, SURVEYS (Notes 7 & 8) | CONCESSIONARY FARES | PASSENGER TRANSPORT | | TOTAL |
|---|---|---|---|---|---|---|
| | | | | LA (detailed on LFR 22) | NON-LA (detailed below) | |
| 4 | 5 | 6 | 7 | 8 | 9 | 10 |
| | | | | | | |
| | | | | | | |
| | | | | | | |
| | | | | | | |
| | | | | | | |
| | | | | | | |
| | | | | | | |
| | | | | | | |
| | | | | | | |
| | | | | | | |
| | | | | | | |
| | | | | | | |
| | | | | | | |
| | | | | | | |
| | | | | | | |
| | | | | | | |

ADD: Proportion of central administrative costs allocated to these services

TOTAL

LOCAL FINANCIAL RETURNS 1981-82

ROADS AND TRANSPORT

| CAPITAL ACCOUNT OUT-TURN (Notes 6 and 16) | ITEM | ROADS | ROAD LIGHTING |
|---|---|---|---|
| EXPENDITURE | | 1 | 2 |
| Acquisition of land and existing buildings | 0600 | | |
| New works (Note 17): dwellings | 0601 | | |
| other | 0602 | | |
| Vehicles and Vessels | 0603 | | |
| Plant, machinery and equipment | 0604 | | |
| Maintenance from capital | 0605 | | |
| Other capital expenditure on goods and services | 0606 | | |
| TOTAL EXPENDITURE ON ASSETS* | 0607 | | |
| EXPENDITURE ON GRANTS AND LOANS | | | |
| Payments on capital accounts to other local authorities, Central Government and abroad | 0609 | | |
| Other grants | 0610 | | |
| Other loans | 0612 | | |
| TOTAL EXPENDITURE ON GRANTS AND LOANS | 0613 | | |
| TOTAL CAPITAL EXPENDITURE (0607 + 0613) | 0614 | | |
| INCOME | | | |
| Specific grants | 0615 | | |
| Revenue contributions (from 0360) | 0616 | | |
| Loans | 0617 | | |
| Transfers from special funds | 0618 | | |
| From other local authorities | 0619 | | |
| Sale of assets | 0620 | | |
| Other capital income | 0621 | | |
| TOTAL INCOME | 0622 | | |
| DEBT | | | |
| Capital debt o/s at start of year | 0630 | | |
| New borrowing (from 0617) | 0631 | | |
| Repayments (inc. any repayments out of capital) | 0632 | | |
| Capital debt o/s at end of year | 0633 | | |
| LOAN CHARGES | | | |
| Repayments | 0640 | | |
| Interest | 0641 | | |
| Loans fund expenses | 0642 | | |
| TOTAL LOAN CHARGES (from 0380) | 0643 | | |

* Of which expenditure on assets bought from authorities

| | | | |
|---|---|---|---|
| | £000 | | |

RE 88458/12 TEL

LFR 5 (1982)

£000

| PARKING | ROAD SAVETY: TRAFFIC MANAGEMENT AND SURVEYS | NON-LA TRANSPORT UNDERTAKINGS | TOTAL |
|---|---|---|---|
| 3 | 4 | 5 | 6 |
| | | | |
| | | | |
| | | | |
| | | | |
| | | | |
| | | | |
| | | | |
| | | | |
| | | | |
| | | | |
| | | | |
| | | | |
| | | | |
| | | | |
| | | | |
| | | | |
| | | | |
| | | | |
| | | | |
| | | | |
| | | | |
| | | | |
| | | | |
| | | | |
| | | | |
| | | | |
| | | | |
| | | | |
| | | | |
| | | | |

GENERAL NOTES (cont.) LFR 5 (1982)

8. TRAFFIC MANAGEMENT: Should include only thos schemes specifically designed for traffic management purposes. The traffic management element in road schemes should not be separately apportioned.

9. ENQUIRIES: All enquiries in connection with this form should be made to Mrs Stewart, Scottish Office Finance Division (031-556 8400 ext. 5549).

REVENUE ACCOUNT NOTES

10. STAFF COSTS: Staff costs should include salaries and wages, national insurance contributions and superannuation contributions (under "pay") and other staff costs directly chargeable to the service.

11. RUNNING EXPENSES: Running expenses should include property costs, supplies and services, transport costs, payments to persons and other bodies and direct administration expenses.

12. PAYMENTS TO BODIES AND PERSONS: This section covers a heterogeneous group of payments. Subsidies comprise payments to commercial and other undertakings to provide public benefit. Payments of grant on revenue account include (a) grants to individual persons not employed by the local authorities and (b) payments to private voluntary bodies such as charities, universities, churches etc.

13. LOAN CHARGES: Loan charges arising from road construction works should be recorded in column 1.

14. SPECIFIC GRANTS: The information on grant income (other than rate support grants) has now been requested in 2 parts. Item 0441 should cover income from those specific grants relating to RSG relevant expenditure (ie those grants listed in Appendix E of the Report on the Rate Support Grant (Scotland) Order 1980); Item 0444 should show other grant income which will include grants towards job creation schemes, EEC grants, SDA revenue grants etc.

15. INTEREST: Only interest receipts from lending should be shown here. Interest receipts deriving from revenue balances should not be credited to a specific service but should be shown in total in the appropriate section of Form LFR 10.

CAPITAL ACCOUNT NOTES

16. Expenditure on professional services charged to capital should be included in expenditure on the asset.

17. Including site preparation and demolition work.

CERTIFICATE OF DIRECTOR OF FINANCE

I certify that, to the best of my knowledge and belief, the entries given in Form LFR 5 (1982) are correct.

Date ________________ Signature ________________
Director of Finance

Name and telephone number of local authority officer to whom enquiries may be made:-

Name ________________ Telephone No. ________________

# APPENDIX 7

### Budget Estimates

Prior to the commencement of each financial year, local authorities in Great Britain provide central government with budget estimates (see paragraph 4.2.5). At the same time, and on the same form, they provide revised expenditure estimates for the current year. In England and Wales, these figures are returned not only to DOE/Welsh Office but also to CIPFA/SIS which is responsible for the publication of budget statistics in [QRL.4]. In Scotland, a parallel return on form DO/BE is made to the Scottish Office which does not publish the results (see paragraph 4.2.6). But CIPFA in Scotland does publish the estimates in [QRL.5] having obtained the data indirectly i.e by questionnaire from the authorities themselves.

This Appendix includes a specimen copy of the England and Wales return (RER) plus Annexes 1, 2 and 3 plus Appendices A and B which contain definitions of value to the user of local authority statistics.

Department of the Environment/Welsh Office
Chartered Institute of Public Finance and Accountancy

**Ref. RER 83**
**PAGE 1**

**RETURN OF EXPENDITURE AND RATES FOR THE YEAR ENDING 31 MARCH 1983**

Local Authority.............................................................................................

*PLEASE READ THE SUPPORTING NOTES TO THIS RETURN BEFORE COMPLETION*

Inquiries concerning completion of the return should be made to the Department of the Environment: telephone 01-212 4475

NB. Changes in drafting or content from RER82/82A form are shown thus: †

**FINAL CHECKS**

Before you sign and return this form could you please check the following points which are prone to error

1. Units:
   - (i) rate poundages should be in pence to two places of decimals (eg 89.00p, NOT 89p) – lines 81 to 87, Cols (c) (d) & (e)
   - (ii) all other money items should be in £000s.
2. At lines 81 to 87 the maximum rate Col (e) should be greater than the minimum rate Col (c), with average rate Col (d) being between.
3. That the form is complete and totalled correctly.

Please return completed forms, not later than 26 March 1982, as follows:

| FIRST COPY to | Authorities in England | OR | Authorities in Wales |
|---|---|---|---|
| | Department of the Environment<br>FLAS Division, Room N4/11<br>2 Marsham Street, LONDON SW1P 3EB | | Mrs B Wilson<br>Welsh Office<br>Economic Services & Statistical Division (ES8)<br>New Crown Building, Cathays Park, CARDIFF CF1 3NQ |

SECOND COPY to CIPFA (Statistical Information Service)
c/o J E Barton Director of Finance & Administration
PO Box 1 Cambridge CB3 OAS
Tel 0223 358811 Ext 284 (or Ext 259)

Please indicate the name of the person who should be contacted if further information is required relating to this return.

Signed.............................................................................................

Name............................................................................................. Status.............................................................

Telephone (STD Code)......................................(No)......................................(Ext)..........................

Address to which all correspondence should be sent.

.............................................................................................................................................................

.............................................................................................................(Postcode)...........................................

**GENERAL NOTES**

Report in Cols 3 and 6 of lines 1 to 34 (on page 2 of this form) Net Revenue Expenditure, which is defined as gross expenditure less income specific to the service (eg Sales, Fees, Charges, Reimbursement, those Specific Grants listed in Annex 3, and Interest Receipts)

† Include in cols 1 & 4 (debt charges) leasing payments for assets counting as prescribed expenditure for capital control purposes. If you are unable to give a service breakdown include the total amount of such leasing payments within line 33 (Unallocated Contingencies) AND line 34 (Totals)of cols 1 & 4.

The assets affected by this treatment are: (i) leases of property for a period of more than 20 years; (ii) leases of property for a period of more than one year where a permanent building has been erected at the request of the authority (eg in sale/leaseback or lease/leaseback schemes); (iii) leases of vehicles where acquired for a period exceeding one year; (iv) plant and equipment acquired by finance leases

**AGENCY AND JOINT ARRANGEMENTS BETWEEN LOCAL AUTHORITIES**

(a) District Councils as Agents of the County Council

Where a District Council has expenditure and income on functions delegated by the county council, the district council should NOT report this expenditure. Full details of the net expenditure (obtained by consultation with the district council as necessary) should be included on the return made by the county council. The same procedure should be used for expenditure by the GLC and the London boroughs on metropolitan roads.

b. Joint Police Authorities

Constituent authorities of combined police authorities (other than the Metropolitan Police authority) should obtain from the Treasurer of the combined police authority an analysis of their net precept to show the relevant proportions of the following items for inclusion in the county authority's return:–

Line 7 Police – net current expenditure at November, 1981 prices (deducting interest receipts)
– debt charges
– total net expenditure at November, 1981
33 Unallocated contingencies (not pay and price changes)
35 Allowance for pay and price changes
56 Revenue contribution to capital outlay
57 Unallocated contributions to special and capital funds
63 Specific government grant.

c. Joint Probation Areas

Authorities which are part of a joint probation area should include the proportion of expenditure (gross of specific grant – see 63) attributable to their area.

d. Joint Committees

Where a service is provided by a joint committee each constituent authority should include that proportion of net expenditure of the committee attributable to its own area.

e. Contributions to and receipts from national education pools

Authorities are asked to record their latest estimates of net contributions to or receipts from national education pools, showing at line 38 the best estimates of the cash amount in each year. Accordingly no element of pooling contribution or receipt should be included in line 1 (education) or line 35 (allowance for pay and price increases).

f. Inter-authority recoupment adjustments - education

Authorities should enter at line 39 for each year their best estimates of the net payments to or receipts from other authorities in respect of extra-district recoupment. As in the case of the pooling transaction, no element of the net payment or receipt should be included in line 1 (education) or line 35 (allowance for pay and price changes).

In most cases the best estimates of the cash amounts for the year will be informed by the recoupment rates for 1981/82 given by the Local Education Authorities' Advisory Committee on Inter-Authority Payments in IAPC letter 7 May 1981 and the rates for 1982/83 if those are notified before the due date for return of this form.

**DOE 12508**

| | | | Column 1 1981/82 REVISED ESTIMATE £000 | Column 2 1982/83 BUDGET £000 |
|---|---|---|---|---|
| | 49 | "Current Expenditure" at Outturn Prices (In Col 1 repeat line 48 Col 2; In Col 2 repeat line 48 Col 5B) | | |
| | 50 | Rate Fund Contribution to Housing Revenue Avvount | | |
| | 51 | Net Expenditure on Standard Amount Rent Rebates | | |
| | 52 | Net Expenditure on Standard Amount Rent Allowances | | |
| | 53 | Net Expenditure on Rate Rebates | | |
| † | 54 | Debt Charges met from Rate Fund (excluding leasing payments). See note | | |
| † | 55 | Leasing Payments met from Rate Fund | | |
| | 56 | Revenue Contributions to Capital Outlay | | |
| | 57 | Unallocated Contributions to Special and Capital Funds | | |
| † | 58 | Unallocated Contributions to or from Direct Labour organisations | | |
| | 59 | Expenditure on Mandatory Student Awards | | |
| † | 60 | Interest Receipts on Revenue and other Balances – see notes | – | – |
| | 61 | Total Local Rate & Grant Borne Expenditure | | |
| | 62 | Specific Grant on Mandatory Student Awards | – | – |
| | 63 | Specific Grants as Defined for RSG Purposes | – | – |
| | 64 | Transport Supplementary Grant taken to Revenue | – | – |
| † | 65 | "Total" Expenditure | 0 | |
| | 66 | London Equalisation | | |
| | 67 | Block Grant : Anticipated Entitlement | – | – |
| † | 68 | Riots : net amount of qualifying expenditure after deducting Government assistance received | | 0 |
| † | 69 | Steel compensation etc : see note | – | – |
| | 70 | Variation in Balances | | |
| | 71 | Total Local Rate/Precept | | |

| | | | | |
|---|---|---|---|---|
| | 72 | Domestic Element of Rate Support Grant | | |
| | 73 | Balance of Transport Supplementary Grant | | |
| | 74 | Rate Fund Balance at 31 March 1982 | | 0 |
| | 75 | Net Debt at 1 April 1982 | 0 | |

| | | | | |
|---|---|---|---|---|
| † | 76 | Total Uncapitalised Capital Expenditure within cols 2 & 5 on page 2 | | |
| † | 77 | Total Interest Receipts on services within cols 2 & 5 on page 2 | | |

78 Price Base of 1982/83 Estimates (Lines 1 to 34)

1 = Estimates at Nov 81 prices : See Annex 1 for details
2 = Estimates at Outturn Prices.
3 = Some other Price Base (please specify)

| Debt charges | Other expenditure | Income |
|---|---|---|
| | | |

£000

79 Estimate Product of 1p Rate in 1982/83 on Actual RV

| |
|---|
| |

80 Estimated Population at June 1982

| |
|---|
| |

RATE LEVIES

| | | | | 1981/82 Amount (a) £000 | 1982/83 Amount (b) £000 | 1982/83 MINIMUM (c) p | 1982/83 AVERAGE (d) p | 1982/83 MAXIMUM (e) p |
|---|---|---|---|---|---|---|---|---|
| | 81 | District/Borough Rate | Rating Authorities ONLY | | | | | |
| † | 82 | Special District expenses | Rating Authorities ONLY | | | | | |
| | 83 | Parish/Community Precepts | Rating Authorities ONLY | | | | | |
| | 84 | Total Local Rate/Precept (as LINE 71) | ALL Authorities | | | | | |
| | 85 | County/GLC Precept | Rating Authorities ONLY | | | | | |
| | 86 | Metropolitan Police Precept | Rating Authorities ONLY | | | | | |
| | 87 | Total Rate/Precept | ALL Authorities | | | | | |

SUPPLEMENTARY RATES

| | | | £000 |
|---|---|---|---|
| † | 88 | Own Supplementary Rate (a) due to changes in rate & grant borne expenditure | |
| † | 89 | Own Supplementary Rate (b) due to alternative use of balances/loss of grant | |
| † | 90 | Total own supplementary rate | |
| † | 91 | Total Supplementary Precepts collected | |

1

**NOTES FOR GUIDANCE** PAGE 3

**ALL ITEMS ON THIS PAGE SHOULD BE REPORTED AT OUTTURN PRICES**

| | | |
|---|---|---|
| | 49 | Totals carried forward from Page 2 |
| | 50 | Expected net contribution to or from HRA net of reimbursements by the Welfare Authority for Welfare activities. |
| | 51 | Exclude administrative costs and expenditure above the standard (include in line 50): Report expenditure net of Rent Rebate Grant. |
| | 52 | Exclude Administration costs and expenditure above the standard (include in line 18): Report expenditure net of specific grant. |
| | 53 | Include Rate Rebates to the disabled under rating (Disabled Persons) Act 1978: |
| † | 54 | Lines 54 & 55 together should equate to the totals of cols 1 & 4 respectively on page 2: Note price base : outturn prices. |
| † | 55 | |
| | 56 | Note Price Base (outturn prices). Follow same practice as used for outturn forms If distinction between lines 56 and 57 is uncertain, follow convention used in own budget. |
| | 57 | Exclude revenue contributions from HRA, Trading or PTE. Accounts and contributions to Superannuation Funds. Do not include in line 57 repayments of principal or interest relating to loans from capital funds (show these as debt charges in Cols 2 and 4 on page 2). All contributions should be net: For JPAs see note on page 1. |
| † | 58 | Show here only direct subsidies to or from DLO s from the rate fund. Contributions to DLO reserve funds from the Rate Fund (Revenue Account) should be included in line 57. Charges by DLOs to rate fund services should be included with the running expenses of those services. |
| | 59 | Anticipated (Gross) expenditure at outturn prices. |
| † | 60 | Net of interest payments on revenue borrowing and discounts for prompt payment for rates. Exclude receipts arising from particular local services where these would normally be credited direct to those services. Include interest on revenue and other balances where this is credited direct to the Rate Fund revenue account. |
| | 61 | Total of lines 49 to 60 |
| | 62 | Should be 90% of line 59 |
| | 63 | See list at Annex 2 Exclude: Grants listed at Annex 3 : Grants at 62 and 64. Include National Parks Supplementary Grant: Report estimated total grants receivable: For JPAs see note on page 1. |
| | 64 | Counties and GLC only: TSG credited to Rate Fund Account only (see also line 73): Total TSG is thus lines 64 + 73. Report estimate total grants receivable allowing for increase orders. |
| † | 65 | Total of lines 61 to 64 (Note : definition was different for 1981/82). Equates to line 1 of Grant Claim Form BG 1982/83-3 |
| | 66 | |
| | 67 | London boroughs should include here grant payable in respect of GLC and ILEA; authorities in the Metropolitan Police area should include grant payable in respect of the Metropolitan Police. The GLC, ILEA and Metropolitan Police should leave this box **blank**. |
| † | 68 | Authorities qualifying for special Government financial assistance |
| † | 69 | Compensation for sharp losses of Rateable Value as the result of a closure of a single industrial plant or works. |
| † | 70 | Increase (+) or decrease (–). Include here Needs element of Rate Support Grant: adjustments under 2nd Increase Order for 1980/81. |
| | 71 | For 1981/82: Total of lines 61 to 70: For 1982/83: Total of lines 65 to 70. |

| | |
|---|---|
| 72 | Rating Authorities only. |
| 73 | Counties and GLC only: Balance of TSG not credited to Rate Fund (see 64) |
| 74 | Exclude Transport Board Payments by DOE in lieu of Rates. Allow for increase orders on which the Variation at line 70 is calculated. |
| 75 | Total debt for which authority is responsible and which relates to its own functions, whether the debt is administered by the authority or not. Include advances from Capital Funds. |

| | |
|---|---|
| 78 | Enter a single number in each box according to the price base at which estimates have been calculated |

| | |
|---|---|
| 79 | Exclusive 1p product as defined by the rate product rules. Rate rebates and the cost of rate collection are no longer met from the rate income account – see DOE letter dated 5 December 1980. |

| | |
|---|---|
| 80 | Authority's own estimate on a basis equivalent to that of the Registrar General: District Councils/London Boroughs should ensure consistency with their counties. |

| | |
|---|---|
| 81 | |
| 82 | |
| 83 | |
| 84 | |
| 85 | Include any special levies or precepts. Land drainage precepts to GLC. |
| 86 | |
| 87 | Cols (a) and (b) total lines 71 + 85 + 86 : ALL AUTHORITIES |

IN LINES 81 TO 87

Col (d) equals Col (b) divided by the penny rate product (line 79)

Cols (c) and (e) are the actual minimum and maximum rates levied.

| | |
|---|---|
| 88 | |
| 89 | |
| 90 | Line 88 + line 89 |
| 91 | RATING AUTHORITIES ONLY |

**ER83**

**RETURN OF EXPENDITURE AND RATES FOR THE YEAR ENDING 31 MARCH 1983 (RER 83)**

**Please use black ink to complete the return**

| Name of local authority | 1981/82 REVISED ESTIMATE (AT OUTTURN PRICES) | | | 1982/83 BUDGET (SEE NOTES FOR PRICE BASE) | | |
|---|---|---|---|---|---|---|
| | Debt charges (1) £000 | Other expenditure (2) £000 | Total (3) £000 | Debt charges (4) £000 (See Annex 2) | Other expenditure (5) £000 | Total (6) £000 |
| 1 Education | | | | | | |
| 2 School Meals and Milk | | | | | | |
| 3 Libraries | | | | | | |
| 4 Museums & Art Galleries | | | | | | |
| 5 Personal Social Services | | | | | | |
| 6 Port Health | | | | | | |
| 7 Police | | | | | | |
| 8 Fire Service | | | | | | |
| 9 Magistrates Courts | | | | | | |
| 10 Other Courts | | | | | | |
| 11 Probation & After-care | | | | | | |
| 12 School Crossing Patrols | | | | | | |
| 13 Civil Defence | | | | | | |
| 14 Registration of Electors | | | | | | |
| † 15 Highways, Lighting, Road Safety | | | | | | |
| † 16 Transport Admin. and Car Parking | | | | | | |
| 17 Net cost of Housing Act Advances | 0 | | | 0 | | |
| 18 Other (non HRA) Housing | | | | | | |
| 19 Careers Service | | | | | | |
| 20 Sheltered Employment and Workshops | | | | | | |
| 21 Consumer Protection | | | | | | |
| 22 Refuse Collection | | | | | | |
| 23 Refuse Disposal | | | | | | |
| 24 Environmental Health | | | | | | |
| 25 Swimming Pools, Sports and Recreation Centres | | | | | | |
| 26 Parks and Open Spaces | | | | | | |
| 27 Town and Country Planning | | | | | | |
| 28 Cemeteries | | | | | | |
| 29 Crematoria | | | | | | |
| 30 Agriculture and Fisheries | | | | | | |
| 31 General Administration (Unallocated) | | | | | | |
| 32 Other Services | | | | | | |
| 33 Unallocated contingencies | | | | | | |
| 34 Total of rows 1 to 33 | | | | | | |
| 35 Provision for Pay & Price Changes | 0 | 0 | 0 | | | 0 |
| 36 Total of rows 34 & 35 | 0 | | 0 | | | 0 |

| | Col 2 | Col 5A | Col 5B | Col 5C |
|---|---|---|---|---|
| | 1981/82 OUTTURN PRICES | NOV 81 PRICES | 1982/83 OUTTURN PRICES | REPRICING FACTORS |
| 37 In Col 2 repeat figure from line 36 Col 2; in Col 5B repeat figure from line 36 Col 5; for Cols 5A & 5C see note opposite | | | | |
| 38 National Education Pools | | | | 1.0500 |
| 39 Inter-Authority Education Recoupments | | | | |
| 40 Concessionary Fares (Elderly, blind, and disabled) | | | | |
| † 41 Rate Fund Contribution to Passenger Transport | | | | |
| 42 Cost of Rate Collection | | | | |
| 43 Rate Fund Contributions to Trading Services | | | | |
| 44 Parish Precepts | | | | |
| 45 Land Drainage Precepts | | | | |
| 46 Other Precepts | | | | |
| 47 Other Adjustments | | | | |
| 48 Total "Current" Expenditure (Total of lines 37 to 47) | | | | |

| NOTES FOR GUIDANCE | | PAGE 2 |
|---|---|---|
| | ALL 1982/83 BUDGET FIGURES AT LINES 1–33 ON THIS PAGE SHOULD BE GIVEN AT NOV 81 PRICES – SEE ANNEX 1 FOR DETAILS. [SEE PAGE 1 FOR EXPENDITURE DEFINITIONS] | EQUIVALENT LINE ON 1980/81 OUTTURN FORMS (WHERE APPLICABLE) |
| 1 | Exclude: Mandatory Student Awards (see 59): Pooling and Recoupment (see 38 & 39) | RO1 COL 15 (PART) |
| 2 | | RO1 COL 16 + 17 |
| 3 | | RO2 LINE 1 |
| 4 | | RO2 LINE 2 |
| 5 | Exclude: Bed & Breakfast and Temporary Accommodation (see 18): Structural Adaptations to Local Authority/New Town Corporation/Housing Association Homes (see 18). | RO3E LINE 14 |
| 6 | Include Port Health Precept Payments | RO3E LINE 10 |
| 7 | For Joint Police Authorities see note on page 1; | RO7 LINE 3 |
| 8 | | RO9 |
| 9 | | RO10 LINE 1 |
| 10 | | RO10 LINE 5 |
| 11 | Exclude Expenditure on Probation and Bail Hostels, except flat rate payments by Probation Committees. For Joint Probation Areas see note on page 1. | RO11 LINE 5 |
| 12 | | RO8 LINE 1 |
| 13 | | RO10 LINE 6 |
| 14 | | RO8 LINE 2 |
| † 15 | Do not deduct TSG taken to Revenue } Include in Cols 1/4 total estimated (separately identified) debt charges on public passenger transport | RO18 LINE 8+RO19 LINE 4 + RO19 LINE 12 COL 3 |
| † 16 | Exclude Expenditure incurred on behalf of other Authorities under Agency arrangements. | RO19 LINES 3 & 5 |
| 17 | Show net cost to rate fund – ie All Expenditure less all Income. | |
| 18 | Exclude: Contributions to HRA : Include non HRA Expenditure on Temporary Accommodation (see line 5). Exclude residual expenditure on standard amount Rent Allowances (see line 52). | RO4 |
| 19 | | RO5 LINE 1 |
| 20 | | RO5 LINE 2 |
| 21 | | RO12 LINE 5 |
| 22 | | RO14 LINE 1 |
| 23 | | RO14 LINE 2 |

| | | |
|---|---|---|
| 24 | Include Public Baths and Laundries | RO14 LINE 9 |
| 25 | Exclude: Public Baths & Laundries: Civic Halls and Theatres defined for RSG purposes as Trading Services. | RO15 LINES 3, 7, 8 |
| 26 | Include National and Country Parks. | RO15 LINES 1, 2, 4, 5, 6 |
| 27 | | RO15 LINE 15 |
| 28 | | RO14 LINE 10 |
| 29 | | RO14 LINE 11 |
| 30 | Include Smallholdings, Land Drainage (non precept), Flood Prevention, Sea Fisheries (non precept), Agricultural Pest Control etc. | RO12 LINE 4 |
| 31 | Net all recharges from this item (even where they are technically recharges of debt charges). Follow same recharging practice as at outturn, estimating if necessary. Exclude: Cost of Rate Collection (see 42). Registration of Electors (see 14). Do not deduct interest receipts (see 60). | RO13A LINE 1 |
| 32 | | RO17 LINE 9 |
| 33 | Not pay and price changes (see 35) : for JPAs see note on page 1 | |

In Col 5B lines 37 to 48 should be at Outturn Prices: For JPAs see note on page 1
In Col 5A lines 36 to 48 should be at November 81 prices. See Annex 1 for details

In Col 5C Please show repricing factors to 4 Decimal Places (ie 1.xxxx) – See below for suggested method

| | NOTES | |
|---|---|---|
| 37 | | |
| 38 | Anticipated net contribution (+) or receipt (–) | RS1B LINE 5 |
| 39 | Anticipated net contribution (+) or receipt (–) | |
| 40 | Relates to expenditure incurred by the Authority | RO19 LINE 15 |
| † 41 | Include both support for municipal undertakings and contribtuions to PTEs. Include direct expenditure on public passenger transport except debt charges. Do not deduct TSG taken to revenue | RO19 LINE 12 |
| 42 | Include cost to rate fund of agency arrangements for collecting regional water authorities' charges | RO13A LINE 2 |
| 43 | Report net transfer to (–) or from (+) rate fund. Include all trading accounts other than transport undertakings (see 41) | RS1B LINE 2 |
| 44 | | RS1B LINE 18 |
| 45 | Exclude: charges collected on behalf of RWAs: precepts to GLC (see 85) | RS1B LINES 19 + 20 |
| 46 | See examples at Annex 5 | RS1B LINES 8 – 10 (PART) |
| 47 | Please specify. Credits should be shown as negative. | |

**SUGGESTED METHOD**

a. First complete col 5B with figures at outturn prices as shown in your budget.

b. Then calculate the repricing factors in col 5C as follows:

FOR LINE 37: If your entries at lines 1 to 34 are at
(i) November 1981 prices –
divide Col 5 line 36 by Col 5 line 34
or (ii) Outturn or mixed prices –
enter an appropriate factor to represent the allowance you have made for inflation on your gross expenditure in col 5.

FOR LINE 38: the factor is as preprinted on the form.

FOR LINES 39 to 47: either use the factor at LINE 37 OR, if available, individual repricing factors for each service. IN EITHER CASE PLEASE ENTER THE FACTOR YOU USE IN COL 5C LINES 39 TO 47

c. Finally complete col 5A using the repricing factors in col 5C (ie for each line divide col 5B by col 5C).

NOTE: If you have used the November 1981 price base in lines 1 to 34, Line 37 Col 5A should equal Line 34 Col 5.

**Chartered Institute of Public Finance & Accountancy**
**Department of the Environment/Welsh Office**

**RER83 ANNEX 1**

**RETURN OF EXPENDITURE AND RATES FOR THE YEAR ENDING 31 MARCH 1983**

**NOTES ON THE NOVEMBER 1981 PRICE BASE**

**INTRODUCTION**

1. In completing lines 1 to 34 columns 4 and 5, and lines 37 to 48 column 5A, of the RER form, you are asked to follow the price base as defined in Mr Kidgell's letter of 26 November 1981. This price base is referred to as "November 1981 prices", but the conventions used mean that the price base does not simply imply the prices actually paid in November 1981.

**EXPENDITURE**

2. Authorities have already been sent details of the increases included or excluded from the price base (Mr Kidgell's letter of 26 November 1981), but for convenience these details are repeated in Appendix A to these notes.

**INCOME (OTHER THAN GRANTS AND RATES, FOR WHICH SEE NOTES ON THE FORM ITSELF)**

3. The general principle underlying the treatment of local income in the price base is that the 'volume' of income (measured at constant prices) remains constant if and only if it is increased at the same rate as the costs relating to the gross expenditure of the service for which the charges are made. Thus if you increase your charges, rents for properties etc by less than the increase in your gross costs, then in terms of the price base your income falls in real terms, and hence your net expenditure volume will increase (the user is bearing a smaller share of the costs and so your authority is bearing a greater share). Conversely, increases in charges greater than the expected increase in costs relating to your gross expenditure will reduce your net expenditure.

4. In arithmetic terms this convention requires income at November 1981 prices to be calculated by first estimating anticipated income at outturn prices (taking account of planned increases in charges and anticipated changes in demand etc) and then deflating by the appropriate gross expenditure deflator. This is algebraically equivalent to calculating the percentage of income to expenditure which is expected at outturn prices, and then applying this percentage to gross expenditure at November prices.

5. <u>You are asked therefore if possible to estimate income at the November 1981 price base when calculating net expenditure at lines 1 to 34 column 5 of the new RER form</u>. A worked example is given at Appendix B.

6. For many authorities, of course, the difference between their own budget figures for income volume and those calculated on the above methodology will be insignificant, either because they themselves follow an analogous method of allocating increases in income between services concerned and their overall contingency allowance, or because their local income from fees and charges is proportionally small. We do not expect authorities to make detailed calculations where the affect is "de minimus" (but please see paragraph 9 below).

**ADJUSTMENT OF THE CONTINGENCY ALLOWANCE**

7. Note that, where the expenditure plans reported at lines 1 to 34 (columns 4 and 5) differ from those shown in your own budget estimates because you have made adjustments to bring them to the November 1981 price base, you will need to make a corresponding (converse) adjustment to the contingency allowance reported at line 33 columns 4 and 5, so that the outturn price figures at line 36 reflect those underlying your own budget plans. This will ensure that the expenditure reported at line 71 is that used in setting your own rate or precept.

**AUTHORITIES UNABLE TO CONFORM TO THE NOVEMBER 1981 PRICE BASE**

8. You are asked to follow the November 1981 price base (as explained above) in reporting your expenditure at column 5 lines 1 to 34 of the new form, since your own estimates for each service will be more reliable than any made centrally. However, if you are unable to do this, and consider that the expenditure reported in these lines differs significantly from the November 1981 price base (being at mixed or outturn prices) the form design allows you to report, at line 37 column 5A, your own estimate of your overall budgeted expenditure on the services at lines 1 to 34, and you are asked to do this (details of the methodology are given in the notes printed on the form itself). (In this case the figure you reported at line 37 column 5A will differ from that at line 34 column 5.)

**INDICATION OF PRICE BASE OF ESTIMATES**

9. Please complete the memorandum item (line 78) to show the price base at which the expenditure at lines 1 to 34 (columns 4 and 5) has been reported. If you have used an 'other' price base (code 3) it would be helpful if you could give brief details, either on the form if space permits, or in a covering letter.

PTO

**ANNEX 1 Contd**

**FURTHER GUIDANCE ON REPRICING OF SUBSIDIES AND PRECEPTS, ETC**

10. Subsidies, such as Rate Fund Contributions to Passenger Transport; parish precepts; and other items reported at lines 40 to 47, need to be revalued in a similar fashion as any other item of expenditure or income, to be consistent with the general methodology used for public expenditure purposes, ie the value of a subsidy at a constant price base such as November 1981 is different from the value of the anticipated cash subsidy in 1982/83. Authorities who are able to obtain information on the revaluation factor (November 1981 to 1982/83) appropriate to the gross expenditure of the body concerned should of course use this factor. However an easier (but approximate) method of revaluing is to calculate the overall deflator derived from the relationship between an authority's expenditure at outturn prices and the same expenditure converted to the November 1981 base, and to apply this deflator to the planned cash subsidy (or surplus) in outturn prices. (Note that, for consistency with the treatment of income explained above, this factor must relate to gross not net costs.)

**Chartered Institute of Public Finance & Accountancy** **RER 83 Annex 2**
**Department of the Environment/Welsh Office**

**Specific grant as defined for RSG purposes**

This should only include those specific grants which are covered by the rate support grant order for 1982/83. All other specific grant (except those in respect of mandatory student awards – item **62** - and Transport Supplementary Grant – items **64** and **73**) should be deducted in arriving at net expenditure. NB above.
The following list shows those **grants to be included at row 63.**

**Police**: grants towards approved net expenditure of county and combined police authorities, the Metropolitan Police and the City of London Police.

**Magistrates' Courts**: grants to responsible authorities towards approved net expenditure in discharging their functions under the Justices of the Peace Act.

**Probation and after-care**: grants towards the expenses of probation and after-care committees required to be defrayed by county councils, etc.

**Commonwealth** immigrants: grants towards approved expenditure on employment of staff specially provided in the exercise of their functions in consequence of the presence within their areas of substantial numbers of Commonwealth immigrants.

**Civil defence**: grants towards net approved expenditure incurred in discharging civil defence functions.

**Urban programme**: grants towards expenditure incurred by reason of the existence in any urban area of special social need.

**Open space and derelict land**: grants in respect of expenditure incurred in the acquisition of land and carrying out of works for the reclamation or improvement of derelict, neglected or unsightly land. Grants payable to local authorities in respect of expenditure incurred in, or in connection with, the acquisition of land for use as public open space.

**Clean air**: grants towards expenses for adapting dwellings in smoke control areas. Report here grant taken to the revenue account only.

**Urban development**: grants paid for the acquisition or appropriation, clearing and preliminary development of land required for development or redevelopment as a whole, or for the relocation of population or industry or replacement of public open space in the course of such development.

**Transitional Town Development Subsidy**: based on the amount of Town Development subsidy under the Housing Finance Act 1972 paid to sending authorities (payable under S5, Housing Rents and Subsidies Act 1975).

**Improvements grants and area improvement grants**: grants to local authorities for the improvement of predominantly residential areas to ameliorate living conditions by improvement of amenities of the area – General Improvement Areas – Grants to local authorities for improving or converting their non-HRA dwellings. Grants to local authorities in respect of grants they have made for improvement, conversion and repair of private sector accommodation.
Grants to local authorities in respect of environmental works carried on in a Housing Action Area.

**Slum clearance**: slum clearance subsidy payable to local authorities in respect of losses incurred in the exercise of their slum clearance functions.

**Smallholdings**: contributions towards losses incurred by smallholdings authorities.

**Sheltered employment**: training and employment of approved adult blind, partially sighted and severely disabled people in special workshops, and employment of approved adult blind in home-workers schemes.

**National Parks supplementary grant**: supplementary grant to county councils in respect of expenditure in connection with National Parks.

**DOE 12508D**

**RER 83 Annex 3**

The following list shows those **grants to be excluded from row 63**, and therefore deducted in arriving at net revenue expenditure at rows 1 to 33 and 38 to 47.

**Job creation:** reimbursements by Manpower Services Commission of costs incurred under the Job Creation Programme, Special Temporary Employment Programme (STEP) and Community Enterprise Programme (CEP)

**Rent allowance subsidy:** payable to local authorities in respect of the rent allowance which they are required to grant to qualifying tenants of privately-owned accommodation including tenants of housing associations. See also notes to lines 18 and 51.

**Consumer advice grant scheme and price comparison surveys:** grants paid to selected local authorities for approved projects for the provision of new consumer advice facilities, and for local price comparison surveys.

**Careers service:** payable to local education authorities for support to the careers service in respect of staff appointed to deal with temporary high unemployment.

**TOPS:** grants in respect of the training of approved persons under the "TOPS" scheme at existing educational institutes.

**Youth Opportunities Programme:** YOP grants in respect of approved persons under "YOP" at existing educational institutes and at training workshops.

**Civil defence (care of the homeless):** grants towards expenditure incurred before 1 April 1974 in respect of planning on billeting and care of the homeless.

**Oil pollution:** grants towards expenditure incurred on measures taken against oil and chemical pollution round the coast.

**Privately owned agricultural dwellings:** Grants in respect of payments made by local authorities for privately-owned agricultural dwellings.

**Rural water supplies:** grants towards guarantee payments by local authorities to water companies for provision of water supplies in rural localities prior to 1 April 1974.

**Former defence works:** grant paid to cover the cost of removal of former defence works (including air raid shelters).

**Health service building:** compensation towards the cost of rehousing families following the acquisition of land for health service building.

**Port health.** grants in respect of health control at certain airports, and medical examination of aliens and Commonwealth immigrants.

**European Development Fund** assistance and similar EEC subsidies.

**Home Insulation Grant:** the excess of actual administration costs over the set administration fee should be treated as revenue expenditure.

**Countryside Commission Grants.**

**Coast Protection Grants.**

**AMHB Grants:** Ancient Monuments & Historic Buildings maintenance grants.

**Joint finance:** payments received from Health Authorities towards jointly sponsored personal social services projects, where treated as revenue income, should be deducted in arriving at net expenditure at row 3.1.

**Annex 4**

**Row 46 Other Precepts**

National Parks
Country Parks
Sea Fisheries
Sea Defence Commissioners/Harbour Boards
Joint Municipal Airports
Cemetery/Crematoria/Burial Boards
Archives/Charter Trustees
Joint Parks/Sports Centres/Community Councils

Note: Precepts payable to Drainage Boards should be reported at row 45

*RER82*/**Annexes 3/4**

**DEFINITION OF THE NOVEMBER 1981 PRICE BASE** **APPENDIX A**

## A. PAY

(1) The November 1981 price base includes all pay settlements effective on or before 31 October 1981, except that the 1981 London Weighting awards for the following groups are excluded.

(a) teachers and related education groups;

(b) APTC and related groups (including Chief Officers and Executives, Justices' Clerks' Assistants and Probation Officers);

(c) Fire Officers and men;

(d) Manual workers and craftsmen.

The November awards for Fire Officers and men (including control-room staff), manual workers, craftsmen, and other November groups are excluded. The effects of adjustments to grade maxima for Probation Officers on 1 November 1981 are also excluded.

(2) The price base includes the costs of the Pensions Increase (Review) Order 1981 (with effect from 23 November), whether borne directly by the Rate Fund or by increased contributions to the Superannuation Fund.

## B. RUNNING EXPENSES

(1) The price base includes all price increases up to 1 July 1981.

(2) For price increases between 1 July and 1 November, the following items are included:

Police and Fire communications;
Examination fees;
Officers' travelling and subsistence allowances;
Members' and Justices' allowances;
Discretionary awards to further education students;
Officers' car allowances (increases with effect from 1 April 1981 and 1 August 1981).

The following increases are excluded:

Supplementary Rates, 1981-82;
Autumn/Winter Postal and telephone charges increases;
Electricity: any increase after 1 July 1981;

(3) For items not mentioned in (2), an inflation allowance at the rate of 11% pa between 1 July and 1 November, equivalent to 3.54% in the four months, is included.

## C. INTEREST ELEMENT OF DEBT CHARGES

Use the interest rate of 13.0% in calculating debt charges.

**APPENDIX B**

**NUMERICAL EXAMPLE OF THE ADJUSTMENT OF REPORTED NET EXPENDITURE AT NOVEMBER 1981 PRICES TO FOLLOW THE RSG CONVENTION FOR INCOME VOLUME**

Suppose the following figures appear in your own budget plans (here, "budget" prices will be used to denote the price base conventions followed in your own budget whereas "November" prices denotes the November 1981 price base convention):

Example 1: Income increased at a greater rate than gross costs (eg libraries)

Example 2: Income increased less than gross costs (eg property rents)

| | | Examples (1) £000 | (2) £000 | Comments |
|---|---|---|---|---|
| A | Net expenditure at "budget" prices | 100 | 20 | as shown in own budget |
| B | Contingency allowance for pay and price increases on net expenditure (10% say for example 1, 15% for example 2) (including adjustments for budgeted items not at "budget" prices, as usual for the RER) | 10 | 3 | spread back any overall contingency allowance (eg for pay increases) notionally between services if necessary |
| Hence: | | | | |
| C | Net expenditure at outturn prices | 110 | 23 | |
| D | Income at "budget" prices | 20 | 5 | As shown in own budget |
| E | Income at outturn prices eg example 1: budget plans are for a 40% increase in fees to help meet budget cuts: example 2: no rent review this year | 28 | 5 | |
| Hence: | | | | |
| F | gross expenditure at "budget" prices (equals gross expenditure at "November" prices) | 120 | 25 | (A + D) |
| G | gross expenditure at outturn prices | 138 | 28 | (C + E) |
| H | Hence contingency provision on gross | 18 | 3 | (G − F) |
| J | Factor for deflating income | 1.15 | 1.12 | (G ÷ F) |
| K | Hence income at "November" prices | 24 | 4 | (E ÷ J) |
| L | Hence net expenditure at "November" prices | 96 | 21 | (F − K), which is equal to (C ÷ J) |

If your authority had only these 2 services, its planned net expenditure volume at "November" prices would be £117,000 [L(1) + (2)] whereas "budget" prices would show planned net expenditure of £120,000 [A(1) + (2)]

# APPENDIX 8

## Rate Fund Services

*Summaries of Outturn Revenue Expenditure and Income*

Form RS1B (England and Wales) shown in this Appendix is referred to in paragraphs 4.4.9, 5.2.5, 6.3.2, 6.4.5 and 8.2.1. It supplements the Rate Fund RO forms (see Appendix 4). Form RS1A which is referred to in RS1B, simply contains summaries of the Rate Fund Revenue Outturn forms and is not reproduced here.

Form FO 85 (Scotland) relates to the final paragraph of Appendix 12 and is also one of the principal sources of the statistics in [QRL.65]. When issued by the Scottish Office it is accompanied by a very full set of notes for guidance on completion. These include important definitions of terms used in the relevant tabulations.

**GENERAL NOTES TO FORM RS1B**

Form RS1B comprises the total expenditure and Income included on RO forms (Form RS1A) together with those items of expenditure and income within the rate or county fund for which no entry is made on the RO forms. As a general rule, any direct revenue expenditure or income on services for which the authority alone is responsible – as opposed to transfers to and from, or expenditure on behalf of, other accounts, funds, authorities or miscellaneous bodies – should be included under the appropriate heading on the RO forms (and therefore also on RS1A).

Columns 1 and 2 are included at the request of Audit Inspectorate. In columns 3 and 4 please report Net expenditure (+) or income (–), as appropriate: except where otherwise indicated these columns will show the difference between columns 1 and 2.

The notes are substantially unaltered from last year. Changes and additions are side lined in the margin for ease of reference.

**SPECIFIC NOTES TO FORM RS1B**

Line 2: Exclude contributions to passenger transport undertakings which should be incorporated on form RO19 and not here. Contributions to or from other trading accounts of the authority or to a joint trading account of which the authority completing RS1B also keeps the accounts, should be entered at line 2 to agree with the totals of columns 14 and 4 respectively on form TSR1 (see TSR1 notes for guidance). Authorities having contributions to or from trading undertakings (other than public passenger transport) the accounts of which are held by another authority or miscellaneous body should record the amounts at line 11 RS1B.

Line 5: Report here additions to or deductions from Block Grant in respect of pooled services, which should not therefore be included on RO forms.
This will include all pooling adjustments made in 1981-82 irrespective of the year to which they relate.

**JOINT COMMITTEES, BOARDS, ETC**

Lines 6 to 11: Joint arrangements
Include contributions to and receipts from local authorities and other bodies . acting as constituent members of joint committees, boards, or under other joint arrangements. All contributions should be shown against the appropriate headings on form RS1B and not as part of the service expenditure on the RO forms, with the following exceptions:

Port Health (form RO3E)
Passenger Transport (form RO19) )
Burial boards cemetary and crematoria ) see notes thereon
committees etc (form RO14) )

Where the accounts of the joint body are kept by a local authority, this authority should enter the total contributions received as income, and the authority's own contributions as expenditure (but see line 2 note concerning joint trading accounts), all expenditure and income (other than contributions) of the joint committee being included on the relevant RO forms of the authority.

Lines 6, 7, 8, 9 and 10: Please name the joint board involved in the space provided.

Line 9. Includes contributions to National Park Planning Boards and the Lee Valley Regional Park Authority.

Line 11: See line 2 note above.
For the purpose of this section, police authorities should be treated as local authorities.

Line 12 to 15: Only the agent authority should enter the expenditure incurred and offsetting reimbursement on form RS1B. The principal/sponsor authority should report the full subjective and objective detail of such expenditure on the RO forms concerned in accordance with the general notes for guidance. (This differs from the practice on capital outturn forms).

Line 16: Include expenditure and income incurred as agents of Regional Water Authorities, Regional Health Authorities and similar bodies. Agency Arrangements for Government Departments should include motorways, trunk roads, reception centres, youth employment, rent officers, vehicle licence administration, teaching in prisons and borstals etc. However, transactions in probation (after care) should be shown on RO11 and not on RS1B. Refer to the notes for guidance in completing CO forms concerning the treatment of transferred debt. **EXCLUDE** expenses on parliamentary elections – these should not appear anywhere on forms RS1A or RS1B.

Line 17: Record here all expenditure and income incurred including reimbursements from the Manpower Services Commission in respect of the Community Enterprise Programme (CEP) and the Youth Opportunity Programme (YOP). Such expenditure and income should not be recorded on RO Forms.

Line 18: Report here net precept paid

Lines 19 & 20: Include (i) land drainage – payments by County Councils to Regional Water Authorities or Internal Drainage boards
(ii) payments to Regional Water Authorities under section 81 of the Land Drainage Act 1976

exclude general environmental services charge; water charge; precept payments to GLC (see notes on lines 35–37

Amounts deducted from precept payments for cost of collection should be recorded as income against "cost of rate collection" on RO13A & RS1A

Line 21: Report here rateable contributions refunded from County Councils to District Councils whose areas are exempt from contributions for certain services. These contributions should thus not be deducted from county precepts (line 36).

Line 23: Report here (constituent authorities of JPAs only) the authority's share of the JPA's (gross, before deduction of grant) current expenditure (as recorded at line 26 of their RS1B form.

Line 24: CONSTITUENT AUTHORITIES of JPAs (only): Report in col 1 the authority's actual contribution to the JPA. This will be net of the specific grant paid to the JPA.
JOINT POLICE AUTHORITES (only) : Report in col 2 Contribution income

Line 25: This net expenditure should be calculated from RO1 Col 18 (lines 32 + 33 minus line 67)

Line 26: This net expenditure should be calculated from RO4 line 4(a) (col 2 minus col 6)

Line 28: Report here, as applicable, the share of any expenditure met from balances (+), or the share of any contributions to balances (–), of any joint committee or board etc for which contributions are reported at lines 6 to 11 and 24 above. This adjustment is necessary because it is the share of the expenditure of a joint body, rather than the precept payable, which is included within the "total" expenditure of an authority.

| RS1B | Name of Local Authority |
|---|---|

**RSIB RATE FUND SERVICES SUMMARY FORM 1981–82**
**ALSO PROVIDES FINAL GRANT CLAIM FORM**
**Expenditure and Income not included on RO/CO forms**

**Authorities in England**

Forms RS1A and RS1B should be completed using the latest available figures and 1 copy of each returned to the Department of the Environment, FLAS Division, Room N4/11, 2 Marsham Street, London SW1P 3EB, by 31 August following the end of the financial year.

Form RS1B will also act as a final grant claim form. If certificated OR audited figures are not available by the due date please submit your best available estimates and then provide your final **certificated figures** as soon as possible thereafter. An extra copy of form RS1B is included for this purpose.

Telephone enquiries about completion of this form may be made to the Department of the Environment 01-212 4487 or 01-212 4475.

**Authorities in Wales**

Forms RS1A and RS1B should be completed using the latest available figures and 2 copies of each returned to the Welsh Office, Economic & Statistical Services Division, (ESS8) Room 2004, New Crown Building, Cathays Park, Cardiff, CF1 3NQ, by 31 August following the end of the financial year.

Telephone enquiries about completion of this form may be made to the Welsh Office, Cardiff 0222 825067 or 823963.

Name, telephone number and extension of council officer to whom any enquiries may be made

.......................................................

.......................................................

Telephone (STD Code) (No) (Ext)

**Entries should be in whole pounds only**

**NB: Grant for Mandatory Student Awards should be shown as a Non-relevant grant in Col 7 of form RS1A.**

| | EXPENDITURE (1) £ | INCOME (2) £ | NET REVENUE EXPENDITURE (3) £ | NET "CURRENT" EXPENDITURE (4) £ |
|---|---|---|---|---|
| TOTALS BF FROM LINE 39 FORM RS1A | | | | 0 |
| 1. RS1A Col 1, in Col 1: RS1A Col 2, in Col 2: RS1A Col 1 – Col 2, in Col 3<br>1A. RS1A Col 1 – Col 2 – Col 3 – Col 4 – Col 5 + Col 6, in Col 4 | 0 | 0 | 0 | |
| 2. CONTRIBUTIONS TO TRADING ACCOUNTS OF THE AUTHORITY (OTHER THAN PASSENGER TRANSPORT) | | | | |
| 3. TOWN DEVELOPMENT Transactions between authority in connection with Housing Rents and Subsidies Act 1975 – Transitional Town Development Subsidy | | | | |
| 4. Other overspill | | | | |
| 5. NATIONAL EDUCATION POOLING (Net) | | | | |
| JOINT COMMITTEES, BOARDS, ETC<br>6. Education (Name . . . . | | | | |
| 7. Probation (Name . . . | | | | |
| 8. Sea fisheries (Name . . . | | | | |
| 9. Recreation Parks and Swimming Pools (Name . . . | | | | |
| 10. Other Non-Trading Services (Name . . . | | | | |
| 11. Trading Services (other than Passenger Transport) | | | | |
| AGENCY ARRANGEMENTS WITH OTHER LOCAL AUTHORITIES<br>12. Education (including school health service operated by Inner London boroughs) | | | | |
| 13. Highways (including metropolitan roads) | | | | |
| 14. Personal Social Services | | | | |
| 15. Other | | | | |
| 16. AGENCY ARRANGEMENTS FOR GOVERNMENT DEPARTMENTS, REGIONAL WATER AND REGIONAL HEALTH AUTHORITIES | | | | |
| 17. JOB CREATION SCHEMES | | | | |
| 18. Parish Councils and Parish Meetings (PRECEPTS) | | | | |
| 19. Land Drainage payment to Regional Water Authorities or | | | | |
| 20. " " " Internal drainage boards | | | | |
| 21. Rateable contributions refunded by County Councils | | | | |
| 22. Housing Act Advances (from form RO4 line 7) | | | | |
| 23. Share of current expenditure of Joint Police Authority | 0 | 0 | 0 | |
| 24. Contribution to Joint Police Authority/Contribution income | | | | 0 |
| 25. Deduct net expenditure on mandatory student awards (ie net of specific grants already deducted at line 1A above) | 0 | 0 | 0 | |
| 26. Deduct net expenditure on standard amount rent allowances | 0 | 0 | 0 | |
| 27. Net "current" expenditure: Sub-total of lines 1A, 2 to 22, 23, 25 & 26 | 0 | 0 | 0 | |
| 28. Adjustment for share of contributions to or from balances of joint bodies | 0 | 0 | | 0 |
| 29. SUB TOTAL FOR TRANSFER TO LINE 30 (page 2): Sub-total of lines 1, 2 to 22, 24 & 28 | 0 | 0 | | 0 |

| | | EXPENDITURE (1) £ | INCOME (2) £ | NET REVENUE EXPENDITURE (3) £ | |
|---|---|---|---|---|---|
| | 30. SUB-TOTAL TRANSFERRED Repeat entry at line 29 from page 1 (Col 3) | 0 | 0 | | |
| | RATE FUND CONTRIBUTIONS TO OR FROM HOUSING REVENUE ACCOUNT | | | | |
| + | 31A. Net cost of Standard amount rent rebates | | | | |
| + | 31B. All other rate fund contributions (see note ) | | | | |
| | 32A. Unallocated contributions to or from capital funds/capital reserve funds | | | | ↑ |
| | 32B. Unallocated contributions to or from other special funds | | | | |
| + | 33. Unallocated direct contributions to or from Direct Labour Organisations | | | | |
| + | 34. Interest on other balances credited to the rate fund | 0 | | | |
| | 35. Rate Rebates and Rate Rebates Grant | | | | ↑ |
| | 36. Transport Supplementary Grant taken to REVENUE account (County Councils and GLC only) | | | | |
| | 37. "Total" expenditure (as defined for block grant purposes) total of lines 30 to 36 | 0 | 0 | | |
| | 38. Rate income gross of precepts including contributions in lieu (N.B. exclude "Water rates" – see note) | | | | |
| | 39. PRECEPTS: Greater London Council | | | | ↑ |
| | 40. County Councils | | | | |
| | 41. Receiver for Metropolitan Police | | | | |
| | 42. Interest on Revenue Balances | | | | |
| | 43. Domestic Rate Relief Grant | 0 | | | ↑ |
| + | 44. Block Grant | | | | |
| + | 45. Steel Compensation etc (see note) | | | | |
| + | 46. Riots: expenditure and grant (see note) | | | | |
| | 47. London Equalisation payments | | | | ↑ |
| | 48. Adjustments, in respect of 1980/81 Second Increase Order, to Needs and Resources element | 0 | | | |
| | 49. Local Taxation Licence Duties | | | | |
| | 50. Other adjustments: Transition of financial adjustments | | | | ↑ |
| | 51. Holding Accounts | | | | |
| + | 52. Report here an equal and opposite sum to that shown at line 28. | 0 | 0 | | |
| | 53. BALANCE brought forward on 1 April 1981 | | | | ↑ |
| | 54. BALANCE carried forward on 31 March 1982 | | | | |
| | 55. TOTAL of lines 37 to 54 | | | 0 | |

1 ↑

THIS SECTION TO BE COMPLETED ONLY WHEN THE FIGURES ARE FINAL

CERTIFICATE OF CHIEF FINANCIAL OFFICER

I certify that to the best of my knowledge and belief the figures relating to land drainage payment at lines 19 and 20, and "Total" expenditure at line 37 above are correct, and claim the related block grant.

Chief Financial Officer

Date

CERTIFICATE OF DISTRICT AUDITOR

I have examined the entries in this form and certify that (subject to the observations in the report dated ..................)* the entries at lines 19, 20 and 37 are to the best of my knowledge and belief correct

Assistant District Auditor

Date

**delete as appropriate*

In lines 31A and 31B record all rate fund contributions to or from the Housing Revenue Account
Do not include payments by the welfare authority to the HRA (which should be reported by the welfare authority on form RO3C as a direct payment to the HRA of the housing authority).

Line 31A: Record here rent rebates (10% of standard amount) (rent rebate subsidy from central government is payable direct to the HRA and should not therefore be included on RS1B)

Line 31B: Record here (a) rent rebates – all allowances in excess of the standard amount (b) the cost of administering the rent rebate scheme (c) other Rate fund contributions

Line 32A: Record here only those contributions to capital funds that have not been allocated to particular rate fund services (allocated contributions should appear as running expenses on the appropriate RO form). The sum of such allocated and unallocated contributions to capital funds should equal the amount recorded in column 5 of form SF2B. All contributions that represent repayments of advances or borrowing from capital funds for capital purposes should be included under debt charges on RO forms against the service concerned.

Line 32B: Following the practice on capital funds contributions (see line 30A note), record here only those contributions that have not been allocated by particular services and are not repayments of borrowing on advances from special funds. Include here any contributions from the Rate Fund (Revenue Account) to DLO reserve funds, or vice versa.

Line 33: Show here only direct subsidies to or from DLOs from the rate fund. Contributions to DLO reserve funds from the Rate Fund (Revenue Account) or vice versa should be included in line 32B. Charges by DLOs to rate fund services should be included with the running expenses of those services.

Line 34 : Report here any interest on balances other than rate fund revenue balances (for which see line 42) – eg capital balances – WHICH IS CREDITED DIRECT TO THE RATE FUND REVENUE ACCOUNT. (Interest on investments by service accounts should continue to be reported on RO forms)

Line 35: Report in Col 1 the cost of rate rebates, including rate rebates to the disabled under the Rating (Disabled Persons) Act 1978. Report rate rebates grant in Col 2. EXCLUDE Administration costs: these should be shown on form RO13A line 2.

Line 36: County Councils and the GLC should report the entirety of TSG receivable in the manner specified below: Other authorities should not attempt to distinguish on the returns an allowance for TSG accruing to them through transactions with the county authority. TSG taken to the revenue account should be entered as such on forms RS1B, and RS1C, but not on form RO19. In addition, if any TSG is directed to the revenue account in the first instance for expenditure of a capital nature which is then capitalised, the entry for Revenue contributions to capital outlay on RO18/19 and CO18/19/20 forms should include this part of TSG. Grant taken direct to the capital account should be recorded as grant income on CO18/19/20B and as a memorandum item on RS1C; the latter should also show TSG taken to a suspense account or accounted for in some other manner.

Line 38:

**INCLUDE:** Transport Board Payments (by DOE, in lieu of rates); other receipts in lieu of rates in respect of Crown Property and under para 4 of part I of schedule 5 of the General Rate Act 1967; other contributions in lieu; central Government grants for rate relief in Enterprice Zones.

**EXCLUDE:** General and environmental service charges levied on behalf of Regional Water Authorities: these are NOT part of rate income. Also exclude all other charges (eg for water) collected on behalf of a body other than a local authority – these should not appear anywhere on forms RS1A or RS1B.

**DO NOT DEDUCT FROM INCOME OR ADD TO EXPENDITURE:**

Rate rebates granted to ratepayers. (see line 35): other allowances and discounts to ratepayers (for prompt payment) – see line 42: cost of rate collection (recorded as a service expenditure on RO13A and RS1A).

**DO NOT ADD TO INCOME:** Rate rebates grant from central government, (which should be shown on line 35 of this form); rate support grant/block grant (lines 44/47/48)

Line 39 to 41: All precept payments other than by London Borough Councils (see below) should represent the actual amounts collected, even though as a matter of convenience certain sums due to or from the recipient authority may have been added on or deducted from the actual amounts collected.

GLC Precepts: London Boroughs' precept payments to the Greater London Council should represent the actual amounts collected as adjusted by the Greater London Rate equalisation scheme amounts, and should include land drainage precepts paid to the GLC (No 3) account.

County precepts. See line 21 note.

Line 42: Interest receipts on revenue balances held in the General Rate Fund ie including interest receipts not allocated to specific services – see also line 34. Report interest payments on temporary borrowing for revenue purposes (eg revenue bills), and the cost of discounts for prompt payment of rates in col 1.

Line 44: Exclude London Equalisation payments.

Line 45: Compensation for sharp losses of Rateable Value as the result of a closure of a single industrial plant or works.

Line 46: Authorities qualifying for special Government financial assistance in respect of additional expenditure arising from the civil disturbances in April and July 1981 should enter their qualifying expenditure and the amount of Government assistance receivable by the authority. Record here qualifying expenditure on clearing and making safe streets and public areas, emergency repairs to roads and pavements, and the repair or replacement of damaged street furniture as well as expenditure on claims under the Riot (Damages) Act 1886 and additional police expenditure (eg on overtime and mutual aid).

Line 49: Only the cost of collecting the dog licence fee should be recorded as expenditure (Col 1) – other expenditure in connection with dogs (eg Wardens) should be reported on RO17 line 8. Report all gross income from dog licence fees in Col 2.

Line 52: Since this expenditure is not shown in the authority's own accounts.

Line 53 and 54: The balances brought and carried forward at lines 53 and 54 of RS1B should be the balances on the rate or County fund at the beginning and end of the financial year. Where the expenditure and income of a joint committee has been included in the RO forms by the authority keeping the accounts of the Committee, in accordance with the general notes, any balances of the joint committee should be included with the balances of that authority. Balances brought forward should be shown as "income" ( – ) and balances carried forwarded as "expenditure" ( + ), except in the case of overdrawn balances (when the reverse applies).

FORM F.O. 85

## SUMMARY OF FINAL OUT-TURN INFORMATION ON RELEVANT EXPENDITURE FOR 1984-85

All information recorded on the return should reflect authorities' abstracts of accounts or, in the absence of accounts, the latest financial information available at the due date for submission.
**The information should be reconcilable with the detailed LFR forms which authorities will submit in due course, and with Rating Review Actual Expenditure for 1984-85 returns.**

A copy of this return, duly completed and signed, should be sent to your Council's auditor with copies to (1) Scottish Office Finance Division, Room 7/22, New St Andrew's House, Edinburgh EH1 3TB and (2) Mr AV Tait, Convention of Scottish Local Authorities, 16 Moray Place, Edinburgh EH 3 6BL

The return should be submitted **AS SOON AS POSSIBLE** but, in any event, not later than **16 October 1985**. If an estimated return is made, this should be marked clearly on the form, and submitted to the Scottish Office direct (copy to COSLA). A further return based on final accounts should be submitted as soon as the final information is available. This should be directed to the auditor and copied as above.

Enquiries about completion of this return should be made to: Mr I J C Howie, Room 7/26, New St Andrew's House, (tel: 031-556 8400 extn: 4747 or 5449).

**Please complete and sign the declaration below.**

I certify that the entries in the form are correct

.......................... Council

Signed ..........................
(Director of Finance)

Date ......................

Auditors Report

We have examined the books and records of the above authority and have obtained such explanations and carried out such tests as we considered necessary.

On the basis of our examination and of the explanations given to us we report that *(subject to the reservations set out in our letter dated ) the certificate relating to final out-turn information in respect of 1984-85 by the Director of Finance is in agreement with the underlying records and in our opinion is in accordance with the relevant statutory requirements.

Signed ..........................

Date ..........................

* delete if inappropriate

**RATING REVIEW**

Additional information:

| | £,000 |
|---|---|
| **EXPENDITURE** | |
| 1. Expenditure per item 43 | |
| - Net Expenditure | |
| - Loans and Leasing Charges | |
| 2. Non-relevant expenditure | |
| 3. Deduct Cost of Rate Collection *(Note 21)* | |
| 4. **TOTAL** | |
| **INCOME** | |
| 5. Specific grants | |
| 6. RSG - Needs | |
| - Resources | |
| 7. Movement in balances (DR/CR) | |
| 8. Rates (including RSG domestic element) | |
| 9. **TOTAL** | |

**Totals at lines 4 and 9 should agree.**

## FINAL OUT-TURNS 1984-85

| ITEM | SERVICE (*Note 3 and appropriate service note in the accompanying notes for guidance*) | | NET EXPENDITURE (excluding loan & leasing charges) (*Note 1*) £000 | LOAN & LEASING CHARGES (*Note 2*) £000 |
|---|---|---|---|---|
| 1 | Education | | | |
| 2 | Libraries | | | |
| 3 | Museums and Art Galleries | | | |
| 4 | Social Work Services | | | |
| 5 | School Crossing Patrols | | | |
| 6 | Police | *Note 4* | | |
| 7 | Fire | *Note 4* | | |
| 8 | Civil Defence | | | |
| 9 | District Courts | | | |
| 10 | Roads and Road Lighting | *Note 5* | | |
| 11 | Concessionary Fares | *Note 6* | | |
| 12 | Transport subsidies Bus | *Note 6* | | |
| | Non-Bus Transport | *Note 6* | | |
| 13 | Public Water (Rate Fund Contribution) | *Note 6* | | |
| 14 | Sewerage | | | |
| 15 | Coast protection | | | |
| 16 | Flood prevention and land drainage | | | |
| 17 | River Purification (Requisition) | *Note 6* | | |
| 18 | Cleansing (including street cleansing by Districts) | *Note 7* | | |
| 19 | Public Analyst | | | |
| 20 | Environmental Health | *Note 8* | | |
| 21 | Planning | *Note 9* | | |
| 22 | Leisure and Recreation | *Note 10* | | |
| 23 | Careers Service | | | |
| 24 | Sheltered Workshops | | | |
| 25 | Improvement etc Grants | *Note 11* | | |
| 26 | Loans for House Purchase | *Note 12* | | |
| 27 | Administration of Housing Benefits | *Note 13* | | |
| 28 | Registration of Electors | | | |
| 29 | Lands Valuation | | | |
| 30 | Registration of Births, Deaths and Marriages | | | |
| 31 | Collection of Rates (Regional & Island Authorities) | *Note 14* | | |
| 32 | Building Control | | | |
| 33 | Consumer Protection | | | |
| 34 | Non-Road Lighting | *Note 15* | | |
| 35 | Urban Programme | *Note 16* | | |
| 36 | Burial Grounds and Crematoria | | | |
| 37 | Rate Fund Contributions a. Slaughter-houses | *Note 6* | | |
| | to Trading Services b. Other Trading Services | *Note 6* | | |
| 38 | Unallocated Transfers to and from Special Funds | *Note 17* | | |
| 39 | Contributions to and from DLOs | *Note 18* | | |
| 40 | Miscellaneous (including Licensing, Elections and Community Councils) | | | |
| 41 | Central Administration Costs | *Note 19* | | |
| 42 | Interest on Revenue Balances - Indicate CR or DR | *Note 19* | | |
| 43 | **TOTAL RELEVANT EXPENDITURE** (Items 1-42) | | | |
| 44 | **LESS DISREGARDS 1984-85** | *Note 20* | | |
| | (i) Urban Programme | *Note 20(i)* | | |
| | (ii) Police Pay Additional Costs | *Note 20 (ii)* | | |
| | (iii) Police Costs on Coal Dispute | *Note 20 (iii)* | | |
| | (iv) Civil Defence Expenditure in Excess of 1983-4 | *Note 20 (iv)* | | |
| | (v) Commonwealth Games Appeal Fund | *Note 20(v)* | | |
| 45 | **TOTAL RELEVANT EXPENDITURE FOR GUIDELINE COMPARISONS** (Item 43 less Items 44(i)-(v)) | | | |

**FOR INFORMATION COULD YOU PLEASE PROVIDE THE FOLLOWING:** £000

1. The amount of capital from revenue contained in Item 43, total relevant expenditure excluding loan charges: *(Note 1)*
2. Your authority's rating balance at the end of 1984-85: *(State credit or debit.)*

# APPENDIX 9

**Quarterly Financial Returns to Central Government**

Form CPA 4 is referred to in paragraphs 4.3.3, 4.3.4 and 4.4.3. Note that matching forms are in use in Wales and Scotland.

Form QRR 2 is referred to in paragraphs 4.4.9 and 4.4.10. This form is in use throughout the United Kingdom.

| NAME OF LOCAL AUTHORITY |
|---|
| |

**CPA 4** 82/3

DEPARTMENT OF THE ENVIRONMENT
CAPITAL PAYMENTS RETURN

1 April 1982 to 31 March 1983

TABLE I: CAPITAL PAYMENTS AND CAPITAL RECEIPTS All figures in £000's

| | FIXED ASSETS | Acquisition of land and existing building and works (Note 4a) (1) | New Construction and conversion (Note 4b) (2) | Vehicles, Plant equipment & machinery (Note 4c) (3) | TOTAL PAYMENTS (Cols 1 to 3) (Note 4d) (4) | RECEIPTS (Note 4e) (5) | Urban Programme Payments (Note 4f) (6) | Coverage in terms of 1981/82 Outturn Forms | Row Code |
|---|---|---|---|---|---|---|---|---|---|
| 1 | Nursery education | | | | | | | CO1A row 1 | 1 |
| 2 | Schools: other | | | | | | | CO1A rows 2 to 4 | 2 |
| 3 | Further education | | | | | | | CO1A row 5 | 3 |
| 4 | Youth service | | | | | | | CO1A row 6 | 4 |
| 5 | Other education | | | | | | | CO1A row 7 | 5 |
| 6 | Total education (Rows 1 to 5) | | | | | | | | 6 |
| 7 | Personal social services. (Note 5a) | | | | | | | CO3 row 3 | 7 |
| 8 | Highways & lighting | | | | | | | CO18A rows 4 and 5 | 8 |
| 9 | Parking | | | | | | | CO19A row 3 | 9 |
| 10 | Public passenger transport (Note 5b) | | | | | | | CO20A row 6 | 10 |
| 11 | PTE's/LTE (Note 5c) | | | | | | | | 11 |
| 12 | Road safety | | | | | | | CO19A row 4 | 12 |
| 13 | Ports and piers | | | | | | | CO21A row 5 | 13 |
| 14 | Tolled road bridges, tunnels and ferries | | | | | | | CO18A row 6 | 14 |
| 15 | Total roads and transport (Rows 8 to 14) | | | | | | | | 15 |
| 16 | Housing including HRA (Note 5d) | | | | | | | CO4A rows 2, 3, 4 & 7 | 16 |
| 17 | Libraries, museums and art galleries | | | | | | | CO2A row 3 | 17 |
| 18 | Fire Service | | | | | | | CO7A row 2 | 18 |
| 19 | Civil defence | | | | | | | CO7A row 6 | 19 |
| 20 | Aerodromes | | | | | | | CO21A row 6 | 20 |
| 21 | Careers service and sheltered workshop | | | | | | | CO5A row 3 | 21 |
| 22 | Agriculture & fisheries | | | | | | | CO12A row 4, CO21A row 4, and CO22A row 4 & 7 | 22 |
| 23 | Refuse collection and disposal | | | | | | | CO13A row 4 | 23 |
| 24 | Recreation parks and swimming pools | | | | | | | CO15A row 8 | 24 |
| 25 | General administration | | | | | | | CO13A row 1 | 25 |
| 26 | Town & country planning | | | | | | | CO16A row 6 and CO21A row 3 | 26 |
| 27 | Other local services | | | | | | | CO12A row 5, CO14A row 9, CO17A row 6 & CO22A rows 3, 5, 8 & 9 | 27 |
| 28 | Total other services (Rows 17 to 27) | | | | | | | | 28 |
| 29 | TOTAL ALL SERVICES (Rows 6 + 7 + 15 + 16 + 28) | | | | | | | | 29 |
| | GRANTS AND ADVANCES | | | | | | | | |
| 30 | Housing | | | | | | | | 30 |
| 31 | Other | | | | | | | | 31 |
| 32 | Total (Rows 30 + 31) | | | | | | | | 32 |

MEMORANDA

£000s

| | | Acquisitions | Disposals |
|---|---|---|---|
| 33 | **Capital value of leased assets (note 6a)** (a) property leases | | |
| 34 | (b) vehicles, plant and machinery | | |
| 35 | **DE MINIMIS EXPENDITURE (note 6b)** | | 1 |
| 36 | **Direct Labour Organisations (note 6c)** (a) net capital payments for fixed assets | | (b) rate of return provision |

DOE 12507

**CPB 4** 82/3

NAME OF LOCAL AUTHORITY

Name, telephone number and extension of Council Officer to whom any enquiries may be made

TABLE II: ALLOCATIONS, PRESCRIBED EXPENDITURE, AND RECEIPTS — All figures in £000's

| | CAPITAL EXPENDITURE ALLOCATIONS | 1 April 1982 to 31 March 1983 (1) | | |
|---|---|---|---|---|
| 1 | Education | | | Enter total 1982/3 allocations (including by 31 March 1983 in column 1, |
| 2 | Personal Social Services | | | |
| 3 | Transport | | | |
| 4 | Housing | | | |
| 5 | Other Services | | | |
| 6 | Urban Programme | | | |
| 7 | Total | | ↑ | Total of rows 1 to 6 |
| 8 | Allocations carried forward from previous year | | | A negative figure should be shown if any 1982/3 |
| 9 | Transfers from other local authorities | | | See note 7 |
| 10 | Trading profits | | | Include the estimated entitlement of profits of a of a joint committee of local authorities |
| 11 | Transfers to other local authorities | | | See note 7 |
| 12 | Total capital allocation at the end of reporting period | | ↑ | Rows 7 + 8 + 9 + 10 minus 11 |
| | **PRESCRIBED EXPENDITURE** | **1 April 1982 to 31 March 1983** | | |
| 13 | Capital Payments on fixed assets | | | Table I row 29 column 4, minus non-prescribed Memorandum box 36(a), minus Table I Memorandum |
| 14 | Capital Grants and Advances | | | Table I row 32 column 4 minus a) amounts of |
| 15 | Leasing of land and buildings | | | Table I Memorandum row 33 acquisitions plus back or for more than 20 years that were entered |
| 16 | Agency work carried out by other local authorities | | | Include payments made by one local authority to basis. Row 16 should record payments to agency |
| 17 | Agency work carried out for other local authorities | | | |
| 18 | De Minimis expenditure | | ↑ | Table I Memorandom box 35 |
| 19 | Expenditure met from insurance payments | | | Expenditure of any description insofar as it is met damage or destruction of that item, or is defrayed |
| 20 | Total expenditure counting against allocations | | | Rows 13 + 14 + 15 + 16 − 17 − 18 − 19 |
| | **CAPITAL RECEIPTS** | | | |
| 21 | Receipts from sales of fixed assets (excluding housing) | | | Table I row 29 column 5 minus row 16 column 5 |
| 22 | Receipts from repayment of grants and advances (excluding housing) | | ↑ | Table I row 31 column 5 minus repayments of car loans |
| 23 | Housing sales of fixed assets (prescribed proportion only) | | | The prescribed proportion of Table I row 16 column 5 (see note 8) |
| 24 | Repayments of housing grants and advances (prescribed proportion only) | | | The prescribed proportion of Table I row 30 column 5 (see note 8) plus the prescribed proportion of HAG |
| 25 | Leasing of land and buildings | | | Table I memorandum row 33 'disposals' |
| 26 | Receipts from district or area health authorities | | | See note 9 |
| 27 | Other capital receipts | | | See note 10 |
| 28 | Total capital receipts | | | Row 21 + 22 + 23 + 24 + 25 + 26 + 27 |

**One copy of the completed form should be returned to the Secretary, Department of the Environment, FLAS, N4/11, 2 Marsham Street, London SW1P 3EB. Any queries concerning the completion of this form can be addressed to Mr Norman (01-212 4613)**

**NOTES**

any additions or reductions) as notified to your authority

allocation was applied in 1981/82.

trat'ing undertaking owned by the authority individually or as a member

expenditure on capitalised repairs (see note 11), plus Table I box 36(b).

interest capitalised on lending to housing associations, and b) car loans

leasing charges for land or building leases which were either lease/lease into between 1 April 1980 and 31 March 1981

another local authority for capital expenditure undertaken on an agency authorities and row 17 receipts by agent authorities (see also note 2)

'a ,a payment received under a contract of insurance against the loss from an insurance or contingency fund maintained by that authority

**NOTES FOR GUIDANCE IN COMPLETING TABLE II**

7. The authority transferring a capital allocation to another should record the amount transferred in row 11, and the authority receiving the allocation should record the amount in row 9. See also note 2.

8. The prescribed proportions for housing capital receipts are as follows:–

| Receipts arising from | Proportion |
|---|---|
| i. A dwelling house normally let or available for letting under Part V of the Housing Act 1957 (other than a dwelling house let on a shared ownership tenancy, or sold under a homesteading agreement). | 50% |
| ii. An advance granted to enable a person:– a) to acquire other than from an interest in a dwelling house, or b) to improve a dwelling house | 50% |
| iii. A grant or advance made to a housing association | 1% |
| iv Other | 100% |

9. Any payment to an authority by a district or an area health authority to the extent that such a payment is used to finance a payment in respect of prescribed expenditure.

10. INCLUDE here as sums to be treated as net capital receipts:–

i. Any payment to an authority under an agreement made under section 52 of the Town and Country Planning Act 1971 or Section 37 of the Town and Country Planning Act 1962.

ii. Any payment to an authority under Part IX of the Highways Act 1959 or Part XI of the Highways Act 1980.

iii. Any payment to an authority in accordance with an agreement made under Section 60 of the Highways Act 1971 or Section 278 of the Highways Act 1980.

iv. Any payment to an authority of damages or compensation pursuant to an order of judgement of a court or an agreement for the compromise of any legal proceedings or by way of settlement of any claim for expenses against a person by an authority under any default power.

v. The proceeds of any lottery promoted by a local authority under the Lotteries and Amusements Act 1976 after the deduction of the expenses of promoting it and the sums required for prizes.

vi. Any payment to an authority by way of gift or bequest.

vii. Any payment made by a displacing authority to a rehousing authority, as respectively defined in Section 42 of the Land Compensation Act 1973, in accordance with that section of that Act.

viii. From 1 April 1982, grants of a capital nature made to local authorities by the following centrally funded bodies:–

The Manpower Services Commission
The Countryside Commission
The Sports Council
The Development Commission
The English Tourist Board
The Arts Council of Great Britain.
The Museums and Galleries Commission
The Council for Educational Technology
Urban Development Corporations
Water Authorities
Research Councils
Area Museum Councils

DO NOT INCLUDE the following capital receipts as the related expenditure is excluded from this return (see note 3(b)):–

i. Any payment to a highway authority by the Minister of Transport or the Secretary of State under any agency, arrangement whereby the authority discharges any functions of the Minister or Secretary of State as a highway authority for the purposes of Section 1(1) of the Highways Act 1959 or the Highways Act 1980, to the extent that the payment is used to finance a payment in respect of prescribed expenditure.

ii. Any payment by an undertaker to a highway authority, as respectively defined for the purpose of the Public Utilities Street works Act 1950, under any provision of that Act.

11. Capital expenditure on repairing and maintaining a property is prescribed only if the property is a dwelling house or is a property held under Part V of the Housing Act 1957 and the expenditure is defrayed by the raising of a loan.

**NOTES FOR GUIDANCE IN COMPLETING TABLE 1**

1. The coverage of this form and the definitions used are generally the same as for the Capital Outturn (CO) forms. Differences between CPR and CO, such as the exclusion of the Police, Courts and Probation sector, from the CPR are shown in the CPR notes in italics.

2. Duplication.

*i. Where the local authority executies work for another on any agency basis or makes contributions to the expenses of another total capital payments should be included in Table 1 of the authority executing the work and excluded from Table 1 of the paying authority. Inter-authority payments and receipts in respect of such work should be included in Table II rows 16 and 17.*

*ii. The payments of joint committees should be included in Table 1 of the constituent authority which keeps the account of the joint committee. This local authority should also record in Table II receipts from other local authorities in the Committee. Other constituent authorities should not include any payments made to the joint committee in Table 1 of the form. These payments should be included in Table II rows 16 and 17 as agency transactions with other local authorities.*

*iii. All dealings with the Consolidated Loans Fund should be excluded from all parts of the form.*

3. EXCLUDE FROM TABLE 1.

(a) The capital value of leased assets. These should be recorded in memoranda boxes 33 and 34.

(b) The cost of work executed by a local authority as the agent of a Government department, Regional Water Authority, or a public utility. Also exclude the receipts of monies to fund this expenditure.

(c) Payments for work on 100% grant-aided road schemes.

(d) County contributions to new town roads.

(e) Payments by an authority to a water authority insofar as it is met by a grant to the authority for that purpose from the Land Settlement Association Limited.

(f) Expenditure on the Youth Opportunities Programme (YOP) and the Community Enterprise Programme (CEP), even if some capital appreciation may arise.

(g) Expenditure on any grant or advance to the owner of a historic building insofar as it is met by a grant or advance from the Secretary of State under Section 4 or 6 of the Historic Buildings & Ancient Monuments Act 1953 or Section 10 or 10B of the Town and Country Planning (Amendment) Act 1972.

(h) Amounts of interest capitalised that arise after an asset is completed or open for public use eg capitalised interest on tolled bridges and tunnels.

(i) Expenditure by the Common Council of the City of London which is defrayed out of the following funds:–

(i) The Bridge House Estates

(ii) City's Cash, except expenditure by the Common Council as a local authority or port health authority.

(j) Expenditure on the improvement of any highway where the improvement does not have the effect of:–

(i) widening the carriageway of any road by more than 0.25 metres,

(ii) significantly changing the alignment of the carriageway of any road,

(iii) substantially improving the standard of the road, or

(iv) providing new traffic signals other than by the replacement of existing traffic signals.

(k) Payments for the appropriation of capital assets to other services within the authority; these are book keeping arrangements concerned with transfers of assets purchased in earlier years between services and are not the subject of this return.

*(l) The repayments of loans or the investment of cash.*

*(m) Grants and Advances to other local authorities or Passenger Transport Executives.*

*(n) Receipts of Housing Association Grants (HAG), which are excluded here for cash limit monitoring purposes, although the prescribed proportion should be included in Table II as a capital receipt.*

*(o) Expenditure and receipts on police, courts and probation services.*

4. INCLUDE all payments, including legal fees, architectural and engineering services, whether from capital account or revenue accout, renewals or reserve accounts, or from Government grants for:–

(a) **Column 1. Acquisition of land and existing buildings,** including any road charges. Show payments against the service for which the land is expected to be used, rather than that appropriate to the powers used for acquisition. Interest on purchase money should be excluded.

(b) **Column 2. New Construction and Conversion.** Payments should include the wages and salaries (including the employers' share of national insurance and superannuation contributions) of employees engaged on direct labour schemes, and the cost (including accommodation) of architectural, engineering and other services (including the authority's own professional staff) connected with capital works,. Where there is difficulty in allocating this expenditure a pro rata apportionment over services is recommended. Expenditure on demolition and site clearance associated with the erection of buildings should be included, as well as any costs incurred in repairing or maintaining property provided under the Housing Acts which are met by raising of a loan.

(c) **Column 3. Acquisition, renewal or replacement of vehicles and vessels, plant machinery and equipment,** including furniture and fittings, and the installation of street lighting, road signs, traffic signals and related equipment where the payment for these can be separately identified.

(d) **Column 4. Total Payments** should equal the sum of columns 1 to 3. For Grants and Advances (rows 30 to 32) record gross payments resulting from a flow of finance instigated by your authority, including the passing on of a Central Government grant (excluding historic building grants, see note 3(g)), from the local authority to a person or body. Include advances for car purchase, private sector renovation grants and home insulation grants, gross lending to private persons for house purchase or renovation, and gross amounts advanced to housing associations (including amounts of interest capitalised). Also include contributions to any transport operator (other than Passenger Transport Executives and an authority's own trading undertakings) in respect of capital expenditure incurred by such operators.

(e) **Column 5. Record the total amount received by the authority from the sale of assets of a capital nature,** including those where the original purchase was not capitalised, and receipts from the disposal of community land. Housing sales (row 16) snould include initial receipts from the disposal of council dwellings (or shares in the equity of council dwellings) and other housing assets. These comprise initial deposits on completed sales where the purchase is financed by a mortgage under Part I of the Housing Act 1980 or Section 104 of the Housing Act 1957 (or as amended). For all other sales include the full sale price (net of any discount). Include receipts from the sale of dwellings to Housing Associations.

Receipts from Grants and Advances include repayments of renovation grants, repayments of principal (regular and premature) of loans to private persons, repayments of sums left outstanding on the sale of council dwellings (or shares in the equity of dwellings), and repayments of principal of loans to Housing Associations (but excluding payments of HAG see note 3(n)).

(f) **Column 6. Urban Programme Payments.** Include here amounts included in column 4 that relate to projects approved for inclusion in the urban programme.

5. Services

(a) Personal Social Services should include Port Health. Local authorities should record the payments they incur in respect of jointly financed schemes, including payments made to them by the Area Health Authority where these form part of the local authority payment.

(b) Public Passenger Transport should include only direct expenditure by an authority on its own trading undertaking.

(c) PTE's/LTE should include only direct expenditure by Passenger Transport Executives and London Transport Executive.

(d) Housing. Include payments for new housebuilding, slum clearance, renovation of local authority's own dwellings, expenditure under GIA and HAA powers, acquisition of existing dwellings for retention in housing stock and acquisition and improvement of dwellings for sale.

6. Memoranda. Expenditure recorded in rows 33, 34 and 36(a) should not be included in Table 1. Expenditure in row 35 and 36(b) should be included in Table 1.

(a) Capital Value of leased assets. Include here for those leased assets in which the local authority acquired an interest in present or future use during the reporting period:–

| | |
|---|---|
| Row 33 Acquisitions | (a) the freehold value of property leased for a period of more than 20 years<br>plus (b) the freehold value of any property lease of over 1 year where a permanent building has been erected at the request of the authority (eg sale/leaseback or lease/leaseback schemes) |
| Row 33 Disposals | the corresponding value of such types of assets disposed of during the reporting period. |
| Row 34 Acquisitions | (a) the capital value of vehicles leased for more than 1 year<br>(b) the capital value of plant and machinery acquired by finance leases |
| Row 34 Disposals | the market value of such types of assets disposed of during the reporting period |

(b) De Minimis Expenditure. Include here capital payments on items of plant, vehicles, equipment and machinery that individually cost less than £5500 and that the local authority has decided should not count against its capital allocation.

(c) Direct Labour Organisations

Row 36(a): include payments by the DLO on its own behalf on the acquisition of capital goods net of receipts from the sale of capital goods.

(b) include those parts of DLO charges included in rows 1 to 29 above which are intended to cover the depreciation of capital assets or to secure the rate of return required by Part III of the 1980 Act.

Department of the Environment

**LOCAL AUTHORITIES – RETURN OF RATES, INTEREST RECEIPTS & PAYMENTS, WAGES & SALARIES 1983/84**
**NOTES FOR GUIDANCE IN COMPLETING FORM**

1. **Rate Receipts net of Rate Rebates**

Include
- (i) General rates received from the public, including rates received from another authority;
- (ii) Payments in lieu of rates;
- (iii) Rates on council property;
- (iv) Precepts collected on behalf of other local authorities;
- (v) Regional Water Authority precepts for land drainage (including payments under Section 25, Land Drainage Act 1961 agreements).
- (vi) Payment from DHSS via Housing Benefit for Supplementary Benefit claimants

**Do not include**
- (i) Rates collected on behalf of another local authority (eg from council house tenants in over-spill areas)
- (ii) Grants from central government in respect of standard rate rebates and rate support grant;
- (iii) General and environment service charges collected on behalf of regional water authorities;
- (iv) Share of any rate rebate not reimbursed by Central Government and any rebate above standard funded by the Local Authority.

2. **External Interest receipts**

Include
- (i) Interest on deposits of all types and dividends on financial assets purchased by the authority;
- (ii) Discounts on Treasury Bills;
- (iii) Interest on loans for the erection, purchase and repair of dwellings; for sale of council houses; for private street works; and any other interest received from external sources.

**Do not include**
- (i) Interest received as a result of the investment of balances/funds within the authority;
- (ii) Interest receipts on account of transferred assets;
- (iii) Dividends and interest arising from superannuation fund assets.

3. **External interest payments**

Include
- (i) Interest payments only on debt administered by your authority; authorities administering debt on behalf of other local authorities eg the head of a consortium should include the total interest paid on that debt before deducting receipts from other local authorities;
- (ii) Discounts on stocks, bonds and bills.

**Do not include**
- (i) Internal interest payments eg interest on internal balances

4. **Wages and salaries**

Include
- (i) Gross pay together with payments and allowances in the nature of remuneration, compensation for loss of office, and redundancy payments;
- (ii) Employers share of national insurance contributions plus contributions to superannuation funds, and beneficiaries under pension schemes eg police & firemen;

**Do not include**
- (i) Payments for training expenses, removal expenses etc;
- (ii) Transfer values on superannuation funds,
- (iii) Allowances paid to councillors.

## QRR 2

Name, telephone number and extension of Council Officer of whom any enquiries may be made

............................................................

............................................................

Total Payments and Receipts in the period **1 April 1983 to 30th September 1983**

| | £ thousand |
|---|---|
| Rate receipts (net of refunds) *(see note 1)* | |
| External Interest Receipts *(see note 2)* | |
| External Interest payments *(see note 3)* | |
| Wages and Salaries of all services *(see note 4)* | |

1

**One copy of the completed form should be returned to the Secretary Department of the Environment, FLAS N4/11, 2 Marsham Street, London SW1P 3EB by Monday 17th October.**

**Enquiries concerning the completion of this form may be made to the Department of the Environment 01-212 6236.**

**DOE 12523 (Revised September 1983)**

# APPENDIX 10

**Capital Expenditure and Debt Financing [QRL.8]**

A specimen copy of the CIPFA/SIS questionnaire (including completion instructions) and covering letters are contained in this Appendix. It should be noted that a slightly amended version of the covering letter is sent to authorities in Scotland and Northern Ireland (see paragraphs 7.3.2 and 7.5.2).

The Chartered Institute of Public Finance and Accountancy
(tel: 01-930-3456)

My ref: PR/SC

April, 1985

Dear Colleague,

**CAPITAL EXPENDITURE AND DEBT FINANCING STATISTICS 1984–85 — (Form B85E)**

I enclose two copies of this letter, guidance notes, and a questionnaire for Capital Expenditure and Debt Financing Statistics 1984–85. I shall be grateful if you would arrange for the completion and **return of one copy of the questionnaire as soon as possible and in any event not later than Monday 18th May, 1985 to:**

(1) County Councils (including Met. Counties)

V. F. Phillips, Esq., IPFA
County Treasurer,
Bedfordshire County Council,
County Hall,
Bedford, MK42 9AP.

(2) Other Authorities

The Director (ref STATS),
C.I.P.F.A.,
3 Robert Street,
London, WC2N 6BH.

The format of the survey in England and Wales is changed this year, in that in order to avoid duplication Capital Payments data will be derived from the Department of Environment/Welsh Office returns [see Preface notes overleaf], and will, as normal refer to provisional outturn. I hope these arrangements are convenient and with your assistance I will endeavour to publish the results as early as possible.

Yours sincerely,

N. P. Hepworth
*Director*

To: All Chief Financial Officers of Local Authorities in England and Wales.

PS. County Treasurers who are also Treasurers to Joint Police Authorities have been sent two sets of questionnaires so that a separate return can be submitted in respect of the Joint Police Authority.

**CIPFA — 'CAPITAL BORROWING STATISTICS'**

*Full notes for guidance on completion are set out on page 6.*

**Preface:**

This year a change has occurred in the format of the survey in two respects:

(a) Capital Payments data will be derived from returns made to DOE on form CPR4 — 1984/85, and the Welsh Office on the equivalent form for Wales. However, to ensure that the figures requested in this survey, relating to borrowing against the various service headings, correspond to those definitions for Capital Payments, the appropriate CPR/Welsh Office return references are set out below for your guidance.

(b) An attempt is being made to distinguish internal and external borrowing within the overall analysis of total "outstanding borrowing" made for Capital Purposes. Although the Institute holds the view that the term "debt" should include temporary use of available internal funds, there is merit in making a distinction which can, through the published statistics, provide lay interests with a more balanced portrait of "debt", based on externally funded borrowings outstanding. Thus, although the questionnaire requests that you continue to provide the data in the traditional form (i.e. inclusive of Capital Fund advances outstanding), the publication will only describe external borrowings as "debt".

**Service definitions — Capital Payment's Return References (Welsh Office in brackets):**

| | |
|---|---|
| Education | Table 1 Row 6 (1) |
| Personal Social Services | Table 1 Row 7 (2) |
| Highways and Local Transport | Table 1 Row 16 (5 + 9) |
| Refuse Collection and/or Disposal | Table 1 Row 23 (13) |
| Recreation and Leisure | Table 1 Row 24 (14 part) |
| Housing Revenue Account<br>Housing Act Advances<br>Housing (Other Rate Fund Housing Services) | Table 1 Row 17 (26) |
| Police/Courts/Probation | CPC Row 7 (CPRA Row 7) |
| Trading Services<br>Other Services | Table 1 Rows 28-23-24<br>(15 + 16 - 9 - 13 - 14 part) |

**CIPFA QUESTIONNAIRE — CAPITAL BORROWING —**
**(Capital Expenditure and Debt Financing Statistics 1984/85)**

L.A. Name ..............................................

**B85E**

Amount £'000

**SECTION A. CAPITAL AUTHORISATIONS 1984–85**

1. Prescribed Expenditure (as per CPR Table 3 Row 19 or Welsh Office Form Table 3 Row 24)
2. *Less* Capital Receipts Applied to Supplement Allocations
3. Capital Allocations Utilised (Line 1 less Line 2) see Note 1.
4. Total Capital Allocations for year (including amounts brought forward — see Note 1.)
5. Capital Allocations c/fwd (see Note 2.)

**SECTION B. OUTSTANDING ADVANCES FOR CAPITAL PURPOSES AS AT 31 MARCH 1985 (see Notes 3 to 7.)**

Amount £'000

6. Education
7. Personal Social Services
8. Highways and Local Transport (see Note 4, and Line 24)
9. Refuse collection and/or disposal
10. Recreation and Leisure
11. Housing Revenue Account (see Note 5)
12. Housing Act Advances
13. Housing (Other Rate Fund Housing Services)
14. Police/Courts/Probation Services
15. Trading Services
16. Other Services
17. Total advances outstanding on services for which Authority is responsible (Lines 6 to 16)
18. Cash in hand of CLF and Investment of loan and sinking fund balances
19. Total advances for Authority's own purposes
20. *Less* Outstanding advances for which authority is responsible but which is administered by other authorities

*Add* Outstanding advances administered by the authority for which other authorities are responsible:—

21. — Local Authorities
22. — Other Public Bodies
23. All outstanding advances administered by the local authority (to agree with line 39)

MEMORANDUM ITEM for PTA's only:

24. Amount in respect of PTE's/LTE own advances (see Note 4.)

**SECTION C. ANALYSIS BY TYPE AND MATURITY OF LOAN** — Amount £'000

**BORROWING FOR CAPITAL PURPOSES AS AT 31 MARCH 1985**

**External Advances**

**Temporary Debt for Capital Purposes (see Note 10) maturing:**

25. On or before 7 April 1985
26. On or after 8 April 1985 but before 1 July 1985
27. On or after 1 July 1985 but before 1 April 1986
28. Total Temporary debt for Capital Purposes (lines 25 to 27)

**Long Term Debt (see Note 10)**

29. Public Works Loan Board
30. Stock Issues
31. Local Authority Negotiable Bonds
32. Market Bonds/Mortgages
33. Local Bonds/Mortgages
34. Other Loans (incl. foreign currency loans)
35. Total external long-term debt (lines 29 to 34)

**Internal Borrowing**

36. Capital Fund Advances outstanding
37. Internal advances (see Note 8.)
38. Revenue and other balances temporarily used for capital purposes (see Note 9.)
39. Total Borrowing for Capital Purposes (Lines 28, 35, 36, 37 and 38; to agree with line 23)

**Memorandum Items**

40. Amount of external long term debt which is at variable rate — i.e. in line 35 (e.g. "rollover" loans; variable rate stock; and variable rate PWLB). [see note 11.]

Amount of borrowing from other local authorities for Capital Purposes:

41. — Temporary (i.e. maturing within 12 months)
42. — Long term (see Note 12)

Amount of long term lending by your authority (other than to banks and other monetary sector) (see Note 13):

43. — as Gilt Edged Securities
44. — to Other Local Authorities
45. — Other

**SECTION D. ANALYSIS OF LONG TERM DEBT BY MATURITY (EXCLUDING PWLB) AS AT 31 MARCH 1985**

| (See Notes 14 to 21) | Amount of Debt £'000 (Nearest whole numbers) | Average Interest Rate % at 31.3.85 (To one decimal Place) |
|---|---|---|
| Long Term Debt Maturing: | | |
| 46. Before 1 April 1986 | | |
| 47. On or after 1 April 1986 but before 1 April 1987 | | |
| 48. On or after 1 April 1987 but before 1 April 1988 | | |
| 49. On or after 1 April 1988 but before 1 April 1989 | | |
| 50. On or after 1 April 1989 but before 1 April 1990 | | |
| 51. On or after 1 April 1990 but before 1 April 1995 | | |
| 52. On or after 1 April 1995 but before 1 April 2000 | | |
| 53. On or after 1 April 2000 | | |
| Total Non-PWLB long term debt: 54. Sum of lines 46 to 53 — to agree with line 35 less line 29. | | |

**SECTION E. CALCULATION OF AVERAGE RATES OF INTEREST 1984/85 (See Notes 22 to 25)**

| | | [Whole Pounds!] | |
|---|---|---|---|
| 55. Interest on external debt (See Note 23) | £ | | |
| 56. Interest on internal borrowing (See Note 24) | £ | | |
| 57. Total Interest (Lines 55 and 56) | £ | | |
| 58. Equated Advances (See Note 25) | £ | | |
| 59. Average Rate of Interest (See Note 26) $\left(\frac{\text{Line 57}}{\text{Line 58}} \times \frac{100}{1}\right)$ | | • | % to Two Decimal Places |
| 60. Total Debt Management Expenses 1984/85 | £ | | |
| 61. Percentage of Debt Management Expenses to Equated Advances $\left(\frac{\text{Line 60}}{\text{Line 58}} \times \frac{100}{1}\right)$ | | • | % to Two Decimal Places |

1

In the event of the compiler requiring further information please show the name of the person to be contacted:

Mr./Mrs./Miss ........................................................ Tel. No. .............................. Extn. ..............

Address to which all correspondence may be sent: ..........................................................................

..........................................................................................................................................

..........................................................................................................................................

*Notes for guidance*

**SECTION A. CAPITAL AUTHORISATIONS 1984–85**

1. Capital Allocations Utilised. Include here all prescribed expenditure funded by means other than Capital Receipts. At item 4 include: amounts specified for the 1984–85 year net of transfers to/from other authorities; Trading profits; and allocations under/over utilised in the 1983–84 financial year.

2. Capital Allocations c/fwd will be line 4, less line 3, less any adjustment applied for tolerance subject to the 10% limitation.

**SECTION B. OUTSTANDING ADVANCES FOR CAPITAL PURPOSES AS AT 31 MARCH 1985**

3. Services for which the Authority is responsible whether or not the debt is administered by the authority should be entered at lines 6 to 16. These lines should also include Capital Fund advances outstanding.

4. PTE's/LTE. Exclude from line 8 outstanding advances of the PTE/LTE for which your authority is responsible. BUT, show that amount which you have excluded, at line 24. However, please include, at line 8, the "debt" of the PTA.

5. Transferred Housing. Include at line 11, all outstanding advances on housing which have been transferred to your authority. Similarly, exclude the outstanding advances on housing which you have transferred to other authorities, even though you may still be administering this "debt". (Instead, include this amount at either line 21 or 22, as appropriate.)

6. Outstanding advances serviced on an agency basis should be included in lines 21 and 22.

7. Joint Police Authorities. Whether or not a separate loans pool is maintained for the police authority, please return a separate questionnaire for the police authority, showing all police debt at line 14 and the debt administered by constituent or former police authorities at line 20. Non-administering constituent and former police authorities should include any police debt they service in line 21.

**SECTION C. ANALYSIS BY TYPE AND MATURITY OF LOAN**

8. Internal advances — this includes balances used temporarily for Capital Purposes from Repairs and Renewal Funds, Capital Funds, Insurance Funds and unapplied Capital Receipts etc. It should *not* include advances from Superannuation Funds which should be included in lines 25 to 34.

9. Revenue and other balances used temporarily for capital purposes. This arises due to such factors as cash received in advance of payments made (i.e. Cash in hand) plus unpresented cheques etc., when it is usual for this money to be invested temporarily in the loans pool or CLF.

10. The categories of borrowing included under 'temporary Loans' and 'long term debt' are the same as those classified in the DoE monthly 'Borrowing and Lending' enquiries.

*MEMORANDUM ITEMS*

11. "Rollover" loans are those where the interest rate is fixed at a premium over the rate of some predetermined money market rate. The rate changes at agreed intervals in line with the movement of the agreed determining figure. The interval between changes is usually related to the indicator used e.g. a rollover based on the money market's three month price would be changed every three months.

12. Borrowing from, and lending to, other authorities — Long Term borrowing is defined as outstanding advances where the initial term for repayment was *more than* twelve months.

13. Lending — Please show all lending to others *except* to Banks and other monetary sector institutions.

**SECTION D. ANALYSIS OF LONG TERM DEBT BY MATURITY (EXCLUDING PWLB) AS AT 31 MARCH 1985**

14. The analysis of debt required in this section is in relation to the remaining period of maturity measured from the 31 March of the indebtedness of the Authority to its external lenders.

15. Loans repayable by instalments should be classified by the expected payment dates. Loans repayable on maturity should be classified by maturity date.

16. Stocks with a range of redemption dates should be classified according to the final redemption date.

17. Loans running on at notice should be classified according to the first date on which notice can be exercised. Any longer term loans running on at notice should not be reclassified as temporary borrowing.

18. The interest rate relating to "rollover" loans should be the appropriate rate current at 31 March 1985. "Rollover" loans should be analysed by reference to the Voluntary Code of Borrowing.

19. Loans repayable at lenders or mutual option in advance of the final redemption date should be classified according to the first date on which the lender's option or mutual option can be exercised.

20. Loans with an unspecified redemption date should be classified as maturing on or after 1 April 2000 (line 53).

21. Average Interest Rates — for each of lines 46 to 53 should be calculated as:-

$$\text{The sum for each outstanding loan for that line of } \frac{(\text{interest rates} \times \text{principal repayable in that interest period})}{(\text{The sum of the principal repayable for that line})}$$

This provides a weighted average interest rate. Do not use the amount of interest expected to be paid. For further guidance on the calculation of average interest rate you are referred to Volume 6 of the Financial Information Service, Chapter 13, paragraphs 27 and following.

**SECTION E. CALCULATION OF AVERAGE RATE OF INTEREST**

22. This section relates to *all* outstanding advances administered by the Local Authority included at Line 23.

23. Total of interest on external debt — include interest due for the year on all debt included at Lines 28 and 35. Interest due should be included even if not accrued in the Local Authority's own accounts.

24. Total of interest on internal borrowing — for the purpose of this calculation include interest at notional rates on ALL internal borrowing included at Lines 36, 37 and 38. This should include funds to which an Authority may not normally allocate a credit e.g. Capital Fund, Rate, or County Fund balances used temporarily for capital purposes, capital receipts unapplied etc. Refer to FIS. 6, Chapter 7, paragraphs 11 to 19.

25. The equated advances are calculated either by adding outstanding advances at 1st April to half the annual capital advances, or by adding outstanding advances at 1st April to a different proportion of annual capital advances which have been arrived at either by examination of past trends or in the light of definite information regarding the year in question.

26. The average interest rate is calculated by ascertaining the equated advances and applying the formula:

$$\text{Estimated average rate} = \left(\frac{\text{Total Interest}}{\text{Equated advances}}\right) \times \left(\frac{100}{1}\right)$$

# APPENDIX 11

The form Local Financial Return (LFR 11) contained in this Appendix is Scottish central government's source of the following statistics of actual income or expenditure published in [QRL.77].

1) Rate Income (see paragraph 5.2.5)
2) Domestic element of Rate Support Grant (see paragraph 6.4.12)
3) Rate rebates paid (see paragraph 6.3.2)
4) Rate rebate grant from central government to individual authorities (see paragraph 5.5.4)

Note: There is not a matching source document in use in England and Wales (See Appendix 8). Item 2 above is not payable in England and Wales.

LFR 11 (1982)

RATE INCOME 1981-82

Regional/Islands Council

The information requested in this form will be used to prepare the published Local Financial Returns. One copy of the form should be completed and returned to the Scottish Office Finance Division, Room 7/23, New St Andrew's House, Edinburgh EH1 2TB as soon as possible and, in any event, not later than 31 August 1982. All enquiries about completion of this form should be made to Mr Morris Scottish Office Finance Division (Tel 031 556 8400 Ext 5761)

| | | £000 |
|---|---|---|
| 1. (a) TOTAL RATEABLE VALUE AT 1 APRIL 1981 including effective rateable value from running roll (Note 1) | £000 | |
| (b) LESS: Remissions, reliefs, exemptions etc (Note 2) | | |
| (c) RATEABLE VALUE ON WHICH RATES WERE LEVIABLE | | |
| 2. (a) GROSS RATE INCOME FROM REGIONAL/GENERAL RATE (1(c) x regional/general rate) | | |
| (b) GROSS RATE INCOME FROM ALL DISTRICTS RATES (sum of appropriate proportions of (c) x each district rate). | | |
| (c) ADD: Contributions in aid of rates (Note 3) and recoveries of rates previously written off | | |
| TOTAL | £000 | |
| (d) LESS: i) Total amount of rate rebates paid (Note 4)<br>ii) Other reductions in rates due (Note 5)<br>iii) Domestic element of rate support grant | | |
| (e) NET INCOME FROM RATEPAYERS | | |
| (f) ADD BACK: Rate Rebate grant<br>Domestic element of rate support grant | | |
| (g) NET RATE INCOME | | |

Please provide the name and telephone number of the Council official to whom enquiries about the information contained in this form can be made.

Name ____________________ Tel No ____________________

Signature ____________________ Designation ____________________

Date ____________________

RE 88459/2 TBL

NOTES FOR COMPLETION LFR 11 (1982)

1. Excluding the rateable value of properties for which contributions are received in aid of rates (Crown etc property) but including the rateable value of subjects valued by the Assessor of Public Undertakings.

2. The remissions, reliefs etc shown here (including remissions, reliefs etc from running roll) should be shown net of any rates to be collected in respect of unoccupied property under Section 24 of the Local Government (Scotland) Act 1966 and should include reliefs under Sections 4 and 5 of the Local Government (Financial Provisions etc) (Scotland) Act 1962 to charitable and other institutions; relief for poverty (under Section 244 of the Local Government (Scotland) Act 1947); reductions in respect of lands and heritages which the Assessor of Public Undertakings (Scotland) is required to value under Section 215(1) of the Local Government (Scotland) Act 1973 and exemptions under Sections 243 (unlet and unoccupied subjects) and 240 and 241 (rates levied on owners) of the Local Government (Scotland) Act 1947.

3. Contributions by the Crown, by police authorities and by the Commissioner of Northern Lighthouses.

4. Including rebates paid under the Rating (Disabled Persons) Act 1978.

5. Including net cost of collection of rates, repayments of rates paid in error (under Section 20 of the Local Government (Financial Provisions) (Scotland) Act 1963) and rates written off as irrecoverable.

# APPENDIX 12

## Rate Support Grants

This Appendix consists of:

a) an outline of the method of distributing the Block Grant element of RSG in England and Wales;
b) an introduction to the statistics behind the calculation of the cost of providing a standard service in England;
c) the build up of GREA's for each authority in England and Wales;
d) a note on the nature of the resources and needs element of RSG in Scotland and
e) a brief note on the client group method recently introduced as a means of allocating, to individual authorities, the general portion of the needs element of RSG (see paragraphs 6.4.8 and 6.4.13).

*The Statistics of Block Grant Distribution in England and Wales and Resources and Needs Elements of RSG in Scotland*

**1.** The aims of the distribution of block grant are twofold:

a) to enable all authorities within the same class (e.g metropolitan districts, non-metropolitan councils) and therefore providing the same "bundle" of services, to levy the same rate;
b) to enable central government to influence the spending levels of authorities.

**2.** To achieve these ends central government has attempted first of all to calculate what each authority would require to spend to provide a standard service, allowance having been made for the characteristics of each area. The initial step in calculating the cost of providing a standard service is taken by DOE/Welsh Office agreeing with the local authority associations, the national totals in cash to be assigned to each of 64 service components (e.g nursery education, social services for the elderly, refuse disposal, concessionary fares). The agreed figures for England are contained, with accompanying comments, for each component in the DOE Blue Book *Grant Related Expenditure: How the Expenditure Needs of Local Authorities are Assessed For Block Grant* [QRL.28]. (Note: There is no comparable publication for Welsh authorities.) For each component and class of authority, the cost of providing a *unit* of service is assessed and published in the *Rate Support Grant Report (England)* [QRL.60] and the *Rate Support Grant Report (Wales)* [QRL.61].

**3.** In order to arrive at the notional cost to each authority, of providing a standard service, these unit costs have to be factored by indicators which allow for the needs

of each area. There are 53 needs indicators which fall into 5 main groups:
a) people in the area;
b) physical features of area;
c) social and environmental problems;
d) differences in the costs of providing services;
e) special requirements of particular services.
The value of each indicator for each authority in England is published in *Block Grant Indicators* [QRL.6] and for Welsh authorities in *Rate Support Grant (Wales)* [QRL.62]. (Note: Throughout the year, needs indicators data for individual authorities is maintained and updated by DOE/Welsh Office as a computerised data base.)

**4.** The results of the calculations referred to at the beginning of the last paragraph, when added together are called the *grant related expenditure assessments* (GREA) or amounts which each authority would need to spend to provide a standard service.

**5.** The GREA for each authority in England is published in [QRL.60] and for Wales in [QRL.61] and [QRL.92]. A detailed statistical build up of each English authority's GREA is provided by DOE in its Green Book, *Technical Handbook of Grant Related Expenditure* [QRL.84]: a large selection of this Green Book data is reproduced in [QRL.6].

**6.** The Reports [QRL.60] and [QRL.61] publish for each class of authority a *grant related poundage* (GRP) schedule from which any authority may deduce the rate poundage which central government assumes it would charge if it were to spend its GREA. The schedule also enables an authority to compute its GRP for varying levels of expenditure per head *above* or *below* its GREA. Increases in GRP are in line with increases in expenditure up to certain *threshold* expenditures per head beyond which an incremental rise in proposed expenditure would lead to a larger incremental change in GRP. Threshold expenditure per head figures are fixed for a year for each class of authority in England and Wales and are contained in [QRL.60] and [QRL.61].

**7.** Statistics of estimated variations in GRP for individual authorities in England and Wales are published by CIPFA/SIS in *Finance and General Statistics* [QRL.20]. For English and Welsh authorities, the statistics of GRP to be applied to each authority's block grant calculation (see below) are published by CIPFA/SIS in *Block Grant Statistics* [QRL.7] under the heading "GRP before holdback" (for meaning of term "holdback" see paragraph 11 below).

**8.** Before moving towards the calculation of its block grant entitlement, each authority must estimate its *total expenditure* (TE), a quantity which is defined in the Local Government Planning and Land Act, 1980 but which in any year may be redefined marginally in [QRL.60] and [QRL.61]. The term is defined by CIPFA in [QRL.20] as equivalent to "total rate and grant borne expenditure less specific government grants". *Estimates* of this quantity for each authority in England and Wales are contained in [QRL.20] under the heading "expenditure ranking for block grant"; the figures are repeated in [QRL.7]. The source of the data in [QRL.20] is the RER form shown in Appendix 4. At outturn stage, the source of the data on which final calculations of block grant are based is the RS1B (see Appendix 11) which is the grant claim form. However, there is not any national publication of "total expenditure" as such.

**9.** Before finally applying the block grant formula to make its initial estimate of its grant entitlement, an authority requires to identify two further quantities: its *multiplier* (M) and its *rateable value* (RV). The former is determined by central government and is published for each authority in [QRL.60] and [QRL.61]; but the multiplier (known at this stage as a "composite multiplier") can be adjusted in Supplementary Reports to [QRL.60] and [QRL.61]–see paragraph 11 below. The RV's for purposes of block grant calculation differ from the total and actual RV's as defined in chapter 9. Adjustments are made to total RV's to allow for (a) unoccupied non-domestic hereditaments and (b) hereditaments subject to charity relief. Statistics of RV's for block grant purposes are published in [QRL.20] and repeated in [QRL.7]. The figures are computed by DOE/Welsh Office and transmitted to CIPFA and individual authorities.

**10.** The block grant formula is as follows: BG = TE (RV x GRP x M). But the application of the formula in itself does not yet enable an authority to calculate its block grant entitlement because, as mentioned in paragraph 1 above, central government uses block grant distribution to achieve a second objective i.e to influence the spending levels of authorities. To this end it has devised a means of penalising individual authorities by reducing their block grant entitlements if their spending exceeds expenditure *targets* or their GREA, which ever is the greater. Target expenditure figures for each authority in England and Wales are listed in [QRL.20] as "expenditure ranking for target". The data is repeated for England and Welsh authorities in [QRL.7] and for Welsh authorities only in [QRL.62]. The source of the target data is of course, central government.

**11.** The amount of block grant reduction which an authority may suffer is referred to as *holdback, abatement* or *penalty*. The method by which the reduction is effected is to increase the multiplier. This amended multiplier (called the *effective multiplier*) is the product of the composite multiplier, already referred to, and a quantity known as the *guideline related multiplier*, the final value of which cannot be determined for each authority until outturn expenditure is known. The basis of calculating guideline related multipliers is set out in the RSG Reports [QRL.60] and [QRL.61], in the form of a schedule showing the incremental increases for each 1% overspend for each class of authority. (Note: any increase in the value of a guidance related multiplier has the effect of increasing GRP in the block grant formula and thus of reducing grant entitlement.)

**12.** *Estimates* of abatement (or penalty) of block grant for each authority in England and Wales are published in [QRL.20] and repeated in [QRL.7]. The estimates are made by the CIPFA/SIS compiler by applying the block grant calculations rules to the RER figures provided by authorities (see Appendix 4). Statistics of *actual* holdback/abatement plus effective multipliers and guidance related multipliers in respect of each authority are published in Supplementary RSG Reports [QRL.60] and [QRL.61].

**13.** In *Scotland*, the aims of Rate Support Grant distribution (i.e the resources element plus the needs element) are similar to those set out for block grant distribution in England and Wales (see paragraph 1 above). However, the basis of distributing the two elements in Scotland is quite different from that employed in allocating block grant. But the Scottish means of distributing the needs element does now begin to resemble the block grant method.

**14.** Not all authorities receive a share of *resources element* of RSG. The amount of resources element payable to an authority is calculated by reference to the amount by which its standard penny rate product exceeds the actual product of a rate of 1p in the £. The standard penny rate product is arrived at by multiplying the population of the authority by a national standard amount per head of population. This national standard is shown in [QRL.72], the step by step calculation being set out in [QRL.71]. Since the *total* size of the resources element is fixed in advance, that is to say, before actual rate poundages are determined, it becomes necessary in the light of up to date information to revise the national standard amount per head of population and publish this in a Supplementary version of [QRL.72]. It is worth noting that the statistics of individual authority calculation are not published.

**15.** As to the distribution of the *needs element* of RSG: up to and including the year 1983/84 this was effected by means of a three stage method. First, there was an allocation to each regional and islands *area*. Second, an amount was allocated to each region and to the *sum* of districts within each region. The statistical rules of each stage in the allocation which was wholly demographic were set out in detail in [QRL.71]. But the values of the demographic variables used in making the allocation to individual authorities have not been published.

**16.** From the year 1984/85, the method of allocating the general portion of needs element, known as the client group approach, begins to move towards the England/Wales block grant allocation method in that an attempt is made to estimate the level of expenditure required to provide the services for which an authority is responsible–rather similar to a GREA. To this end 64 service components have been identified (e.g further education, home helps, road lighting and fire staff). For each component a primary indicator of expected expenditure is identified. (e.g "students" in respect of further education; "population aged 65 and over" for home helps; "Km's of urban road lane" for road lighting.) Also, for some service components, secondary indicators have been identified (e.g for home helps the "numbers of low income pensioners"). For each indicator–both primary and secondary–an expenditure per unit is estimated (e.g x pounds per student in further education; y pounds per person aged 65 and over for home help provision; z pounds for Km of urban lane for road lighting). By summing the products of these and the values of the indicators for each authority, an estimated expenditure per head of population for individual authorities is computed and published in [QRL.72]. (Details of authority-by-authority calculations, which are carried out by the Scottish Office, are not published. In other words there is no Scottish equivalent to the DOE's Green Book)–see paragraph 5 above. The estimated expenditure per head per authority are used as weights for calculating national grant shares for each authority. These shares (expressed as needs grant per head) are also set out in [QRL.72]. Because the application of the client group approach has resulted in significant reductions in the grant paid to some authorities and as the total of needs element is smaller than that of the previous year, the actual distribution of grant was modified for the first year to which the new approach applied. The numerical basis of the modification was explained in [QRL.72]–House of Commons paper 171 of 1983/84 Session.

**17.** With the object of influencing the spending levels of individual authorities (see paragraph 1), the Secretary of State has powers to penalise authorities spending, or planning to spend, above provision. For individual authorities provision takes the form of a guideline issued to every authority prior to each financial year. Statistics of those guidelines are published in [QRL.65]. Should an authority fail to limit its total expenditure to the guideline level, the Secretary of State may reduce its RSG needs element allocation. He also has power to take "selective action" to reduce the rates and/or rate support grant of any particular authority planning to spend at a level considered to be "excessive and unreasonable". These selective powers are contained in the 1981 (Miscellaneous Provisions) (Scotland) Act 1981 and the Local Government and Planning (Scotland) Act 1982, both of which amended the 1966 Local Government (Scotland) Act. Use of the selective action powers are subject to the approval of Parliament. Finally, the Rating and Valuation (Amendment) (Scotland) Act 1984 gave the Secretary of State powers to set a limit on the rate level to be determined by any class of authority. Again use of these "rate capping" powers is subject to the approval of Parliament.

**18.** In order to decide whether an authority should be penalised, the Secretary of State (i.e the Scottish Office) receives from each authority a return which shows final outturn expenditure relevant to RSG. A specimen of the form used (Form FO 85) is contained in Appendix 8.

# APPENDIX 13

**Special Funds**

The following Special Funds are identified in the data source forms referred to in Section 8.5.

*Form SF 1 –England and Wales*

Sinking or Redemption Funds
Trading Reserve Funds
Reserve Funds of Direct Labour Organisations
Insurance Funds
Repair Renewals and other Reserve Funds
Other Special Funds
Lottery Holding (or operating) Accounts
Lottery Funds

*Form SF 2 –England and Wales*

Capital Funds and Capital Reserve Funds

*Form LFR 23 –Scotland*

Reserve Funds (a) Renewal and Repairs (b) other Reserve Funds
Insurance Funds
Trust or Charity Funds
Capital Funds
Other Special Funds

*Form LRF 25 –Scotland*

Common Good Fund

# APPENDIX 14

## Rateable Values

As explained in section 9.1, the authority responsible for the valuation of properties for rating purposes in England and Wales is the Inland Revenue Valuation Office; in Scotland, the corresponding authorities are the Land Valuation Assessors employed by the Regional and Islands Councils. This Appendix contains specimen copies of the forms on which these authorities provide annual statistical data of the numbers and rateable values of properties in each rating area; form V07140 in the case of England and Wales, form RRV/ASS in the case of Scotland.

**INLAND REVENUE VALUATION OFFICE**

**STATISTICAL ANALYSIS**

**OF**

**THE NUMBERS AND RATEABLE VALUES OF ALL HEREDITAMENTS**

**SHOWN IN THE VALUATION LIST FOR THE**

**RATING AREA**

**OF**

..............................................................................

**ON**

**1 APRIL 19......**

STATISTICAL ANALYSIS OF NUMBERS AND RATEABLE VALUES OF ALL HEREDITAMENTS SHOWN IN THE VALUATION LIST ON 1 APRIL 19 ......

| Classification | | Code | Numbers | RV £ |
|---|---|---|---|---|
| DOMESTIC (CATEGORY 1) | Houses, Bungalows, Flats, Maisonettes, Caravans, etc. (other than Agricultural Dwelling Houses) with Rateable Values:- | | | |
| | Not exceeding £75 ... ... ... ... | 1010 | .......... | .......... |
| | Exceeding £75 but not exceeding £100 ... | 1020 | .......... | .......... |
| | " £100 " £125 ... | 1030 | .......... | .......... |
| | " £125 " £150 ... | 1040 | .......... | .......... |
| | " £150 " £400 ... | 1050 | .......... | .......... |
| | " £400 ... ... ... ... | 1060 | .......... | .......... |
| | TOTALS ... ... ... ... | 1099 | ---------- | ---------- |
| | Agricultural Dwelling Houses (of all Rateable Values) ... ... ... ... ... | 1110 | .......... | .......... |
| | Crown Dwelling Houses ... ... ... | 1120 | .......... | .......... |
| | Separately assessed Single Caravan Sites ... | 1210 | .......... | .......... |
| | **TOTALS OF CATEGORY 1** ... ... | **1999** | | |
| COMMERCIAL (CATEGORY 2) | Shops assessed with Private Dwelling Accommodation ... ... ... ... | 2010 | .......... | .......... |
| | Shops (including Banks in Shopping Areas) and Cafes ... ... ... ... ... | 2020 | .......... | .......... |
| | Offices (including Banks in Office Areas) | 2030 | .......... | .......... |
| | Public Houses including Beerhouses ... | 2040 | .......... | .......... |
| | Hotels and Boarding Houses ... ... | 2050 | .......... | .......... |
| | Restaurants ... ... ... ... ... | 2060 | .......... | .......... |
| | Holiday Camps and Caravan Fields ... | 2070 | .......... | .......... |
| | Warehouses, Stores and Workshops ... | 2080 | .......... | .......... |
| | Garages (commercial), Petrol Filling Stations and Car Parks ... ... ... ... | 2090 | .......... | .......... |
| | Lock-up Garages (separately assessed) ... | 2100 | .......... | .......... |
| | Advertising Stations and Rights ... ... | 2110 | .......... | .......... |
| | Markets ... ... ... ... ... | 2120 | .......... | .......... |
| | **TOTALS OF CATEGORY 2** ... ... | **2999** | | |
| INDUSTRIAL UNDERTAKINGS (CATEGORY 3) | Factories, Mills and other premises of a similar character ... ... ... ... ... | 3010 | .......... | .......... |
| | Mineral Producing hereditaments (other than National Coal Board hereditaments assessed under S.19 Local Government Act 1974) | 3020 | .......... | .......... |
| | **TOTALS OF CATEGORY 3** ... ... | **3999** | | |
| **TOTALS OF CATEGORIES 1, 2 & 3 CARRIED FORWARD** ... ... | | | | |

**I certify that the information on this form is correct**

........................................................................................ **Valuation Officer** ....................................19
**(Date)**

**VO 7140** ........................................................................................ **Valuation Area**

FOLD HERE

County ........................................................................ THIS SPACE FOR USE BY DOE ONLY

Rating Area ........................................................................

| Classification | | Code | Numbers | RV £ |
|---|---|---|---|---|
| TOTALS OF CATEGORIES 1, 2 & 3 BROUGHT FORWARD | | | | |
| OTHER UNDERTAKINGS (CATEGORY 4) | Gas ... ... ... ... | 4010 | | |
| | Electricity ... ... ... | 4020 | | |
| | National Coal Board (hereditaments referred to in S.19 Local Government Act 1974) | 4030 | | |
| | Water ... ... ... | 4040 | | |
| | Post Office ... ... ... | 4050 | | |
| | British Telecommunications ... | 4055 | | |
| | Docks, Harbours, Canals, etc. ... | 4060 | | |
| | Other Transport (including Ferries, Toll Bridges, Road Transport Depots and Yards, etc) | 4070 | | |
| | **TOTALS OF CATEGORY 4** ... | **4999** | | |
| ENTERTAINMENT AND RECREATIONAL (CATEGORY 5) | Cinemas ... ... ... | 5010 | | |
| | Theatres and Music Halls ... ... | 5020 | | |
| | Sports and Recreational Grounds, Golf Courses and Race Courses ... ... | 5030 | | |
| | Social Clubs, Community Centres and Public Halls ... ... ... | 5040 | | |
| | Residential Clubs ... ... ... | 5050 | | |
| | Other Places of Entertainment or Recreation (such as Billiards Saloons, Bowling Alleys, Dance Halls, Pleasure Parks, Foreshores, Swimming Pools, Fairgrounds and the like) | 5060 | | |
| | Radio and Television Relay Undertakings | 5070 | | |
| | **TOTALS OF CATEGORY 5** ... | **5999** | | |
| EDUCATIONAL AND CULTURAL (CATEGORY 6) | Libraries, Museums, Art Galleries, etc | 6010 | | |
| | Schools and Colleges (Local Authority) | 6020 | | |
| | Schools and Colleges (Non Local Authority) | 6030 | | |
| | Universities and University Colleges ... | 6040 | | |
| | **TOTALS OF CATEGORY 6** ... | **6999** | | |
| MISCELLANEOUS (CATEGORY 7) | Crown occupations (including National Health Hospitals, Clinics, etc. but excluding Dwellings) | 7010 | | |
| | Other Hospitals, Clinics, Nursing Homes, etc | 7020 | | |
| | Public Wash-Houses and Baths ... | 7030 | | |
| | Cemeteries, Crematoria and Columbaria | 7040 | | |
| | Sewage Disposal Works and Refuse Disposal Works ... ... ... | 7050 | | |
| | Town Halls and Municipal Offices ... | 7060 | | |
| | Hereditaments not otherwise classified | 7070 | | |
| | **TOTALS OF CATEGORY 7** ... | **7999** | | |
| **GRAND TOTALS OF ALL CATEGORIES IN VALUATION LIST** | | **9999** | | |

FOLD HERE

RRV/ASS/1983

RATEABLE VALUES IN SCOTLAND 1983 - 84

NUMBERS AND RATEABLE VALUES OF SUBJECTS IN THE VALUATION ROLL AS AT 1 APRIL 1983

REGION/ISLAND AREA

Notes for completion of form:

| | |
|---|---|
| 1. DOMESTIC SUBJECTS: | Include the domestic parts of mixed subjects in terms of subsections (3) and (4) of section 7 of the Local Government (Scotland) Act 1966 <u>except</u> the domestic part of caravan site valuations determined under section 3(1) of the Rating (Caravan Sites) Act 1976. The total single unit valuations for caravan sites should be included with Commercial subjects. Include croft houses and separately assessed caravans. |
| 2. BANDS OF DOMESTIC RATEABLE VALUES: | Croft houses and separately assessed caravans should be excluded from columns 11 - 22 and shown in columns 23 and 24. Properties with a NIL rateable value should be included in columns 11 and 12 and the number of such properties should be shown in this box. [ ] |
| 3. INDUSTRIAL AND FREIGHT-TRANSPORT SUBJECTS: | Include all de-rated factories, workshops, docks etc (including non-industrial parts not valued by formula) as well as intensive livestock subjects de-rated under the Valuation for Rating (Scotland) Act 1970. |
| 4. COMMERCIAL SUBJECTS: OTHER (Columns 31 and 32) | Include total single unit valuations under section 3(1) of the Rating (Caravan Sites) Act 1976. |
| 5. PUBLIC UTILITIES ETC: | Include in columns 38 and 39 all lands and heritages the values of which are entered in your roll by virtue of a direction from the Assessor of Public Undertakings (Scotland). Include in columns 40 and 41 any other undertakings (eg statutory docks and harbours) which you value according to statutory formulae. |

RATEABLE VALUES IN SCOTLAND 1983 - 84 RRV/ASS/1983

NUMBERS AND RATEABLE VALUES OF SUBJECTS IN THE VALUATION ROLL AS AT 1 APRIL 1983

REGION/ISLAND AREA

| DISTRICT | DOMESTIC SUBJECTS (NOTE 1) | | | | | | | | | |
|---|---|---|---|---|---|---|---|---|---|---|
| | LOCAL AUTHORITY | | SCOTTISH SPECIAL HOUSING ASSOCIATION | | NEW TOWN | | PRIVATELY OWNED (INC. OTHER HOUSING ASSOCS. except SSHA) | | TOTAL | |
| | No | RV(£) | No | RV(£) | No | RV(£) | No | RV(£) | No | RV(£) |
| | 1 | 2 | 3 | 4 | 5 | 6 | 7 | 8 | 9 | 10 |
| | | | | | | | | | | |
| | | | | | | | | | | |
| | | | | | | | | | | |
| | | | | | | | | | | |
| | | | | | | | | | | |
| | | | | | | | | | | |
| | | | | | | | | | | |
| | | | | | | | | | | |
| | | | | | | | | | | |
| | | | | | | | | | | |
| | | | | | | | | | | |
| | | | | | | | | | | |
| | | | | | | | | | | |
| | | | | | | | | | | |
| | | | | | | | | | | |
| | | | | | | | | | | |
| | | | | | | | | | | |
| | | | | | | | | | | |
| | | | | | | | | | | |
| TOTAL | | | | | | | | | | |

RRV/ASS/1983

RATEABLE VALUES IN SCOTLAND 1983 - 84

NUMBERS AND RATEABLE VALUES OF SUBJECTS IN THE VALUATION ROLL AS AT 1 APRIL 1983

REGION/ISLAND AREA

| DISTRICT | DOMESTIC SUBJECTS : BANDS OF RATEABLE VALUES (NOTE 2) | | | | | |
|---|---|---|---|---|---|---|
| | £0 - £100 | | £101 - £200 | | £201 - £300 | |
| | No | RV | No | RV | No | RV |
| | 11 | 12 | 13 | 14 | 15 | 16 |
| | | | | | | |
| | | | | | | |
| | | | | | | |
| | | | | | | |
| | | | | | | |
| | | | | | | |
| | | | | | | |
| | | | | | | |
| | | | | | | |
| | | | | | | |
| | | | | | | |
| | | | | | | |
| | | | | | | |
| | | | | | | |
| | | | | | | |
| | | | | | | |
| | | | | | | |
| | | | | | | |
| | | | | | | |
| | | | | | | |
| TOTAL | | | | | | |

RATEABLE VALUES IN SCOTLAND 1983 - 84 RRV/ASS/1983

NUMBERS AND RATEABLE VALUES OF SUBJECTS IN THE VALUATION ROLL AS AT 1 APRIL 1983

REGION/ISLAND AREA

| | DOMESTIC SUBJECTS : BANDS OF RATEABLE VALUES (NOTE 2 ) | | | | | | | |
|---|---|---|---|---|---|---|---|---|
| DISTRICT | £301 - £400 | | £401 - £500 | | Over £500 | | Croft Houses and separately assessed caravans | |
| | No | RV | No | RV | No | RV | No | RV |
| | 17 | 18 | 19 | 20 | 21 | 22 | 23 | 24 |
| | | | | | | | | |
| | | | | | | | | |
| | | | | | | | | |
| | | | | | | | | |
| | | | | | | | | |
| | | | | | | | | |
| | | | | | | | | |
| | | | | | | | | |
| | | | | | | | | |
| | | | | | | | | |
| | | | | | | | | |
| | | | | | | | | |
| | | | | | | | | |
| | | | | | | | | |
| | | | | | | | | |
| | | | | | | | | |
| | | | | | | | | |
| | | | | | | | | |
| | | | | | | | | |
| TOTAL | | | | | | | | |

RATEABLE VALUES IN SCOTLAND 1983 - 84 RRV/ASS/1983

NUMBERS AND RATEABLE VALUES OF SUBJECTS IN THE VALUATION ROLL AS AT 1 APRIL 1983

| DISTRICT | COMMERCIAL SUBJECTS | | | | | | | | | |
|---|---|---|---|---|---|---|---|---|---|---|
| | SHOPS (inc.restaurants) | | OFFICES | | HOTELS, BOARDING HOUSES etc. | | OTHER (Note 4) | | TOTAL | |
| | No | RV(£) | No | RV(£) | No | RV(£) | No | RV(£) | No | RV(£) |
| | 25 | 26 | 27 | 28 | 29 | 30 | 31 | 32 | 33 | 34 |
| | | | | | | | | | | |
| | | | | | | | | | | |
| | | | | | | | | | | |
| | | | | | | | | | | |
| | | | | | | | | | | |
| | | | | | | | | | | |
| | | | | | | | | | | |
| | | | | | | | | | | |
| | | | | | | | | | | |
| | | | | | | | | | | |
| | | | | | | | | | | |
| | | | | | | | | | | |
| | | | | | | | | | | |
| | | | | | | | | | | |
| | | | | | | | | | | |
| | | | | | | | | | | |
| | | | | | | | | | | |
| | | | | | | | | | | |
| | | | | | | | | | | |
| TOTAL | | | | | | | | | | |

RATEABLE VALUES IN SCOTLAND 1983 - 84 RRV/ASS/1983

NUMBERS AND RATEABLE VALUES OF SUBJECTS IN THE VALUATION ROLL AS AT 1 APRIL 1983

| DISTRICT | INDUSTRIAL AND FREIGHT TRANSPORT SUBJECTS (Note 3) | | | PUBLIC UTILITIES ( NOTE 5) | | | | | |
|---|---|---|---|---|---|---|---|---|---|
| | | | | A.P.U. VALUED | | OTHERS | | TOTAL | |
| | No | NAV (£) | RV(£) | No | RV(£) | No | RV(£) | No | RV(£) |
| | 35 | 36 | 37 | 38 | 39 | 40 | 41 | 42 | 43 |
| | | | | | | | | | |
| | | | | | | | | | |
| | | | | | | | | | |
| | | | | | | | | | |
| | | | | | | | | | |
| | | | | | | | | | |
| | | | | | | | | | |
| | | | | | | | | | |
| | | | | | | | | | |
| | | | | | | | | | |
| | | | | | | | | | |
| | | | | | | | | | |
| | | | | | | | | | |
| | | | | | | | | | |
| | | | | | | | | | |
| | | | | | | | | | |
| | | | | | | | | | |
| | | | | | | | | | |
| | | | | | | | | | |
| TOTAL | | | | | | | | | |

RRV/ASS/1983

RATEABLE VALUES IN SCOTLAND 1983 - 84

NUMBERS AND RATEABLE VALUES OF SUBJECTS IN THE VALUATION ROLL AS AT 1 APRIL 1983

| DISTRICT | MISCELLANEOUS SUBJECTS | | TOTAL ALL SUBJECTS | |
|---|---|---|---|---|
| | No | RV(£) | No | RV(£) |
| | 44 | 45 | 46 | 47 |
| | | | | |
| | | | | |
| | | | | |
| | | | | |
| | | | | |
| | | | | |
| | | | | |
| | | | | |
| | | | | |
| | | | | |
| | | | | |
| | | | | |
| | | | | |
| | | | | |
| | | | | |
| | | | | |
| | | | | |
| | | | | |
| | | | | |
| | | | | |
| TOTAL | | | | |

ANNEX A

ADDITIONAL INFORMATION

| RATING AUTHORITY<br>.................... REGION<br>............ ISLANDS AREA | Cumulo value at 1 April 1983 of domestic part of caravan site valuations under section 3(1) of Rating (Caravan Sites) Act 1976 | TOTAL Rateable Value at 31 March 1983 | |
|---|---|---|---|
| | | DOMESTIC Subjects | ALL Subjects |
| DISTRICTS:<br>ISLANDS AREA | 1.<br>£ | 2.<br>£ | 3.<br>£ |
| | | | |
| | | | |
| | | | |
| | | | |
| | | | |
| | | | |
| | | | |
| | | | |
| | | | |
| | | | |
| | | | |
| | | | |
| | | | |
| | | | |
| | | | |
| | | | |
| | | | |
| | | | |
| | | | |
| | | | |
| TOTAL | | | |

Signature ______________________

Assessor for ______________________

Date ______________________

# APPENDIX 15

### Manpower Statistics

The LACSAB Quarterly Joint Manpower Watch form contained in this Appendix is referred to in section 10.1 –but see especially paragraph 10.1.8. The Scottish Local Authorities Negotiating Bodies JMW form is referred to in the same parts of the text.

The note addressed to "newcomers" to the survey, is issued by the JMW Group in England and Wales (see paragraph 10.1.3).

## QUARTERLY JOINT MANPOWER WATCH

### NOTES OF GUIDANCE - DECEMBER 1981

The attached Notes of Guidance for the completion of the Manpower Watch survey form are an update of the previous set of notes, issued in 1978. They contain very little new material but hopefully the changed layout and removal of outdated information will assist authorities in their completion of the form.

For newcomers to the survey the following points may be of interest.

(1) The Quarterly Joint Manpower Watch is conducted by LACSAB (on behalf of the Local Authority Associations and entral overnment) every three months amongst the 456 authorities in England and Wales. A similar survey is conducted in Scotland. The survey is the most comprehensive count of local authority employment available and has been conducted regularly since March 1975. Previous to this date an annual (mid-year) count of local authority employment was conducted by the Department of Employment (L42 format). This latter survey is still conducted at the request of the DE as it gives an important male/female full-time/part-time breakdown of the quarterly national employment figures.

(2) Manpower Watch data are published regularly in a number of publications. The most readily available of these for authorities is <u>Local Government Manpower</u>, published by LACSAB. It contains tables A, B and C from the press release (these are national tables - England and Wales). Other publications which publish the data in different forms include the <u>Department of Employment Gazette, Monthly Digest of Statistics, Social Trends and Local Government Financial Statistics</u>.

(3) Feedback of Information to Authorities. LACSAB regularly sends Joint Manpower Watch information to authorities. This is of two principal sorts:

(a) <u>Quarterly</u> Every three months authorities receive: (1) a copy of tables D-M of the press release a week in advance of the press release date. This is given to authorities <u>primarily</u> for them to check their own figures and alert us if changes need to be made (although obviously the data on other authorities will be of interest), (2) a selection of the national tables. These detail employment by service category in the different types of authorities. Change tables (for the same grouping of authorities) showing changes in employment over the last quarter and year are also included.

(b) <u>Half-yearly</u> (March and September). Actual employment figures for authorities of the same type are circulated. In the case of the Non Metropolitan Districts the authorities are split into 11 bands by population; each authority received the forms of all authorities in its band, and summary tables for that band, and the bands immediately above and below it.

(4) Inquiries relating to the Codes of Practice on annual reports and the quarterly publication of manpower should (apart from Manpower Watch definitions) be directed in the first instance to the Department of the Environment.

-2-

(5) Revisions to Part Quarters'(Survey Forms)

A number of authorities who have notified LACSAB of changes to previous quarters' figures have asked about the procedures LACSAB adopts in incorporating them in the national figures and the National Press Release which is issued by the DOE/Welsh Office. The procedure adopted by LACSAB takes account of the tight dead-lines within which the JMW is conducted and the computer and clerical work which has to be undertaken.

Briefly the procedure is as follows:-

Revisions are taken back one year. The national tables are revised, taking account of individual authority revisions, twice after their original publication.

First, they are revised with the publication of the next quarter's survey results.

Second, they are revised with the publication of the results for the same quarter next year.

For example, when the December 1981 figures are compiled, revised national tables are produced for September 1981 (first revision) and December 1980 (second and final revision). The figures for March and June 1981 (which are not published in the National Press Release) are not revised at this stage; they are retained in a holding file until the appropriate quarter . The annual and quarterly change tables published in the December 1981 National Press Release therefore take account of the revisions referred to above. Consequently, the the individual authority figures for December 1980 published in the December 1981 National Press Release take account of revisions notified by authorities to LACSAB provided they have been received by the time the December 1981 draft press release has been sent to authorities to allow them to check the figures. Any amendments received from authorities at this stage are included in the "addendum" to the Press Release; they are not included in the Revised figures for December 1980. (To date, the revisions received in respect of past quarters would have no impact on the national figures.

As agreed by the JMW Group revisions are not taken back further than one year.

IMPORTANT : PLEASE READ CREAFULLY BEFORE COMPLETING THE RETURN

NOTES ON COMPLETING THE QUARTERLY JOINT MANPOWER WATCH AND THE L42 FORMS

The revised Notes of Guidance should be read carefully before starting to complete the form.

A. Consistency in completing returns

It is important that once your authority has allocated staff according to the revised notes of guidance, that you should follow a consistent line in allocating staff to service and functions in all future returns. A number of authorities have found it worthwhile to note down how their employment is allocated within the return so that the return can be completed on a consistent basis. This has been found particularly useful when there has been a change in staff responsible for the completion of the return.

If you introduce changes in allocation in the present return, please would you detail any significant changes in a covering letter, and, if possible, indicate the number of staff who should similarly be transferred in the previous returns.

B. Timing and staff to be included:

The figures should relate to the last pay period for which pay has been computed at the date shown on the form. They should include all full-time and part-time employees (with the exception of Police officers, traffic wardens, and civilian staff of police authorities, for whom information will be provided by the Home Office), including those employees temporarily absent through sickness, holidays and other causes including uncertified absence. For example, personnel on maternity leave should be included in the return, as should personnel engaged in temporary work.

Where authorities undertake agency work for other authorities, only the authority which actually employs the employees concerned should include them on its form. For example, if a District undertakes agency work on behalf of a County, the District should include those employees engaged on this work on its return. By definition the County should exclude such employees despite subsequently reimbursing the District for the cost of the service.

Where authorities merely act as paymaster to pay the staff of other local authorities, the employing authority and not the paying authority should include them on its form. Where two authorities jointly undertake a function they should agree as to which authority includes the staff concerned on its return and adhere to the arrangement in future returns.

C. Form L42

Line 6. All otherdepartments and services. The service classification in lines 1-6 corresponds with the classification used in the Quarterly Joint Manpower Watch. All local authority employees (other than Police officers, and traffic wardens and civilian staffs of police authorities) who have not been included in lines 1-5 should be included in line 6 of the L42 form, which should correspond with line 32 of the QJMW return in respect of full-time, part-time and total figures as, of course, should the figures on lines 1-5 of each form.

Line 000. Grand total - all departments and services. The full-time, part-time and total figures on line 000 of the L42 should correspond with line 33 of the QJMW return. This means that Retained Firemen (see line 29), staff employed by outside employment agencies, Magistrates' and other courts' staff and Probation and After Care staff should be excluded from these total figures.

D. Non-manual employees

The figures for non-manual employees should include all those staff not included under the definition of manual workers given below. They should include all employees within scope and paid under the JNCs for Chief Executives and Chief Officers; the NJC for Administrative, Professional, Technical and Clerical Services; all education staff within the scope of the Burnham Committees, the Pelham and Soulbury Committees, and the JNCs for Justices' Clerks and Justices' Clerks' Assistants (see line 35), and the JNC for the Probation Service (see lines 36 and 37), all firemen and fire officers (see lines 29 and 30).

-2-

E. Manual Workers

The figures for manual employees should include those employees within the scope and paid under the NJC for Local Authorities' Services (Manual Workers), the JNC for Building and Civil Engineering, the JNC for Engineering Craftsmen, the NJC for Workshops for the Blind and all other negotiating bodies and rates of pay linked to the NJC (Manual Workers).

F. Part-time Employees

The figures for part-time employees should include those ordinarily employed for up to and including 30 hours per week. Teachers with normal full-time engagements should, however, be included in the column headed "full-time" irrespective of the numbers of hours "teaching time" in the week.

G. Seasonal variations

In order to enable movements in the employment data returned by authorities to be interpreted more effectively, it would be helpful if authorities could also make a note of any distinctly seasonal variations in employment - for example, changes in the employment of manual workers in the recreation area.

H. STEP, YOP, WEP, etc.

100% MSC funded job creation schemes should be excluded from the return, i.e. CES (STEP, YOP, WEP, etc.

I. Allocation of staff between services and departments

There will, inevitably, be problems in allocating staff between the services and departments listed. Authorities are asked to use the following general guidelines for all services and departments.

Employees normally employed for the bulk of their working time during the pay period within a particular service or department should be allocated to that service or department.

Where employees work on a variable basis during the pay period for several services or departments then they should be allocated to the service or department which is considered to be their home base. If their home service is a composite service covering several of the services or departments set out in the form, then they should be allocated to the section of it which is considered to be their home base.

Authorities should follow a consistent line in allocating staff to services or departments from one quarterly enquiry to the next. When there is a significant change of practice it is essential that a record be kept of this and a note be forwarded with the form.

Line 1
Education Department

The figures for lecturers and teachers should include supply teachers. The figures for part-time lecturers and teachers should include only those part-time lecturers and teachers who have been paid and employed in the last claim period for which pay has been computed at the date shown on the form. They should not include numbers on call, on the rolls, on the books, etc. Consistent with previous practice of the majority of authorities, a part-time lecturer or teacher of the authority should only be included once and in the full-time capacity.

Line 2

The figures for "other staff" should include, for example, school transport staff, youth leaders, nursery assistants and administrative, secretarial and clerical staff in the education department, school caretakers, cleaners, canteen and school playing field staff, non-teacher staff employed on careers guidance and library staff employed by libraries either located in or servicing educational establishments.

Line 3
Construction

This item refers to employees (including direct supervisors, side supervisors; resident site technicians/engineers/architects and surveyors) engaged wholly or largely on work similar to that undertaken by contractors in the construction industry. The figures should cover house building by direct labour, maintenance work on housing estates (i.e. the figures should not be included in Housing department - line 11), contruction of roads and bridges together with road repair work such as resurfacing, widening and straightening and all other construction, repair and maintenance work with the following exceptions.

If the authority has only a small maintenance staff employed solely on the up-keep of its offices and other buildings the workers concerned should be included in their home service or department.

Staff engaged in traffic engineering and sweeping, tidying and minor repairs to roads, together with associated administrative staff, should also be included in their home service or department.

Staff who are not wholly or largely employed on construction, repair and maintenance work as defined but who work in a technical, administrative or general managerial capacity to those employed on construction should be counted as part of the appropriate home central services department or service. In most cases this will be Engineering services.

Line 4.
Transport Services

The figures should include employees of all public transport services operated by the local authority including docks, river and harbour services, airports, and associated transport maintenance. Staff employed by Passenger Transport Executives and the London Transport Executive should be excluded.

Staff engaged in the running of authority transport services other than public transport should be excluded from Transport services (line 4) and included under the appropriate service (e.g. refuse disposal and collection; recreation, parks and baths). Staff engaged in a general maintenance section who cannot be allocated out to individual services should be included under Engineering services (line 17). Where the maintenance section supports both public and other transpost services the staff should be allocated to the main activity.

Line 5
Social Services

The figure should include only those staff who have been paid and employed in the last pay period for which pay has been computed at the date shown on the form. They should not include numbers on call, on the rolls, on the books, etc.

They should include all staff engaged in all personal social services covering field work, residential, day and community care. The staff involved include professional social workers, administrative staff, home helps, staff of care centres, staff of sheltered workshops and staff of residential accommodation.

Non-Metropolitan Districts should not include any staff in line 5. These authorities should include wardens and caretakers of old people's homes under the Housing department (line 11) and meals on wheels staff under 'Other Services' (line 31) on the QJMW return and 'All Other Departments and Services' (line 6) on the L42 return.

Line 7
Public Libraries, Museums and Art Galleries

The figures should include all staff engaged in the provision of these services. Archaeological services employees should be placed in line 31.

Line 8
Recreation, parks and baths

The figures should include staff engaged in work on parks and open spaces, national and countryside parks, sports centres, theatres, night clubs, baths, swimming baths and public laundries, and associated staff including, for example, entertainments and hall lettings staff, catering staff and car park attendents of recreational establishments.

Line 9
Environmental health

The staff to be included are those employed on public health, meat inspection, prevention of air pollution, public conveniences and port health, and any other environmental health work undertaken on an agency basis (e.g. for a water authority).

Line 10
Refuse collection and disposal

The figures should include all staff - dustmen, drivers and associated supervisors and administrative staff - engaged on refuse collection and disposal including disposal undertaken on an agency basis.

Line 11
Housing

The staff to be included are those employed in Housing on the administration of all housing matters, including rent collection, improvement grants and the authority's own housing.

Line 12
Town and country planning

The figures should include the staff of the Planning department including administrative support staff. Building control staff should also be included. Any apportionment of staff should be undertaken as suggested in paragraph I.

CENTRAL SERVICES

The figures should include all staff employed in central administrative services departments. The objective is to obtain a total figure for Central Services (line 28) and as consistent as possible allocation of staff to the separate central services which are listed (lines 13-23). Authorities are asked to use these listed services as far as possible and practicable. Authorities should enter other central services in the spaces provided (lines 24-27) where they are unable to allocate staff to the listed services.

Staff within Central services should as practicable be allocated to individual central services as follows:

Line 13
Chief Executive

The figures should include the Chief Executive, immediate support staff and corporate planning staff (excluding departmental or service heads).

Line 14
Financial services

The figures should include staff engaged in the provision of central Financial (Treasurer's) services to other services and departments.

Line 15
Secretarial services

The figures should include staff normally engaged in the secretarial function, including the provision of central general administrative services and the servicing of committees. Staff engaged in central typing pools and printing and purchasing units (excluding central construction purchasing units which should be included under Engineering services - line 17) should also be included as should archives staff engaged in registration of electors, civil defence, emergency planning, and public relations. Central offices' cleaners, porters and catering staff who cannot be allocated to individual services should also be included.

Line 16
Legal services

The figures should include professional and administrative support staff engaged in the provision of Legal services. Prosecuting solicitors and associated support staff employed by authorities should be included in one of the lines 24-26 and excluded from Legal services. Prosecuting solicitors and administrative support staff employed by police authorities should be excluded altogether from the return.

-5-

Lines 17 & 18
Engineering services and Architectural services

Professional and administrative staff should be allocated between the two services on the basis of paragraph I. Authorities unable to allocate technical services staff on this basis should include them on line 27. Staff engaged in sweeping, tidying and minor repairs to roads should be included under Engineering services, as should car parking attendants engaged at general municipal car parks.

Line 19
Estates services

The figures should include staff engaged in valuation and property maintenance with the exception of small holdings and allotments, the staff of which should be included in line 31 (Other Services).

Line 20
Personnel services

The staff to be included are those engaged in central Personnel services, including those engaged in recruitment, training industrial relations and other personnel functions.

Line 21
Management services

The figures should include staff engaged in central Management services, including O&M and Work Study.

Line 22
Computing services

The staff to be included are those engaged in the provision of Computer services to other departments and services of the authority, including systems analysts, programmers, computer operators and associated machine operators.

Line 23
Consumer protection,etc.

This line should include all staff employed in the maintenance of trading standards, weights and measures.

Lines 24-26

These lines should be used to show separately the following categories of staff: prosecuting solicitors and their support staff, rent officers and their support staff, motor taxation, Road Construction Unit and Registration of Births, Deaths and Marriages staff. These lines should also include other central services staff not allocated to the listed central services (lines 13-23) with the exception of 'Technical services' staff who should be allocated to line 27.

Line 27
Technical services

See lines 17 and 18 notes.

Lines 29 & 30
Fire services

Firemen should include all uniformed (including fire prevention officers) and control room staff (except retained firemen who should be excluded altogether from the return) and be classified as non-manual employees. "Other staff" should include non-firemen, for example, administrative, clerical and cleaning staff.

Line 31
Other services

The figures should include all staff in all services and departments not covered above. They should include staff employed on school crossing patrols, even where the police authority acts as an agent outside the London Metropolitan area. The Home Office will provide information on school crossing patrols within the London Metropolitan area. Trading services such as cemeteries, crematoria, markets, catering (except recreational establishment and central offices catering staff which should be allocated to lines 8 and 15 respectively) and slaughterhouses should be included as should staff engaged in agriculture and fisheries (pest control). smallholdings, sea fisheries, and drainage and allotments . Non-Met. districts should include meals on wheels staff.

-6-

Line 34
Staff employed by outside agencies

The figures should include those employed by outside employment agencies but working for local authorities, e.g. secretarial, clerical and typing staff, together with any other employees of employment agencies (please note B also).

Line 35
Staff of magistrates' and other courts

The figures should include those engaged in magistrates' and coroners' courts. The information should be available from the Treasurer's department. Figures for magistrates' courts in Inner London will be provided by the Home Office.

Lines 36 & 37
Probation and After Care staff

The figures for "other staff" should include for example, administrative, clerical and cleaning staff. The information should be available from the Treasurer's deparment. Figures for Inner London will be provided by the Home Office.

LOCAL AUTHORITIES' CONDITIONS OF SERVICE ADVISORY BOARD

QUARTERLY JOINT MANPOWER WATCH

EMPLOYMENT BY LOCAL AUTHORITY

| | |
|---|---|
| Name of authority.................................. <br> Address .................................. <br> .................................. | Name and telephone number .......................... <br> of person to be contacted <br> in the event of a query: .......................... <br> .......................... |
| For office use only <br> 1/1 - 5 [ \| \| \| \| ] 1/6 - 7 [3 \| 8 \| ] | Numbers employed on 16 June 1984 |

Police officers, traffic wardens and the civilian staffs of police authorities to be excluded (see note 1)

| Line No: | Department or Service (see note 2a) | For office use only card/Col.No. | (see notes) Non-Manual Full-time | Non-Manual Part-time | Manual Full-time | Manual Part-time | TOTAL |
|---|---|---|---|---|---|---|---|
| 1 | Education Department (see notes )<br>(a) Lecturers and teachers | 2. 6 - 30 | | | | | |
| 2 | (b) Other staff | 2.31 - 55 | | | | | |
| 3 | Construction (see notes ) | 2.56 - 80 | | | | | |
| 4 | Transport services (see notes ) | 3. 6 - 30 | | | | | |
| 5 | Social services (see notes ) | 3.31 - 55 | | | | | |
| 6 | Total of lines 1 to 5 | 3.56 - 80 | | | | | |

LACSAB

For office use only 1-5

| Line No. | Department or Service (see note 1) | For office use only Card/Cal No. | (see note E) Non-Manual Full-time | Non-Manual Part-time | Manual Full-time | Manual Part-time | TOTAL |
|---|---|---|---|---|---|---|---|
| 7 | Public libraries, museums and art galleries | 4. 6 - 30 | | | | | |
| 8 | Recreation, parks and baths (see notes) | 4.31 - 55 | | | | | |
| 9 | Environmental health (see notes ) | 5.56 - 80 | | | | | |
| 10 | Refuse disposal & Collection (see notes ) | 5. 6 - 30 | | | | | |
| 11 | Housing department (see notes ) | 5.31 - 55 | | | | | |
| 12 | Town and country planning (see notes ) | 5.56 - 80 | | | | | |
| 13 | Central Services (see notes ) Chief Executive | 6. 6 - 30 | | | | | |
| 14 | Financial services | 6.31 - 55 | | | | | |
| 15 | Secretarial services | 6.56 - 80 | | | | | |
| 16 | Legal services | 7. 6 - 30 | | | | | |
| 17 | Engineering services | 7.31 - 55 | | | | | |
| 18 | Architectural services | 7.56 - 80 | | | | | |
| 19 | Estates' services | 8. 6 - 30 | | | | | |
| 20 | Personnel services | 8.31 -55 | | | | | |
| 21 | Management services | 8.56 - 80 | | | | | |
| 22 | Computer services | 9. 6 - 30 | | | | | |
| 23 | Consumer protection, weights and measures and trading standards | 9.31 - 55 | | | | | |

- 3 -

LACSAB

For office use only 1 - 5

| Line No. | Department of Service (see note 1) | For office use only Card/Col No. | (see note E) Non-Manual Full-time | Non-Manual Part-time | Manual Full-time | Manual Part-time | TOTAL |
|---|---|---|---|---|---|---|---|
| 24 | (see notes) | 9.56- 9.80 | | | | | |
| 25 | | 10. 6-10.30 | | | | | |
| 26 | | 10.31-10.55 | | | | | |
| 27 | | 10.56-10.80 | | | | | |
| 28 | Total of central services | 12. 6-30 | | | | | |
| 29 | Fire Service (see notes) (a) regular firemen | 12.31-55 | | | | | |
| 30 | (b) other staff | 12.56-80 | | | | | |
| 31 | Other Services (see notes) | 13. 6-30 | | | | | |
| 32 | Total of lines 7,8,9,10,11,12 and 28,29,30,31 (see notes ) | 13.31-55 | | | | | |
| 33 | Grand total of all departments and services | 13.56-80 | | | | | |
| 34 | Additional items<br>Staff employed by outside employment agencies but working for local authority (see notes ) | 14. 6-30 | | | | | |
| 35 | Staff of magistrates' & other courts (see notes) | 14.31-55 | | | | | |
| 36 | Probation and After Care staff (see notes ) (a) probation officers | 14.56-80 | | | | | |
| 37 | (b) other staff | 15. 6-30 | | | | | |

## SCOTTISH LOCAL AUTHORITIES NEGOTIATING BODIES

## NOTES on COMPLETING the QUARTERLY JOINT MANPOWER WATCH FORM

1. Timing and Staff to be Included

The figures should relate to the last pay period for which pay has been computed at the date shown on the form. They should include all full-time and part-time employees, including those employees temporarily absent through sickness, holidays and other causes. N.B. Temporary employees taken on to cover for holidays, illness, maternity leave, etc., should be included in the main form, but also listed separately on the sheet provided.

All people employed under the Youth Training Scheme (YTS), Community Programmes, W.E.P. etc. government schemes, including supervisors, (i.e. those wholly funded by the MSC) should be excluded from the form but shown separately on the sheet provided. Those people employed under the "URBAN AID" programme should be included in the main form. Those employed on Unified Housing Benefit work, wholly financed by the DHSS, should be included in the form but listed separately on the sheet provided.

Where authorities undertake agency work for other authorities only the central authority which actually and directly pays the employees concerned should include them in its form.

2. Allocation of Staff between Departments and Services

There will, inevitably, be problems in allocating staff between the departments and services listed. Authorities are asked to use the following general guidelines for all departments and services.

Employees normally employed for the bulk of their working time during the pay period within a particular department or service should be allocated to that department or service.

Where employees work on a variable basis during the pay period for several departments or services then they should be allocated to the department or service which is considered to be their home base. If their home department is a composite department covering several of the departments or services set out in the form, then they should be allocated to the section of it which is considered to be their home base. Notes 4 to 18 should assist authorities in allocation of staff to the various services.

Authorities should follow a consistent line in allocating staff to departments or services from one quarterly enquiry to the next. When there is a significant change of practice it is essential that a record be kept of this and a note be forwarded with the form.

3. (a) Non-Manual Employees

The figures for non-manual employees should include all those staff not included under the definition of manual workers given below. They should include all employees within the scope of -

(i) the J.N.C. Chief Officials of Local Authorities (Scotland);

(ii) the N.J.C. for Administrative, Professional, Technical and Clerical Services (Scottish Council);

(iii) the S.J.N.C. for Teaching Staff in School Education and the S.J.N.C. for Teaching Staff in Further Education,

(iv)/

(iv) N.J.C.'s for Local Authorities' Fire Brigades and Chief Officers of Local Authorities Fire Brigades

(v) Police Council for the United Kingdom

(vi) negotiating bodies other than (i) to (v) above

and all teaching staff, whole time or part-time.

(b) Manual Employees

The figures for manual employees should include those employees within the scope of -

(i) the N.J.C. for Local Authorities' Services (Manual Workers) (Scottish Council)

(ii) the S.J.N.C. for Manual Workers in Water Services

(iii) the S.J.N.C. for Local Authorities' Services (Building and Civil Engineering)

(iv) the S.J.N.C. for Local Authorities' Services (Craftsmen) (Plumbers/Electricians)

(v) the S.J.N.C. for Local Authorities' Services (Engineering Craftsmen)

(vi) negotiating bodies other than (i) to (v).

(c) Part-time Employees

The figures for part-time employees should relate to those ordinarily employed for less than 30 hours per week. Teachers with normal full-time engagements should, however, be included in the columns headed "full-time" irrespective of the number of hours of "teaching time" in the week.

4. Education

Lecturers and Teachers (Part-time)

These staff should include only those part-time lecturers and teachers who have been paid and employed at the last pay period for which pay has been computed at the date shown on the form. They should not include numbers on call, on the rolls, on the books, etc.

A part-time lecturer or teacher who is also in the full-time employment of the authority should only be included once and in the full-time capacity.

All part-time evening class teachers (many of whom will be employed in a full-time capacity with the authority) should be excluded from the main form but shown separately on the sheet provided.

Other Staff

The figures for "other staff" should include, for example, administrative and/

3.

and clerical staff in the education department, school caretakers, cleaners and canteen staff, non-teaching staff employed on careers guidance and library staff employed by libraries either located in or servicing educational establishments. Where school transport staff are based in the Education Department they should be included on this line.

Part-time youth and community staff, administrative and clerical staff etc. working in the evenings should be treated in the same way as part-time evening class teachers and excluded from the main form but shown separately on the sheet provided.

5. Transport

The figures should include employees of all public transport services operated by local authorities, including road transport, docks, river and harbour services, airports, transport maintenance and all other forms of public transport service. Also included on this line should be school transport staff where these employees are not based in the Education Department. Staff employed by the Greater Glasgow Passenger Transport Executive should be included.

6. Social Services

The figures should include all staff engaged in all personal social services covering field work, residential day and community care. The staff involved includes (professional) social workers, Reporters to Children's Panels, administrative staff, home helps, staff of care centres and staff of residential accommodation.

This service, similarly to education, should include only those staff who have been paid and employed in the last pay period for which pay has been computed at the date shown on the form. They should not include numbers on call, on the rolls, on the books, etc.

7. Cleansing

The figures should include staff engaged on refuse collection and disposal including disposal undertaken on an agency basis.

8. Environmental Health

The staff to be included are those employed on public health, prevention of air pollution, public conveniences and port health, and any other work undertaken on an agency basis.

9. Housing

The staff to be included are those employed in the Housing Department on the administration of all housing matters, including improvement grants and the renting of the authority's own housing. Employees engaged on the maintenance or construction of the authority's housing should not be included on this line but under Direct Works (line 240).

10./

10. Recreation, Leisure and Tourism

The figures should include staff engaged in work on parks and open spaces, sports centres, baths, swimming baths and public laundries, and associated staff including for example, entertainments and hall letting staff.

11. Planning

(a) Policy Planning

The Policy Planning function should include those staff involved in drawing up the authority's "corporate plan".

Where the authority does not have a separate policy planning department and this function is carried out as part of the general duties of the management team then such staff should be included in their home services department and there should be no entry on this line.

Where an authority has a separate "Industrial Development" Department then these staff should also be included on line 200.

(b) Physical Planning

The main tasks involved are the planning of urban, countryside and industrial development to ensure the optimum use of the physical environment. All town and country planning officers should be included on this line.

12. Technical Services

It is very difficult to define exactly the technical services function. In some authorities the Technical Services Department, where it exists, embraces functions such as architecture, surveying, physical planning, building control, etc. In other authorities the Technical Services Department only embraces one function.

It is hoped that the following notes will assist authorities in the allocation of staff when the Technical Services Department embraces several functions.

If an authority has a Technical Services Department which is responsible for several of the services just mentioned, then staff should be allocated to the various service functions as follows.

(i) Architects, quantity surveyors, architectural technicians, etc. should be listed on line 320. Clerks of works should also be shown on this line except when they are attached whole time to the Works Department, if this exists.

(ii) Staff employed on the maintenance or construction of the authority's property, e.g. houses, etc., should be listed under Direct Works (except of course architects, surveyors, etc. who will have been listed elsewhere).

(iii)/

(iii) Physical planners, as already indicated, should be listed on line 200.

(iv) If possible clerical and administrative staff should be allocated to the various services already mentioned. In smaller authorities this may not be possible since the typist working in a Technical Services Department covering many functions may spend a third of her/his time working for the planning officers, etc. In this case such staff should simply be shown on the Technical Services line 220.

(v) The Director of a Technical Services Department and his Depute and immediate support staff, responsible for all the functions mentioned above, should be shown on line 220.

(vi) Building Control

Building Control being of a regulatory and enforcement nature involves work of inspection similar to that in Environmental Health - site inspections being one of the main tasks. In some authorities, Building Control Officers will be included in the Planning Department, in others they will fall within the Technical Services Department. In any event, Building Control Officers should be included on line 220 - the Technical Services Department.

13. Roads/Highways

Staff employed on the construction and maintenance of roads/highways should be included under this heading and also staff concerned with the installation and maintenance of road lighting and with the lighting of traffic signals and signs.

14. Direct Works

As already stated, staff employed on the maintenance or construction of the authority's property, e.g. schools, houses, etc., should be listed under this heading (except of course architects, surveyors, etc., who will have been listed elsewhere).

15. Central Services

(i) Chief Executive's Department

On this line should be included only the Chief Executive and his immediate support staff, e.g. Depute, Secretary, etc. In smaller authorities the Chief Executive's Department is also responsible for the legal and administrative function. In this case the staff should be separated as suggested - only the Chief Executive and his immediate support staff being listed on line 290, the remainder being listed on line 310.

(ii) Finance

Allocation of staff to this service should be straightforward.

(iii) Legal and Administrative Services/

6.

(iii) Legal and Administrative Services

The scope of legal and administrative services includes conveyancing, legal procedures and standing orders, contracts, interpretation of statutes and their application to the work of the authority, etc. Also included is the task of servicing the Council and committees providing the common services.

(iv) Architecture and related services

The staff to be allocated to this service includes architects, quantity surveyors, technicians and draughtsmen, trainees, clerks of works, etc.

(v) Manpower Services/Personnel

This function includes advice on staff development and training and on salary and wage structures, negotiations with trade unions and staff associations, etc. Medical Referees and Safety Officers should be shown on this line.

(vi) Management Services

Under this heading should be included staff employed on O.& M. work and Work Study.

(vii) Computer Services, Public Relations

No difficulty should arise in allocating staff to these services.

(viii) Estates

Where an authority has a separate estates department the staff should be included on line 370.

(ix) Internal Transport

This service relates to the maintenance of the authorities' limousines and special purpose vehicles such as those used in lighting, parks and cleansing. If employees engaged on such work are in a particular department such as Cleansing, spending the majority of their time on servicing the Cleansing Department's vehicles, then such staff should be included in their home service department. Where an authority has a distinct internal transport division then the staff should be included on line 380.

(x) Other Central Services

On this line should be listed those staff responsible for catering and cleaning the authority's offices (cleaners responsible for cleaning school buildings, sports halls, etc., should of course be included in their home service department, i.e. education, recreation, leisure and tourism). In larger authorities where office cleaners may be confined to the cleaning of one particular department or service then they should be listed under the heading of that department or service.

If an authority has a separate supplies department then the staff should be included here and the staff of the Chemist and Public Analyst's Department should be listed on this line.

16./

7.

16. Fire Service

Firemen/Firewomen should be classified as non-manual employees; the figures should exclude retained firemen who should be shown separately on the sheet provided but include control room staff. "Other staff" should include administrative, clerical and cleaning staff.

17. Police and Related Services

Police officers should be classified as non-manual employees; the figures should exclude special constables who should be shown separately on the sheet provided. Police cadets should also be classified as non-manual employees and be shown separately from police officers on line 421.

"Other staff" should include non-police officers, for example, administrative, technical, clerical and cleaning staff.

18. Other Services

The figures should include all staff in all departments and services not covered above. Services such as cemeteries, crematoria, markets and slaughterhouses, motor taxation staff and road bridge staff should be included here.

19. Staff of District Court

The figures should cover clerical, administrative and other staff, such as porters and messengers servicing the District Court.

20. Construction

This special item, required by the Department of Employment, refers to employees engaged on work similar to that undertaken by contractors in the construction industry. The figures should cover house building by direct labour, maintenance work on housing estates, construction of roads and bridges, together with road repair such as resurfacing, widening, straightening, etc. In other words authorities will have to count those employees shown in lines 100 - 460 involved in construction work and state the number on this line.

If authorities have only a small maintenance staff employed solely in the upkeep of offices and other buildings no figures should be entered on this line. Similarly the figure for roadmen engaged solely on sweeping, tidying and minor repairs should not be entered here.

## SCOTTISH LOCAL AUTHORITIES NEGOTIATING BODIES

### QUARTERLY JOINT MANPOWER WATCH
### EMPLOYMENT BY LOCAL AUTHORITY

Name of Authority ..........................................................

Name and Telephone No. of person to be contacted in event of a query ..........................................................

For Office Use Only

Region Code [1 — 5] District Code [6 — 10] Year [11 — 12] Quarter [13] Population

Number employed on 10th March 1984 (See Notes)

| Line No. | Department or Service | Non-Manual Staff (See Note 3) | | | | Manual Staff (See Note 3) | | | | Total Males and Females | |
|---|---|---|---|---|---|---|---|---|---|---|---|
| | | Males | | Females | | Males | | Females | | | |
| | | FT | PT | FT | PT | FT | PT | FT | PT | FT | PT |
| 14—16 | | 17—21 | 22—26 | 27—31 | 32—36 | 37—41 | 42—46 | 47—51 | 52—56 | | |
| 100 | Education (See Note 4) (a) Lecturers and Teachers | | | | | | | | | | |
| 110 | (b) Other Staff | | | | | | | | | | |
| 120 | Transport Services (See Note 5) | | | | | | | | | | |
| 130 | Social Services (See Note 6) | | | | | | | | | | |
| (140) | Total lines 100 to 130 | | | | | | | | | | |
| 150 | Cleansing (See Note 7) | | | | | | | | | | |
| 160 | Environmental Health (See Note 8) | | | | | | | | | | |
| 170 | Housing (See Note 9) | | | | | | | | | | |
| 180 | Public Libraries, Museums and Art Galleries | | | | | | | | | | |

2.

For Office Use Only

| Line No. | Department or Service | Non-Manual Staff (See Note 3) | | | | Manual Staff (See Note 3) | | | | Total Males and Females | |
|---|---|---|---|---|---|---|---|---|---|---|---|
| | | Males | | Females | | Males | | Females | | | |
| | | FT | PT | FT | PT | FT | PT | FT | PT | FT | PT |
| 14—16 | | 17 — 21 | 22 — 26 | 27 — 31 | 32 — 36 | 37 — 41 | 42 — 46 | 47 — 51 | 52 — 56 | | |
| 190 | Recreation, Leisure and Tourism (See Note 10) | | | | | | | | | | |
| 200 | Planning (See Note 11) (a) Policy Planning | | | | | | | | | | |
| 210 | (b) Physical Planning | | | | | | | | | | |
| 220 | Technical Services (See Note 12) | | | | | | | | | | |
| 230 | Roads/Highways (See Note 13) | | | | | | | | | | |
| 240 | Direct Works (See Note 14) | | | | | | | | | | |
| 250 | Water | | | | | | | | | | |
| 260 | Sewerage | | | | | | | | | | |
| 270 | Assessor and Electoral Registration | | | | | | | | | | |
| 280 | Consumer Protection, Weights and Meas. | | | | | | | | | | |
| 290 | Central Services (See Note 15) Chief Executive's Department | | | | | | | | | | |
| 300 | Finance | | | | | | | | | | |
| 310 | Legal & Admin. Services | | | | | | | | | | |
| 320 | Architecture and related services | | | | | | | | | | |
| 330 | Manpower Services/Personnel | | | | | | | | | | |
| 340 | Management Services | | | | | | | | | | |
| 350 | Computer Services | | | | | | | | | | |
| 360 | Public Relations | | | | | | | | | | |
| 370 | Estates | | | | | | | | | | |
| 380 | Internal Transport | | | | | | | | | | |
| 390 | Other Central Services | | | | | | | | | | |
| | Fire Services/ | | | | | | | | | | |

3.

**For Office Use Only** ☐☐☐☐☐

| Line No. | Department or Service | Non-Manual Staff (See Note 3) | | | | Manual Staff (See Note 3) | | | | Total Males and Females | |
|---|---|---|---|---|---|---|---|---|---|---|---|
| | | Males | | Females | | Males | | Females | | | |
| | | FT | PT | FT | PT | FT | PT | FT | PT | FT | PT |
| 14 — 16 | | 17 — 21 | 22 — 26 | 27 — 31 | 32 — 36 | 37 — 41 | 42 — 46 | 47 — 51 | 52 — 56 | | |
| | Fire Services (See Note 16) | | | | | | | | | | |
| 400 | (a) Regular Firemen/Firewomen | | | | | | | | | | |
| 410 | (b) Other Staff | | | | | | | | | | |
| | Police and Related Services (See Note 17) | | | | | | | | | | |
| 420 | (a) Police Officers | | | | | | | | | | |
| 421 | (b) Police Cadets | | | | | | | | | | |
| 430 | (c) Other Staff | | | | | | | | | | |
| 440 | (d) Traffic Wardens | | | | | | | | | | |
| 450 | Other Services (See Note 18) | | | | | | | | | | |
| 460 | Staff of District Court (See Note 19) | | | | | | | | | | |
| (470) | **Total** of lines 150 to 460 | | | | | | | | | | |
| (480) | **Total** lines 140 and 470 | | | | | | | | | | |
| 490 | Construction (See Note 20) | | | | | | | | | | |
| (500) | **Total** line 470 less lines 420, 421, 440 and 490 | | | | | | | | | | |
| (510) | **Total** lines 140, 490 and 500 | | | | | | | | | | |

QUARTERLY JOINT MANPOWER WATCH

Name of Authority ..................................................

(a) Please indicate below, according to the department or service line numbers on the survey form, the number of employees on YTS, Community Programmes, Work Experience etc. Schemes wholly funded by the MSC, in the employment of your authority as at the survey date, and whether they are non-manual or manual, full-time or part-time. N.B. YTS etc., employees should not be included when completing the form.

YTS, COMMUNITY PROGRAMMES, WORK EXPERIENCE ETC. STAFF

| Line Number | Non-Manual | | | | Manual | | | |
|---|---|---|---|---|---|---|---|---|
| | FT | | PT | | FT | | PT | |
| | Male | Female | Male | Female | Male | Female | Male | Female |
| | | | | | | | | |
| | | | | | | | | |
| | | | | | | | | |
| | | | | | | | | |
| | | | | | | | | |
| | | | | | | | | |
| | | | | | | | | |
| | | | | | | | | |
| | | | | | | | | |
| | | | | | | | | |
| | | | | | | | | |
| | | | | | | | | |
| | | | | | | | | |
| | | | | | | | | |
| | | | | | | | | |

(b) Please indicate the number of part-time staff specified below who should be excluded from the form.

All Part-Time

| | Non-Manual | | Manual | |
|---|---|---|---|---|
| | Male | Female | Male | Female |
| Part-time Evening Class Teacher/ Lecturer | | | | |
| Other Part-time Education Staff e.g. Community Education Workers, Clerical Staff, etc. | | | | |
| Reserve Firemen | | | | |
| Special Constables | | | | |

2.

(c) Please indicate by line number and whether non-manual or manual, full-time or part-time, the number of seasonal workers in the employment of your authority as at the survey date. (These staff should be included in the main form).

SEASONAL EMPLOYEES

| Line Number | Non-Manual | | Manual | |
|---|---|---|---|---|
| | FT | PT | FT | PT |
| | | | | |
| | | | | |
| | | | | |
| | | | | |
| | | | | |
| | | | | |
| | | | | |
| | | | | |
| | | | | |
| | | | | |
| | | | | |

(d) A brief explanation of any substantial increases or decreases in numbers employed apart from increases or decreases due to seasonal factors outlined in (c) above should be given below.

| Line Number | Explanation for increases or decreases |
|---|---|
| | |

(e) It would be very helpful if a forecast of any likely or planned increases or decreases in staff over the next quarter other than those due to seasonal changes, could be indicated. The forecast need not be numerically precise but the department or service should be identified and whether the staff affected are manual or non-manual.

| Department or Service | Forecast of increases or decreases in staff over the next quarter | |
|---|---|---|
| | Non-Manual | Manual |
| | | |

3.

(f) Please indicate below, according to the department or service line numbers on the survey form, the number of temporary employees taken on to cover for holidays, illness, maternity leave, etc., who should be included in the main return.

Temporary Employees

| Line Number | Non-Manual | | Manual | |
|---|---|---|---|---|
| | FT | PT | FT | PT |
| | | | | |

(g) Please indicate below the number of staff employed on Unified Housing Benefit (UHB) work wholly financed by the DHSS*.

| Line Number | Non-Manual | | Manual | |
|---|---|---|---|---|
| | FT | PT | FT | PT |
| | | | | |

* Where an authority has not employed any 'new' staff to undertake UHB work but existing employees are allocating say 10% of the working week to such work, then an approximate full-time equivalent figure should be entered in column 1 or 3 or both.

# APPENDIX 16

**Public Protection –Fire**

Home Office forms FDR1, FDR3 and 44C (see paragraphs 12.2.6 and 12.2.11) are not suitable for reproduction. The following is a brief description of their contents.

*FDR1* is the basic Home Office document containing a report of each fire notified to a fire brigade in Great Britain. It contains a detailed description of each fire incident under the following headings:

1) General (e.g address, time);
2) location (e.g type of property);
3) construction of building or ship at which fire occurred (e.g dimensions, materials of construction);
4) extinction of fire (e.g methods of fighting fire; resources used);
5) damage and spread;
6) supposed cause of fire;
7) life risk (e.g fatalities, number of persons who escaped);
8) explosion and dangerous substances at fire incident.

*FDR3* is a statistical document entitled Miscellaneous calls Report. It is completed monthly and relates to the number of false alarms, chimney fires, secondary fires.

*44C* is also a statistical document which is completed annually (but not in Scotland). It aggregates all data previously reported in FDR1 and FDR3.

As mentioned in paragraph 12.2.11 the form FS(A) used as data source by HM Chief Inspector of Fires in Scotland, is in 13 parts and is not therefore suitable for reproduction. However a list of the titles of the 13 parts is contained in this Appendix.

FORM FS(A) 1983

________________________________ Fire Brigade

For year ended 31st December 1983

CONTENTS


| | | |
|---|---|---|
| Part 1. | Personnel - Actual Strengths. | Page 1. |
| Part 2. | Changes in Wholetime Strength. | Page 2. |
| Part 3. | Personnel qualified for promotion. | Page 2. |
| Part 4. | Qualified Firemen. | Page 2. |
| Part 5. | Sickness and Injury. | Page 2. |
| Part 6. | Discipline. | Page 3. |
| Part 7. | Deaths and Rescues. | Page 3. |
| Part 8. | Road Accidents. | Page 3. |
| Part 9. | Over the Border assistance. | Page 3. |
| Part 10. | Fatalities - Table. | Page 4. |
| Part 11. | Vehicles and Appliances. | Page 5. |
| Part 12. | Fire Prevention - Establishments. | Page 6. |
| Part 12. | F.P. - Certifiable Premises. | Page 7. |
| Part 12. | F.P. - Non-Certifiable Premises. | Page 8. |
| Part 12. | F.P. - FP Act Appeals. | Page 9. |
| Part 12. | F.P. - Work done during the year. | Page 10. |
| Part 13. | Summrary of Fires and Other Incidents. | Page 11. |


Form to be completed and TWO COPIES returned to:

HM Chief Inspector of Fire Services (Scotland)
Scottish Home and Health Department
Room 258
St Andrew's House
Edinburgh
EH1 3DE

not later than Friday 20 January 1984.

# APPENDIX 17

**Electorate and Election Results**

OPCS form RPF 29 is referred to paragraph 11.2.2. In reproducing this form we are required to acknowledge as follows: Source OPCS. Crown copyright. Reproduced with the permission of the controller of Her Majesty's Stationery Office.

Registrar General for Scotland's form Statistical Return of Election Results is referred to in paragraph 11.3.2.

When completed please return before 1 March 1984 to:-

**R P F 29**

POPULATION ESTIMATES UNIT, Office of Population Censuses and Surveys, Segensworth Rd., Titchfield, Fareham, Hants. PO15 5RS
Tel. Titchfield (0329) 42511 Ext.265

# REPRESENTATION OF THE PEOPLE ACTS
## Summary of Register of Electors 1984
(Qualifying date 10 October 1983)

DISTRICT/LONDON BOROUGH OF
I HEREBY CERTIFY that the following is an accurate summary of my electoral register
Date (Signature) Electoral Registration Officer
Office Address
Telephone No.

SPECIMEN

## PART 1 ELECTORATE TOTALS

| Constituencies within District or London Borough (if part only, add 'Part of') | Local Govt. Electors (Total of cols.3 & 4 | Peers* | Parliamentary Electors (including those shown in cols. 8, 9, 10 and 11) | | | | Persons attaining the age of 18 after 16 February 1984 but before 16 February 1985 | | | |
|---|---|---|---|---|---|---|---|---|---|---|
| | | | Total of cols.5,6 and 7 | Ordinary Residence Qualifications | Service Voters Qualifications | Merchant Seamen | Total of Attainers (cols.9,10 and 11) | Ordinary Residence Qualifications | Service Voters Qualifications | Merchant Seamen |
| 1 | 2 | 3 | 4 | 5 | 6 | 7 | 8 | 9 | 10 | 11 |
| Borough/District Total | | | | | | | | | | |
| | 13 | 19 | 25 | 31 | 37 | 43 | 49 | 55 | 61 | 67 |

*Excluding Irish Peers

PS(T)89A 7 83

## PART 2 — NUMBER OF ELECTORS BY WARDS OF DISTRICT OR LONDON BOROUGH

The name of the constituency to which each ward or part of a ward relates should be entered in column 1.

The number of electors given in each line relates to the district or borough ward or part of the district or borough ward and NOT to parishes, wards of parishes or polling districts.

Where only part of a district or borough ward is located in a constituency the figures should relate only to that part and not to the entire ward. If any such part contains no electors the name of the ward should be entered followed by the word NIL in column 4.

NAME OF DISTRICT/LONDON BOROUGH ....................................

| Name of Constituency | Ward name or number (if part only, add 'Part of') | Polling District Code(s) | Total Parliamentary Electorate (including the number in col. 5) | Persons attaining the age of 18 after 16 Feb 1984 but before 16 Feb 1985 | Peers | Name of Constituency | Ward name or number (if part only, add 'Part of') | Polling District Code(s) | Total Parliamentary Electorate (including the number in col. 5) | Persons attaining the age of 18 after 16 Feb 1984 but before 16 Feb 1985 | Peers |
|---|---|---|---|---|---|---|---|---|---|---|---|
| 1 | 2 | 3 | 4 | 5 | 6 | 1 | 2 | 3 | 4 | 5 | 6 |

SPECIMEN

PS(T)898 7

**LOCAL GOVERNMENT ELECTIONS, 1984**
**STATISTICAL RETURN OF ELECTION RESULTS**

........................................................................ **District Council**

**UNCONTESTED ELECTIONS**

1 Number of councillors returned unopposed ..........

**CONTESTED ELECTIONS**

2 Number of councillors returned as a result of contested elections ..........

3 Aggregate number of electors for district wards in which elections were contested ..........

4 Aggregate number of ordinary ballot papers issued and not otherwise accounted for, as shown on Presiding Officer's ballot papers accounts, and of postal ballot papers received before the close of poll (ie Number of votes cast in contested elections) ..........

Please show below the district wards names or numbers of those district wards which were contested and uncontested.

| CONTESTED DISTRICT WARD(S) | UNCONTESTED DISTRICT WARD(S) |
|---|---|
| No(s) | No(s) |
| | |

Signed ..........................................
Returning Officer

Address ..........................................

..........................................

Date ..........................................

To be returned to:-

The Registrar General
Population Statistics Branch
Room 42
New Register House
EDINBURGH
EH1 3YT

# APPENDIX 18

**Environmental Health Statistics**

This Appendix consists of a specimen copy of the CIPFA/IEHO/SIS questionnaire (plus covering letter and completion instructions) which is the source document of the statistics published in [QRL.19] –see Section 14.1.

The Chartered Institute of 1885–1985 Public Finance & Accountancy

3 Robert Street London WC2N 6BH Telephone 01-930 3456

March 1985

Dear Colleague,

**ENVIRONMENTAL HEALTH STATISTICS — 1983/84 ACTUALS**

We enclose three copies of a questionnaire jointly prepared by CIPFA and the IEHO aiming to collect information about the Environmental Health Service. This survey, a form of which was introduced last Summer, has been developed with full consultation within the Statistical Information Services Environmental Health Working Party.

As is the case with all Statistical Information Service publications, every attempt will be made to present the data collected in a representative and objective manner. The aim of the Survey is to collate information which will be made available to enable individual Councils to examine their own operations with those of others in the national context.

Please will you pass a copy of the letter and questionnaire to the Chief Environmental Health Officer, and arrange for a jointly completed copy to be returned by Friday, 19th April to:

K. P. Hunt Esq., MIEH, AICSA
Chief Environmental Health Officer
Arun District Council
3 Maltravers Road
Littlehampton
West Sussex
BN17 5HH

We are grateful to you for your assistance.

Yours truly

K. J. Tyler
Secretary
IEHO

N. P. Hepworth
Director
CIPFA

To: All Chief Financial Officers and Chief Environmental Health Officers of District Councils, and London Boroughs in England and Wales

## CIPFA/IEHO ENVIRONMENTAL HEALTH CORE FUNCTIONS STATISTICS

**Symbols to use when completing questionnaire**

*Figure is too small to register (i.e. *under* 0.5%)
**Information is not available
##The question is not applicable

## NOTES OF GUIDANCE ON THE COMPLETION OF THIS QUESTIONNAIRE

This survey is intended to collect information about environmental health activities. The Core Functions referred to include:-

— Housing — *monitoring overall conditions including basic inspections in connection with* clearance, general improvement areas, multiple occupation, caravans and basement uses. (NB. The functions include inspections in connection with asessments for improvement grants and house insulation grants, but not the administration of payments of such grants).

— Pollution — measurement and control of industrial and domestic atmospheric pollution and noise nuisance including smoke control and planning advice;

— Occupational Health and Safety including enforcement and the investigation of accidents and giving of preventative advice;

— Shops Act Administration;

— Food Control including Food Hygiene and the administration of Food and Drugs legislation including the supervision of slaughterhouse conditions and the carying out of meat inspection and inspection of foodstuffs;

— Public Health including pest control, control of infectious diseases, offensive trades, drainage and sewerage standards, swimming baths and drinking water quality;

— Animal health and welfare including zoos, dangerous wild animals, boarding establishments; pet shops; riding establishments; work of dog wardens etc.

— Port Health including passenger and crew clearance and the fitness of animals and food in transit.

— Health education including home safety.

It is appreciated that many local authorities organise their staff in departments which carry out wider functions that those listed in the above "Core". However, the following activities are *NOT* to be included:-

(a) other functions such as the provision of public conveniences;
(b) trading standards activities;
(c) waste collection and waste disposal;
(d) rent investigation and administration of Rent Acts;
(e) licensing of vehicles for hire, betting, gaming and lotteries etc.;
(f) building control.

The information provided through this questionnaire and drawn from other sources, will be used to compile statistics showing the following information for each council:-

(a) basic characteristics of the area;
(b) environmental health manpower statistics relating to core functions only;
(c) the allocation of time to core functions;
(d) specified types of inspections;
(e) the level of statutory action;
(f) financial details — these to include the level and amount of central establishment charges and other external costs together with a detailed account of all income including any inter-departmental re-charging.

**Specific Notes in relation to each individual question**

**1. Section 1 — Number of Hereditaments at 1st April 1984**

The number of hereditaments used for residential purposes is shown as an extract of part of Appendix 'A' to HIP 1. The Inland Revenue form CVR 140 will assist in providing information on the numbers of non-residential hereditaments.

**2&3 Section 2 — Demand Factors/Hazards**
**Section 3 — Slaughterhouses & Offensive Trades**

The intention is to collect information which describes the various environmental hazards which exist in the area of your authority. Offensive trades are defined as the following:-

| | |
|---|---|
| Blood Boiler | Gut Scraper |
| Blood Drier | Rag and Bone Dealer |
| Bone Boiler | Size Maker |
| Fat Extractor | Soap Boiler |
| Fat Melter | Tallow Melter |
| Fell Monger | Tripe Boiler |
| Glue Maker | |

(Question 5.1 seeks to find out the implied policy of your local authority in coping with these hazards).

**4. Section 4 — Manpower at 31st March 1984 (FTE Engaged on Core Functions)**

**4a** "Qualified" refers to the Diploma of the Institution of Environmental Health Officer's Education Committee or equivalent.

**4b** Include Technical Officers

**4c** Include Poultry Meat Inspectors.

**4d** Include *only* those vets in the full-time employment of your local authority.

**Section 5 — Activity Analysis During 1983/84**

**5a** These categories make up the 'core functions' described at the front of these notes. Please *estimate* the times, to the nearest whole percent of manpower time spent on these activities or matters. Please include administrative/clerical staff time in this estimate, apportioned over the activities as appropriate. Particular care should be taken to ensure that your returns aggregate to 100%.

**5b** These questions seek to obtain information about the extent of informal and formal enforcement action in different core function areas. Enter the number of notices served, and prosecutions taken for contraventions of the relevant statutory provisions relating to those core functions.

**House Renovation Grants**

**5c** Please indicate extent of Environmental Health staff involvement in HRG activities, e.g. full responsibility, "X" percentage responsibility or no responsibility.

**5d** "Completions in grants paid" is the number of paid grants and not the amount paid.

**5e** Exclude the routine baiting of sewers, and confine the number to cases dealt with. Include work which is contracted out.

**Section 6 — Financial Summary (1983/84 Outturn)**

**6.1** *Employees Expenses* includes: Salaries and Wages, National Insurance, Superannuation Charges, Allowances, Training and re-location expenses and other employee expenses such as pensions and compensation for the loss of office etc.

**6.2** (a) i) *Hired-out and Contracted Services* includes: Veterinary and medical services and hired inspectors' fees, various analytical services including consultants' charges.

(a) ii) *Other Supplies and Services* includes: Equipment, tools and materials, medical and scientific supplies, provisions, clothing, laundry and also the recharges made by other Departments for the supply of such items.

(b) *Premises and fixed plant* includes: Expenditure on *area premises* relating to repairs, alterations and maintenance, fuel, lighting, rent and rates and a contribution to renewal and repairs funds in respect of these premises. Include where appropriate an apportionment of the expenses of shared *operational premises and plant. Do not include costs relating to administrative buildings* — see note 6.2 (d)

(c) *Transport and moveable plant* includes: the use of transport, car allowances, travelling expenses and contributions towards repairs and renewals funds.

(d) *Establishment Expenses* includes: An apportionment of expenses of administrative buildings and Central Establishment charges. Debt Management Expenses, insurances, telephone and general office expenses, including stationery, advertising and postages.

(e) *Agency Services* includes: the net cost incurred in respect of agency payments made by your authority to other authorities, associations, government departments in respect of services rendered by them to you. Do not include Contracting Costs which should be shown at line 6.2 (a) i). Also please show income received from other authorities on whose behalf you are acting at line 6.9 (Other Sources of Income).

**6.5- Notes on Income Items:-**
**6.9** Please show against lines 6.5 to 6.9, as appropriate, all income by source. Do not include grant income in respect of housing renovation or roof insulation. Include agency income at line 6.9. (other sources).

**THE CHARTERED INSTITUTE OF PUBLIC FINANCE AND ACCOUNTANCY/ INSTITUTION OF ENVIRONMENTAL HEALTH OFFICERS**

**ENVIRONMENTAL HEALTH CORE FUNCTIONS:**
**(SIS Survey for the 1983/84 financial year)**

*PLEASE REFER TO THE ATTACHED NOTES OF GUIDANCE.* Some information (e.g. population, area of authority) will be collected by the compiler from other sources.

FOR CIPFA USE ONLY

| | | |
|---|---|---|
| | | |

LOCAL AUTHORITY ..................................................................................................................

Persons to be contacted regarding this questionnaire:

A. Finance Department

Name ............................................................ Tel. No. ............................ Ext. ............................

B. Environmental Health Department

Name ............................................................ Tel. No. ............................ Ext. ............................

**SECTION 1 — NUMBER OF HEREDITAMENTS AT 1ST APRIL, 1984**

1.1 Domestic categories [(extract part of Appendix A to HIP 1, 1984) see note 1]

| | Local Authority | Housing Association | Other public sector | Private sector | Total |
|---|---|---|---|---|---|
| (a) Total dwellings (including non-permanent dwellings) | | | | | |
| (b) Unfit dwellings | | | | | |
| (c) Dwellings fit but lacking basic amenities | | | | | |
| (d) Non-substandard dwellings in need of renovation | | | | | |
| (e) Houses in Multiple Occupation | | | | | |

N.B. The above (a:e) groupings are not mutally exclusive!

continues

**1.2—1.15 Number of Non- Domestic Hereditaments**

1.2 Retail shops—food ........ [ ] 1.2
— FTE number of persons working therein [ ]

1.3 Retail shops—non-food ........ [ ] 1.3
— FTE number of persons working therein [ ]

1.4 Offices ........ [ ] 1.4
— FTE number of persons working therein [ ]

1.5 Hotels, Guest Houses and Boarding Houses ........ [ ] 1.5
— FTE number of persons working therein [ ]

1.6 Public Houses and other licensed premises not included above ........ [ ] 1.6

1.7 Premises licensed for public entertainment not included in 1.5, 1.6 & 1.8 ........ [ ] 1.7

1.8 Holiday camps ........ [ ] 1.8

1.9 Licensed residential caravan sites ........ [ ] 1.9
— Number of caravans [ ]

1.10 Licensed non-residential caravan sites ........ [ ] 1.10
— Number of caravans [ ]

1.11 Restaurants, cafes, and "take-away" premises not included above ........ [ ] 1.11

1.12 Warehouses and stores ........ [ ] 1.12

1.13 Premises licensed under Local Government (Miscellaneous Provisions Act 1982) ........ [ ] 1.13

1.14 Other work places involving E.H. inspections ........ [ ] 1.14

1.15 TOTAL NON-DOMESTIC HEREDITAMENTS (Sum of lines 1.2—1.14) ........ [ ] 1.15

**SECTION 2 — DEMAND FACTORS/HAZARDS (see notes 2 & 3)**

Which of the following affect your area (tick boxes as appropriate):

2.1 Airport/Harbour/Marina [ ]

2.2 Heavy industrial or chemical/oil processing [ ]

2.3 Depositing of hazardous wastes [ ]

2.4 Mining or mineral extraction [ ]

2.5 Tourism [ ]
If substantial, give brief account of effect ____________

2.6 Other Demand factors/Hazards (not included above or in section 3) [ ]
please specify ____________

**SECTION 3 — SLAUGHTERHOUSES & OFFENSIVE TRADES**

How many of the following establishments are there within your administrative area?

3.1 Number of Slaughterhouses — (a) Red Meat [ ]
(b) Poultry Meat [ ]
(c) Knackers Yards [ ]
(d) EEC Licence Holders [ ]
(e) TOTAL (Sum of lines 3.1(a) to 3.1(d)) [ ]

3.2 Number of Establishments engaging in Offensive Trades (See Note 2) [ ]
Types of Offensive Trades: ____________

____________

____________

continues

**SECTION 4 — MANPOWER AT 31st MARCH 1984 (FTE ENGAGED ON CORE FUNCTIONS)**

| | | FTE |
|---|---|---|
| 4.1 | Qualified Environmental Health Officers (see note 4a) | |
| 4.2 | Scientific Officers (see note 4b) | |
| 4.3 | Technical Assistants | |
| 4.4 | Authorised Meat Inspectors (see note 4c) | |
| 4.5 | Veterinarians (see note 4d) | |
| 4.6 | Student Environmental Health Officers | |
| 4.7 | Pest Control Operatives | |
| 4.8 | Manual Employees other than Pest Control Operatives | |
| 4.9 | Administrative and Clerical Support Staff | |
| 4.10 | TOTAL FTE STAFF ENGAGED ON CORE FUNCTIONS (Sum of lines 4.1 to 4.9) | |

**SECTION 5 — ACTIVITY ANALYSIS DURING 1983/1984 (see note 5a)**

5.1 Estimated percentage breakdown *to nearest whole percent* of staff time spent dealing with the following functions (include appropriate apportionments for administrative, clerical and typing services):

| | | % |
|---|---|---|
| (a) | Housing standards | |
| (b) | Air Pollution Control | |
| (c) | Noise Control | |
| (d) | Occupational Health, Safety and Welfare and Shops Act | |
| (e) | Meat Inspection | |
| (f) | Food Hygiene, inspection of foodstuffs, and sampling | |
| (g) | Port Health | |
| (h) | Infectious Diseases Control | |
| (i) | Health Education (including home safety) | |
| (j) | Control of other Public Health risks (including drainage, pest control, nuisances, offensive accumulations) | |
| (k) | Animal health and welfare, boarding establishments, animal breeding, etc. | |
| (l) | Public Entertainment Licensing | |
| (m) | Other | |
| | TOTAL | 100% |

5.2 NUMBER OF MEAT INSPECTIONS
Please give number of carcases inspected:

| | | |
|---|---|---|
| (a) | Poultry | |
| (b) | Pigs | |
| (c) | Sheep and lambs | |
| (d) | Cattle including cows and calves | |
| (e) | Other (please specify .......................................................................) | |
| (f) | TOTAL CARCASES INSPECTED (sum of lines 5.2(a) to 5.2(e) | |

continues

**5.3 ENFORCEMENT ACTION (see note 5b)**

(a) Housing Standards (Housing & Public Health Acts)
— informal notices served
— statutory notices served:
i) Housing Acts
ii) Public Health Acts (Housing)
— prosecutions taken
— houses dealt with:-
i) Work in default
ii) Part III, Housing Act 1957
iii) Part IV, Housing Act 1957

(b) Air Pollution Control
— complaints investigated
— informal notices served
— statutory notices served
— prosecutions taken

(c) Noise Control
— complaints investigated
— informal notices served
— statutory notices served
— prosecutions taken

(d) Occupational Health & Safety inc. Shops Act
— informal notices served
— statutory notices served
— prosecutions taken

(e) Food, food hygiene, and meat and food inspection
— informal notices served
— statutory notices served
— prosecutions taken

(f) Public health including Pest Control, offensive trades (drainage)
— informal notices served
— statutory notices served
— prosecutions taken

continues

(g) Animal Health and Welfare —informal notices served

—statutory notices served

—prosecutions taken

(h) House Renovation Grants (see notes 5c and 5d)
Environmental Staff Involvement ______ %

number of enquiries dealt with over all categories

| | | |
|---|---|---|
| Intermediate | —number of grants approved | |
| | —completions in grants paid | |
| Repair | —number of grants approved | |
| | —completions in grants paid | |
| Improvement | —number of grants approved | |
| | —completions in grants paid | |
| Disabled Persons | —number of grants approved | |
| | —completions in grants paid | |
| Loft Insulation | —number of grants approved | |
| | —completions in grants paid | |

**5.4 PEST CONTROL (see note 5e)**
Number of cases dealt with relating to:-

| | | |
|---|---|---|
| Rodents | —commercial | |
| | —domestic | |
| | —agricultural | |
| Insects (incl. wasps) | —commercial | |
| | —domestic | |
| | —agricultural | |
| Other Pests (incl. birds) | —commercial | |
| | —domestic | |
| | —agricultural | |

continues

**SECTION 6 — FINANCIAL SUMMARY (1983/84 OUTTURN)**

Please refer to the notes of guidance before completion of this section (see note 6)

**EXPENDITURE**

| | | |
|---|---|---|
| 6.1 | Employees' Expenses | £ |
| 6.2 | Running Expenses: | |
| | (a) Supplies & Services:- i) Hired out and Contracted Services | £ |
| | ii) Other (including recharge from other Depts) | £ |
| | (b) Premises and fixed plant | £ |
| | (c) Transport and moveable plant | £ |
| | (d) Establishment expenses | £ |
| | (e) Agency services | £ |
| | (f) Miscellaneous running expenses | £ |
| 6.3 | Debt Charges, RCCO, and Leasing Expenses | £ |
| 6.4 | TOTAL EXPENDITURE (Sum of lines 6.1 & 6.2 & 6.3) | £ |

**INCOME**

| | | |
|---|---|---|
| 6.5 | Pest Control | £ |
| 6.6 | Meat and Poultry Inspection | £ |
| 6.7 | Port Health | £ |
| 6.8 | Licensing Income | £ |
| 6.9 | Other Income | £ |
| 6.10 | TOTAL INCOME (Sum of lines 6.5 to 6.9) | £ |
| 6.11 | Net Expenditure (line 6.4 minus line 6.10) | £ |

Please return this completed questionnaire by 19th April 1985 to:

K. P. Hunt Esq., MIEH, AICSA
Chief Environmental Health Officer
Arun District Council
3 Maltravers Road
Littlehampton
West Sussex BN17 5HH

# APPENDIX 19

**Direct Labour Organisation Statistics**

This Appendix contains a specimen copy of the CIPFA/SIS questionnaires for English and Welsh authorities which are completed annually by each local authority concerned (see paragraph 15.2.3). Also included are: notes for the guidance of those completeing the questionnaire and an explanatory letter from the Director of CIPFA. An approximately similar questionnaire is completed by the authorities in Scotland.

The Chartered Institute of Public Finance and Accountancy
3 Robert Street London WC2N 6BH Telephone 01-930 3456

**Ref: PR** **September, 1984**

Dear Colleague,

**Direct Labour Organisations (DLO's) Statistics, 1983-1984 Actuals**

This letter refers to the attached questionnaire in respect of this survey covering the Actual accounts for the 1983-84 financial year. Two copies have been sent to you, and *I shall be grateful if you will complete and return one copy to the Institute by Friday 9th November, 1984.*

If your authority is exempt from the requirements of Part III of the Local Government, Planning and Land Act, 1980, or else you do not operate a DLO please complete the section at the bottom of the notes, on the reverse side of this letter (page 2), and return the otherwise blank questionnaire.

Please will you note that, throughout the questionnaire, figures are sought in £000's except in Section E where, for the purposes of calculating accurate rates of return, full expenditure and income is sought.

**If you have any difficulty in completing this form please contact the Institute where Paul Cook will be pleased to assist you.**

*Yours truly*

N. P. Hepworth

**N. P. Hepworth**
*Director*

To: Treasurers of all English and Welsh Local Authorities

2

## NOTES OF GUIDANCE ON COMPLETION OF BOXES

(a) All financial figures are to be expressed in £'thousands, except in Section E.

(b) If a "negative" figure is returned enter a "–" sign immediately before the figure. For example – 897 will appear as follows:

| | | | | | |
|---|---|---|---|---|---|
| | | – | 8 | 9 | 7 |

(c) Only figures inside boxes will be accepted; and no extra boxes or lines should be added to the questionnaire.

(d) Entries must be right justified, i.e., any blank squares must be on the left of the box. For example 897 will appear as follows:

| | | | | | |
|---|---|---|---|---|---|
| | | | 8 | 9 | 7 |

(e) If a nil figure is returned enter "0" in the extreme right hand box:

| | | | | | |
|---|---|---|---|---|---|
| | | | | | 0 |

(f) If figures are not available for a particular question, the last two boxes should be entered as follows:

| | | | | | |
|---|---|---|---|---|---|
| | | | | • | • |

but indicate clearly in which box this figure is included.

(g) Please complete boxes using blue or black ink.

(h) The figures to be returned on this questionnaire should relate only to the accounts prepared by the authority for work undertaken by a Direct Labour Organisation, and which is required by Part III of the Local Government, Planning and Land Act 1980 and any subsequent Regulations and Directions. The questionnaire is based on the analysis shown in CIPFA's *"Accounting Code of Practice for Direct Labour Organisations* July 1984".

(i) Any interest on the DLO's cash flow, or on the DLO appropriation account balance, should be included under "Miscellaneous income" (line 15). Interest on the DLO reserve fund should *not* be included in Sections B or C.

---

Please show the name of the person to be contacted should the compiler require further information:

Name. . . . . . . . . . . . . . . . . . . . . . . . . . . . . . . Tel. No. . . . . . . . . . . . . . . . . . . . . Extn. . . . . .

---

**TO BE COMPLETED BY "EXEMPT AUTHORITIES" OR ELSE THOSE NOT OPERATING A DLO:–**

I. If you are making a "nil" return does your authority:

(a) NOT carry out any Direct Labour work ☐

(b) qualify for exemption under the "de minimis" rule ☐

(Please tick √ the appropriate box)

II. What is the name of your authority? . . . . . . . . . . . . . . . . . . . . . . . . . . . . . . . . . . . . . . . . .

. . . . . . . . . . . . . . . . . . . . . . . . . . . . . . . . . . . . . . . . . . . . . . . . . . . . . . . . . . . . . . . . . . . .

3

## DIRECT LABOUR ORGANISATIONS STATISTICS 1983-84 ACTUALS

Local Authority ..........................................................................

1 Local authority reference code (please leave blank)

### SECTION A — GENERAL STATISTICS

2 Population (thousands) OPCS estimate at 30 June 1983

3 Area (hectares)

### SECTION B — SUMMARY HISTORICAL COST REVENUE ACCOUNT FOR THE YEAR ENDING 31 MARCH 1984

*Expenditure* £'000

4 Direct Labour

5 Supplies and Services

6 Transport and Plant

7 Sub-contractors

8 Overhead Accounts

9 Change in work-in-progress during year (+ or –)

10 Change in provision for future losses (+ or –)*

11 Total expenditure (4 + 5 + 6 + 7 + 8 + 9 + 10)

*Income* £'000

12 Charges to other accounts of the Council

13 Charges under agency agreements

14 Charges under works contracts

15 Miscellaneous income (see note i)

16 Total income (12 + 13 + 14 + 15)

17 Surplus (+)/Deficit (–) *before* exceptional items (16–11)

18 Exceptional items

Please list any items included at line 18 above:

.........................................................................

.........................................................................

.........................................................................

£'000

19 Surplus (+)/Deficit (–) *after* exceptional items (17–18)

20 *Memorandum item (see line 10):
Provision for future losses at end of year

### SECTION C — DLO APPROPRIATION ACCOUNT

£'000

21 Balance brought forward at 1 April

22 Contribution to (+)/from (–) General Rate Fund

23 Contribution to (+)/from (–) DLO Reserve Fund

24 Balance carried forward at 31 March (19 + 21 – 22 – 23)

4

**SECTION D — ANALYSIS OF SUMMARY REVENUE ACCOUNT BY DESCRIPTION OF WORK**

The four descriptions of work at lines 25 to 28 opposite (DI, DII, DIII, DIV) are those specified in Part III of the Local Government, Planning and Land Act 1980 (as amended by the 1981 Accounts Regulations) for which separate DLO revenue accounts are required as follows:

DI (Line 25) –General highway works (other than on highways connected with other works of new construction), and sewerage works.

DII (Line 26) –Works of new construction (other than highways and sewerage) estimated to cost *more* than £50,000.

DIII (Line 27) –Works of new construction (other than highways and sewerage) estimated to cost *not more* than £50,000.

DIV (Line 28) –Maintenance work (other than highways and sewerage maintenance work) within the meaning of the Local Authority (Goods and Services) Act 1970. The 1970 Act definition includes minor renewals, minor improvements, and minor extensions.

*Where the de minimis option of section 11 of the 1980 Planning and Land Act has meant that two or more categories of work have been amalgamated, the figures should be returned in accordance with the accounts which have been prepared by the authority.*

**SECTION E — RATE OF RETURN ON CAPITAL EMPLOYED (Complete in £'s *not* £'000s)**

The rate of return on capital employed is to be calculated on a current cost accounting basis in accordance with the (Rate of Return on Capital) Directions 1981 for Direct Labour Organisations. See also the Appendices of CIPFA's "Accounting Code of Practice for Direct Labour Organisations (July 1984)".

| | | | DI £ + |
|---|---|---|---|
| | 30 | Surplus/deficit on historical cost basis *after* exceptional items— to agree with lines 25, 26, 27, 28, 29 of Section D, Col. 3. respectively | |
| + | 31 | Interest charged (include *all* interest paid) less interest received. | |
| + | 32 | Capital Replacement Charges in historical accounts (incl. loan principal repayments and Renewals Fund contributions) | |
| – | 33 | Current Cost Depreciation (gross) | |
| – | 34 | Cost of Sales Adjustment | |
| | 35 | Current Cost Operating Surplus/Deficit (30+31+32–33–34) | |
| | 36 | Average Value of Capital Employed | |
| | 37 | % rate of Return on Capital Employed at Line 36 (Line 35/Line 36) | † % • |
| | 38 | % of Line 35 to Total Income at Lines 25, 26, 27, 28, 29 of Section D Col. 2. respectively | † % • |

† Show negative percentages as zero.

5

| Description of work (see note opposite) | *Col. 1* Total Expenditure £'000 | *Col. 2* Total Income £'000 | *Col. 3* Surplus/deficit *after* exceptional items £'000 |
|---|---|---|---|
| 25 DI | | | |
| 26 DII | | | |
| 27 DIII | | | |
| 28 DIV | | | |
| 29 TOTALS* | | | |
| | *To agree with line 11 | *To agree with line 16 | *To agree with line 19 |

DESCRIPTION OF WORK
(see SECTION D above)

| DII £ | + | DIII £ | + | DIV £ | = | ALL TYPES |
|---|---|---|---|---|---|---|
| | | | | | | |
| | | | | | | |
| | | | | | | |
| | | | | | | |
| | | | | | | |
| | | | | | | |
| | | | | | | |
| | | | | | | |
| † % • | | † % • | | † % • | | † % • * |
| † % • | | † % • | | † % • | | † % • |

* If the rate of return quoted here is different from the figure stated in your authority's DLO statutory annual report please state reason:

..........................................................................................

..........................................................................................

..........................................................................................

6

**SECTION F — MISCELLANEOUS STATISTICS**

I **Manpower**

| | Description of work (see SECTION D above) | *Col. 1* Direct employment (see Note below) 1983-1984 Annual Average Full Time Equivalent | *Col. 2* Indirect employment (see Note below) 1983-1984 Annual Average Full Time Equivalent |
|---|---|---|---|
| 39 | DI | | |
| 40 | DII | | |
| 41 | DIII | | |
| 42 | DIV | | |
| 43 | TOTALS (39 + 40 + 41 + 42) | | |

*Notes:*

Col. 1, Direct Employment, covers those employees who would be covered by the definition contained in Section 11 (i) of the 1980 Planning and Land Act (i.e. those employees who rank for the de minimis test)

Col. 2, Indirect Employment, covers all other employees whose financial costs are relevant to the DLO accounts shown in this questionnaire, which would include employees whose costs are recharged from central administrative departments.

II **Methods of Charging for Direct Labour Work**

See Section 4 of CIPFA's Accounting Code of Practice for Direct Labour Organisations. July 1984

| | | £'000 |
|---|---|---|
| 44 | Lump sum contracts | |
| 45 | Schedule contracts | |
| 46 | "Target hours" | |
| 47 | Day works contracts | |
| 48 | "Bulk" offers | |
| 49 | Other Methods — please specify | |
| 50 | TOTAL | |

III **Contractors Proportions 1983-1984**

Total of the tenders or estimated charges for all functional work and works contracts undertaken by the authority's DLO (see note below) for each description of work:

| | | £'000 |
|---|---|---|
| 51 | DI | |
| 52 | DII | |
| 53 | DIII | |
| 54 | DIV | |
| 55 | TOTAL (51 + 52 + 53 + 54) | |

Total value of all construction or maintenance work carried out by placing contracts with contractors other than the authority's DLO (see note below) for each description of work:

| | | £'000 |
|---|---|---|
| 56 | DI | |
| 57 | DII | |
| 58 | DIII | |
| 59 | DIV | |
| 60 | TOTAL (56 + 57 + 58 + 59) | |

*Notes:*

(i) Description of work (see SECTION D above)

(ii) Lines 51 to 55, as defined in para 2(a) (i) and (ii) of The Local Government (Direct Labour Organisations) Annual Reports Directions 1982 and para 8.04(a) of CIPFA's Accounting Code of Practice for Direct Labour Organisations (July 1984).

(iii) Lines 56 to 60, as defined in para 2(b) of The Local Government (Direct Labour Organisations) Annual Reports Directions 1982 and para 8.04(b) of CIPFA's Accounting Code of Practice for Direct Labour Organisations (July 1984).

Assets and liabilities [illegible]
Association of District Council Treasurers [illegible]
Average interest rates [illegible]
Average rate poundage, 5.3.3; [illegible]
Bank of England, 4.2.7
Banks, [illegible]
Bathhouses and laundries, [illegible]
Block grant [illegible]
Block grant indicators [illegible]
Capital and Reserve Funds [illegible]
Capital transfers, [illegible]
Car parking, [illegible]
Car loans, Staff, 8.2.3
Car ownership, 13.1.7; [illegible]
Casualties and fatalities, 12.2.10
Cemeteries and crematoria, 2.5.1; 14.1.1; 14.1.2; 14.5; 14.6.3; 14.6.5; 14.6.16; 14.6.18; 14.7.1
Cemeteries and crematoria finance, [illegible]
Cemeteries and crematoria manpower, [illegible]
Cemeteries and crematoria organisations, [illegible]

# SUBJECT INDEX

Abattoirs, 2.5.1; 3.3.4; 14.1.2; 14.7.1; 14.7.2
Accounting conventions, 3.1.3; 3.3.1; 3.4.1
Accounting practices, 3.1.5
Accounts, 3.1.1
Accounts, National, 3.1.1; 3.4.3; 4.4.1; 4.5.6
Actual Outturn income, 5.2.3; 5.2.4; 5.3.8; 5.6.3; 6.2.5; 6.3.2; 6.4.2; 6.4.7; 6.4.12; 6.5.3; 6.5.4; 8.1.2; 8.4.1; 8.5.1; 8.7.3
Actual outturn data, 3.1.2; 3.2.3; 4.2.4; 4.2.5; 4.3; 4.4; 4.4.4; 4.5.2; 4.5.4; 17.1.4
Actual outurn expenditure, 3.2.3; 4.2.2; 4.2.10; 4.3.3; 4.3.4; 4.3.5; 4.3.6; 4.4.2; 4.4.3; 4.4.4; 4.5.2; 4.5.5
Administration of Justice, 12.4; 12.4.7
Administration of Justice finance, 12.4.7
Administration of Justice manpower, 12.4.7
Aerodromes, 3.3.4; 13.1.1
Aggregate Exchequer Grants Settlement, 6.4.5; 6.4.10
Aggregates, 6.1.1; 6.1.2
Agriculture Act 1970, 15.3.1; 15.3.3
Airports, 13.4.1; 13.4.2; 13.4.3; 13.4.4
Animal health, 14.2.6
Area Health Authorities, 8.2.3
Arts sponsorship, 2.1
Asset sales, 4.3.2
Assets and liabilities, 7
Association of District Council Treasurers, 3.2.2
Audit Commission, 3.2.4
Auditing, 3.1.5
Average interest rates, 7.2.7
Average rate bill, 5.3.1; 5.3.8; 5.6.4; 6.4.11; 17.2.2
Average rate poundage, 5.3.3; 5.3.5

Bank of England, 7.2.7
Banks, 4.4.11
Baths and laundries, 14.2.2; 14.6.5
Belfast, 3.4.2; 3.4.4; 5.6.2; 5.6.4
Block Grant, 1.3.2; 4.1.2; 6.4; 9.1.2; 9.2.3
Block Grant distribution, 6.4.3; 6.4.4; 6.4.6; 6.4.7; 6.4.8; 9.2.1; 17.1.2
Block grant indicators, 3.2.2
Board of Inland Revenue, 9.1.1; 9.1.5
Book stock, 15.1.3
Borrowing, 3.1.5; 7.1.1; 7.1.2; 7.2; 17.2.2
Borrowing and Lending Inquiry, Quarterly, 7.2.4; 7.2.6; 7.3.4; 7.4.1
Borrowing instruments, 7.1.1; 7.2.7
Borrowing Requirement, 7.2.3; 7.2.4; 7.2.5
Borrowing, Source of, 7.2.6; 7.3.1; 7.3.2; 7.3.3; 7.5.2
Borrowing Survey, Sample, 7.2.5
Borrowing, Term of, 7.2.6; 7.3.1; 7.3.3; 7.5.2
Borrowing, Total net, 7.2.2; 7.2.3; 7.2.4; 7.2.5; 7.2.6; 7.5.1; 17.2.2
Bridges, 13.1.1; 13.1.2
British Rail, 4.4.11
Budgeting, 3.1.5
Building regulations, 2.5.1
Burial grounds, 14.1.2; 14.6.1; 14.6.5; 14.6.16; 14.6.18; 14.7.1
Bus undertakings, 8.4.1; 13.1.2; 13.2.1
By-elections, 11.1.5

Capital expenditure, 4.3; 4.4.2; 4.5.1; 4.5.2; 4.5.5; 6.5.3
Capital grants, 4.4.12; 6.5.1; 6.5.4
Capital accounts, 3.1.3; 3.4.1; 3.4.3; 4.4.12; 6.2.1; 6.2.2; 8.1.2; 8.2.3; 8.4.2; 8.5.3; 8.7.3
Capital and Reserve Funds, 3.3.5; 3.4.3
Capital formations, Fixed, 4.4.6; 4.4.9; 4.4.12; 4.4.13
Capital projects (NI), 6.5.1
Capital transfers, 4.4.12
Car parking, 4.3.2; 13.1.1; 13.1.2
Car loans, Staff, 8.2.3
Car ownership, 13.1.7; 13.2.4; 13.2.5
Casualties and fatalities, 12.2.10
Cemeteries and crematoria, 2.5.1; 14.1.1; 14.1.2; 14.5; 14.6.1; 14.6.5; 14.6.16; 14.6.18; 14.7.1
Cemeteries and crematoria finance, 14.5.2; 14.5.4
Cemeteries and crematoria manpower, 14.5.2; 14.5.4
Cemeteries and crematoria resources, 14.5.4
Central Government, 4.4.11; 17

Central Intelligence Unit, 17.1.1
Cesspools, 14.2.2
Changes in balances, 8.2.1; 8.3.1; 8.4.1
CIPFA database, 4.2.8; 9.1.5; 15.1.5; 15.5.2; 17.1.5; 17.1.7
CIPFA/SIS, 1.4.1; 3.2.2; 3.2.3; 3.4.2; 3.4.6; 4.2.6; 4.3.3; 4.4.13; 5.2.1; 5.3.2; 5.3.4; 5.3.5; 5.3.6; 5.3.7; 5.3.8; 5.4.1; 5.5.3; 17.1.3; 17.1.4; 17.1.5; 17.1.6; 17.2.3; 17.3.2; 17.3.3
Civic restaurants, 3.3.4
Civil defence, 12.4.1; 12.4.4; 12.4.9; 12.4.11; 12.4.13
Classification, Property, 9.1.2
Classifications, Manpower, 10.1.5; 10.1.9; 10.2.3
Clean air, 14.1.2; 14.2.2; 14.2.4; 14.6.5; 14.7.1; 14.7.2
Cleansing, 14.1.2; 14.6.1; 14.6.2
Clustering, 14.2.8; 17.1.5; 17.1.6
Collective agreements, 10.2.1; 10.2.2; 10.2.3
Commercial property, 9.1.2
Commissioner for Complaints (NI), 16.2.1
Commissioner for Complaints Act 1969, 16.2.1
Commissioners of Valuations (NI), 9.3.2
Commissions for Local Administration, 16.1.1; 16.1.2; 16.2.1
Common Good Fund, 3.3.1; 4.3.8; 7.2.1; 8.5.3
Community councils, 2.3.1
Community schemes, 6.5
Comparisons, Inter-authority, 3.2.4; 10.1.10; 14.3.8; 14.4.4; 17.1.1; 17.1.2; 17.1.3; 17.1.4; 17.1.6; 17.3.3
Complaints, 12.1.12; 16
Complaints (NI), Commissioner for, 16.2.1
Complaints Act 1969, Commissioner for, 16.2.1
Concessionary fares, 13.1.1; 13.1.2
Consents, 4.3.5; 4.3.7
Construction manpower, 10.1.2; 10.2.1; 15.2.1; 15.2.2
Consumer protection, 2.1; 2.2.1; 2.2.2; 2.5.1; 3.2.2; 3.3.2; 12.5
Coroners, 12.4.2; 12.4.3
Costs, Rate collection, 5.4.1; 5.4.3; 5.6.5
Council dwellings, 9.1.2
County Land Agent, 15.3.1
Courts, 2.1; 6.2.1; 12.4.1; 12.4.9; 12.4.10
Courts finance, 12.4.3; 12.4.10; 12.4.13
Courts manpower, 12.4.13
Criticisms, 17.2.1; 17.2.2
CSO, 3.2.1; 3.4.2; 4.4.10; 4.4.13
Current expenditure, 4.4.2
Deaths in police custody, 12.1.6; 12.1.12
Debt, Estimated, 7.3.1; 7.3.2
Debt interest, 4.4.11; 7.2.7
Debt, Loan, 7.5.2
Debt management, 7.3.5; 7.5.2
Debt, Outstanding, 7.3; 7.5.2; 17.2.2
Department of Economic Development, 6.5.1
Department of Education (NI), 6.5.1
Department of Transport, 13.1.6
Derating, 6.5.1; 9.3.3
Direct labour organisations, 3.2.3; 15.2
Direct labour organisations finance, 15.2.2
Direct labour organisations manpower, 15.2; 15.2.4
Disabled persons, 5.5.4; 6.3.1; 10.1.13
Disabled Persons Act 1978, Rating, 5.5.2; 5.5.3
Disabled Persons Employment Act 1944, 10.1.13
District rate, 5.6.1
District Councils (NI), 3.4.2; 3.4.3; 3.4.4; 5; 5.6; 6.5.3; 11.4.1
DLO's, Scotland, 15.2.1; 15.2.4
DOE, 3.2.1; 3.2.3; 4.2.4; 4.2.7; 5.2.1; 5.3.2; 5.3.6; 5.5.3; 17.2.2
DOE (NI), 3.4.2; 5.6.1; 6.5.1; 6.5.3; 7.5.1
Dog control, 2.5.1
Domestic water rate, 5.1.3; 5.3.5; 5.3.8; 8.4.2; 14.6.9; 14.6.13
Domestic rate relief, 5.1.4; 5.2.3; 5.3.5; 5.3.8; 6.4.1; 6.4.2; 6.4.9; 6.4.14; 6.5.2; 6.5.4; 9.2.4
Drainage, 2.3.1; 14.6.9; 14.7.2
Drugs, 12.1.9

Earnings Survey, New, 10.2.1; 10.2.2; 10.2.3
Earnings, Wages and, 4.4.10; 10.2.1; 10.2.2; 10.3.3; 12.1.6; 12.1.12; 12.2.5; 12.2.10
Education, 2.1; 2.2.1; 2.2.2; 2.2.3; 2.3.1; 2.5.1; 3.2.2; 3.4.4; 4.3.2; 4.4.1; 6.1.2; 6.5.7; 8.2.1; 10.1.2
Education and Library Boards, 10.3.1
Educational grants, 4.4.11
Election results, 11.3
Elections Incidence of, 11.1; 11.4.1; 11.4.3
Electoral registration, 11.2.2; 11.4.2; 12.4.1; 12.4.5; 12.4.9
Electors, 11.2.1; 11.2.2; 11.4.2
Electricity, 2.5.1; 3.4.4; 9.1.2
Employment by service, 10.1.5; 10.1.7
Employment Enquiry (NI), Quarterly, 10.3.2
Employment, New, 17.2.3
Employment Survey, Quarterly, 10.1.4; 10.1.11
Environmental manpower, 14.2.4; 14.2.5;

14.6.3; 14.6.6; 14.6.7; 14.6.8
Environmental health, 2.2.1; 2.2.2; 2.2.3; 2.3.1; 2.5.1; 3.2.2; 14.1; 14.2; 14.6.18; 17.2.2
Environmental services, 14.1; 14.2; 17.2.2
Estimated debt, 7.3.1; 7.3.2
Estimated grants, 6.2.3
Estimates, Expenditure, 3.1.2; 3.2.3; 4.2.5; 4.2.6; 4.2.7; 4.2.8; 4.2.10; 4.3.5; 4.3.6; 4.4.7; 4.4.8; 4.5.3; 6.4.5; 6.4.10; 6.4.11
Estimates, Income, 3.1.2; 6.4.2; 6.4.6; 6.4.11; 6.4.14; 6.5.4; 8.2.2; 8.3.1; 8.7.2; 8.7.3
Estimates, Rate income, 5.2.1; 5.6.3
Estimates, Revised, 4.2.8; 5.2.1
European Regional Development Fund, 8.2.3
Exchequer Grants Settlement, Aggregate, 6.4.5; 6.4.10
Exchequer contribution, General, 4.5.5; 6.5.2
Expenditure, 3.1.3; 3.2.1; 3.4.1; 4
Expenditure Estimates, 3.1.2; 3.2.3; 4.2.5; 4.2.6; 4.2.7; 4.2.8; 4.2.10; 4.3.5; 4.3.6; 4.4.7; 4.4.8; 4.5.3; 6.4.5; 6.4.10; 6.4.11
Expenditure, Actual outurn, 3.2.3; 4.2.2; 4.2.10; 4.3.3; 4.3.4; 4.3.5; 4.3.6; 4.4.2; 4.4.3; 4.4.4; 4.5.2; 4.5.5
Expenditure, Capital, 4.3; 4.4.2; 4.5.1; 4.5.2; 4.5.5; 6.5.3
Expenditure, Current, 4.4.2
Expenditure, Grant related, 9.2.1
Expenditure, Gross, 4.2.2
Expenditure, National, 4.4.6; 4.4.7; 4.4.8; 4.4.10; 4.4.13
Expenditure, Net, 4.2.3
Expenditure, Planned, 4.4.5; 4.5.4
Expenditure, Public, 4.1.3; 4.4.1; 4.4.2
Expenditure, Revenue, 4.2; 4.5.5
Expenditure, Types of, 4.1.2

Factor cost data, 4.4.13
Fares, Concessionary, 13.1.1; 13.1.2
Fatalities, Casualties and, 12.2.10
Fees and charges, 8.1.1; 8.1.2; 8.2.1; 8.3.1; 8.4.1; 8.7.1; 15.3.2
Ferries and tunnels, 13.1.1; 13.1.2
Final consumption, 4.4.6; 4.4.9; 4.4.10
Finance, Administration of Justice, 12.4.7
Finance, Cemeteries and crematoria, 14.5.2; 14.5.4
Finance, Courts, 12.4.3; 12.4.10; 12.4.13
Finance, Direct labour organisations, 15.2.2
Finance, Fire service, 12.2.4; 12.2.9; 12.2.11
Finance, Highways and Roads, 13.1.3; 13.1.4; 13.2.2; 13.2.3; 13.2.5
Finance, Library, 15.1.2; 15.1.3; 15.1.4; 15.1.6; 15.1.8; 15.1.9
Finance, Museums and galleries, 15.1.2; 15.1.3
Finance, Police, 12.1.5; 12.1.7; 12.1.11
Finance, Probation service, 12.3.1; 12.3.4
Finance, Refuse collection/disposal, 14.3.2; 14.3.5; 14.6.2
Finance, Smallholdings, 15.3.2
Finance, Waste disposal, 14.4.2; 14.4.3
Financial control, 3.1.5
Financial flows, Net, 7.1.2; 7.4
Financial institutions, 4.4.11
Financial planning, 4.2.6
Financial transactions, Net, 7.4
Fire data, Miscellaneous, 12.2.5; 12.2.10; 12.2.11
Fire service, 2.1; 2.2.1; 2.2.2; 2.2.3; 2.3.1; 3.2.2; 4.4.1; 10.2.1
Fire service finance, 12.2.4; 12.2.9; 12.2.11
Fire service manpower, 12.2.5
Fire service resources, 12.2.4; 12.2.8; 12.2.11
Fire service workload, 12.2.3; 12.2.8; 12.2.11
Firemen, 3.3.5; 8.6.1; 8.6.3
Fixed capital formations, 4.4.6; 4.4.9; 4.4.12; 4.4.13
Flood prevention, 2.1
Food and Drugs Act, 14.2.6
Food hygiene, 14.2.6
Fund, Common Good, 3.3.1; 4.3.8; 7.2.1; 8.5.3

Galleries finance, Museums and, 15.1.2; 15.1.3
Galleries manpower, Museums and, 15.1.2; 15.1.3
Galleries, Museums and, 2.5.1; 15.1; 15.5.1
Gas, 2.5.1; 3.4.4; 6.5.1; 8.7.1; 9.1.2
Gas Accounts, 3.4.3
General Exchequer contribution, 4.5.5; 6.5.2
General grant, 4.5.1; 5.5.1; 6.5.1; 9.3.3
General rate, 5.1.3
General revaluations, 9.3.4
General Revenue Account, 3.4.4; 4.5.5
Grant, Block, 1.3.2; 4.1.2; 6.4; 9.1.2; 9.2.3
Grant distribution, 4.1.2; 6
Grant distribution, Block, 6.4.3; 6.4.4; 6.4.6; 6.4.7; 6.4.8; 9.2.1; 17.1.2
Grant, General, 4.5.1; 5.5.1; 6.5.1; 9.3.3
Grant related expenditure, 9.2.1
Grant related poundage, 5.3.9
Grant, Transport Supplementary, 6.2.2; 6.2.4
Grants, Capital, 4.4.12; 6.5.1; 6.5.4
Grants, Educational, 4.4.11
Grants, Housing, 2.5.1; 6.1.2
Grants, Planned Supplementary, 6.2.4
Grants Settlement, Aggregate Exchequer, 6.4.5; 6.4.10

Grants, Subsidies and, 6; 15.3.2
Greater London Council, 1.3.1; 2.1; 2.2.3; 11.1.3; 12.3.1; 15.1.1; 17.1.1
Gross expenditure, 4.2.2

Harbours, 2.5.1; 13.1.1; 13.1.2; 13.5.1
Health and welfare services, 2.5.1; 3.4.4; 10.1.2
Health Authorities, Area, 8.2.3
Health care functions, 14.2.6; 14.2.7
Health education, 14.2.6
Health inspection, 14.1.2; 14.2.4
Highways and Roads, 2.1; 2.5.1; 13.1; 15.2.2
Highways and Roads finance, 13.1.3; 13.1.4; 13.2.2; 13.2.3; 13.2.5
Highways and Roads manpower, 13.1.3; 13.2.2
Highways and Roads workload, 13.1.5; 13.2.2; 13.2.4
Highways, Scotland, 13.2
Hours worked, 10.1.5; 10.1.9; 10.2.3; 12.1.6; 12.1.12; 12.2.5; 12.2.10
House of Commons Papers, 3.2.1
Housing, 2.2.1; 2.2.2; 2.2.3; 2.3.1; 3.2.2; 3.3.2; 3.4.4; 4.3.2; 4.3.3; 4.3.5; 4.4.11; 4.4.12; 6.5.2
Housing grants, 2.5.1; 6.1.2
Housing benefits, 1.3.2
Housing Capital Account, 7.2.1
Housing Revenue Account, 3.3.1; 3.3.3; 4.2.10; 6.2.1; 8.3

IDLG, 17.2.2; 17.2.3
ILEA, 2.2.3
Incidence of, Elections, 11.1; 11.4.1; 11.4.3
Income Estimates, 3.1.2; 6.4.2; 6.4.6; 6.4.11; 6.4.14; 6.5.4; 8.2.2; 8.3.1; 8.7.2; 8.7.3
Income, Actual Outturn, 5.2.3; 5.2.4; 5.3.8; 5.6.3; 6.2.5; 6.3.2; 6.4.2; 6.4.7; 6.4.12; 6.5.3; 6.5.4; 8.1.2; 8.4.1; 8.5.1; 8.7.3
Income, Interest, 8.2.1; 8.3.1; 8.4.1
Income, Local, 4.2.4
Income, Local Authority, 3.1.3; 3.2.1; 3.4.1; 5; 5.1; 8; 9.2.1
Income, National, 4.4.6; 4.4.12; 17.2.2
Income raising, 3.1.5; 5.1.2
Inconsistencies, 17.2.2
Indebtedness, Net, 7.5.2
Index-linking, 8.6.2
Industrial estates, 3.3.4
Industrial property, 9.1.2
Inland Revenue, Board of, 9.1.1; 9.1.5
Inter-authority comparisons, 3.2.4; 10.1.10; 14.3.8; 14.4.4; 17.1.1; 17.1.2; 17.1.3; 17.1.4; 17.1.6; 17.3.3
Interest income, 8.2.1; 8.3.1; 8.4.1
Interest rates, 7.2.7
Interest rates, Average, 7.2.7
Investments, Local authority, 7.3.6
Island Councils, 2.3.1; 5.1.2
Isles of Scilly, 2.1.1; 2.4; 5.1.2

Joint Manpower Watch, 10.1.3; 10.1.4; 10.1.5; 10.1.7; 10.1.8; 10.1.9; 10.1.10; 10.1.11; 10.2.1; 12.3.4; 12.4.1; 13.1.3; 13.2.2; 14.2.4; 14.5.2; 14.6.3; 14.6.4; 14.6.12; 14.6.14; 14.6.18

Labour costs, 4.3.1; 5.4.1
LACSAB, 10.1.3
Land Valuation Assessors, 5.3.8; 9.1.1; 9.1.7
Land drainage, 2.1
LARIA, 17.1.1; 17.2.3
Law and order services, 10.1.5; 10.1.6
Leasing, 4.3.1
Leisure services, 2.2.1; 2.2.2; 2.3.1; 3.2.2
Liabilities, Assets and, 7
Libraries, 2.2.2; 2.4.1; 3.3.2; 3.4.2; 4.3.2; 8.2.1; 15.1; 15.5.2
Libraries manpower, 15.1.2; 15.1.3; 15.1.6; 15.1.8; 15.1.9
Libraries resources, 15.1.2; 15.1.4
Libraries workload, 15.1.2; 15.1.4
Libraries, Scotland, 15.1.7; 15.1.9
Library finance, 15.1.2; 15.1.3; 15.1.4; 15.1.6; 15.1.8; 15.1.9
Library Boards, Education and, 10.3.1
Licensing powers, 2.1; 2.5.1
Loan debt, 7.5.2
Loans, New, 7.2.1; 7.2.2; 7.5.1
Local income, 4.2.4
Local Ombudsmen, 16.1.1
Local planning, 2.2.1; 2.2.2; 2.2.3; 2.3.1
Local amenities, 2.2.2
Local Authority Income, 3.1.3; 3.2.1; 3.4.1; 5; 5.1; 8; 9.2.1
Local authority investments, 7.3.6
Local Authority Associations, 4.2.7
Local authority sector, 3.1.1; 8.6.1
Local Government (Misc Prov) Act, 8.5.1
Local Government Act (Northern Ireland) 1972, 2.5.1; 3.4.1; 4.5.2; 6.5.1; 7.5
Local Government Act (Scotland) 1966, 6.4.9
Local Government Act (Scotland) 1973, 2.1.1; 5.5.2
Local Government Act (Scotland) 1975, 16.1.1
Local Government Act 1963, 2.1.1
Local Government Act 1972, 2.1.1; 2.4.1; 5.5.2
Local Government Act 1974, 16.1.1

Local Government Act 1985, 2.1
Local government changes, 1.3; 2.1.2; 2.5.1; 17.1.1
Local Government Finance Working Party, 4.2.9
Local Government Planning and Land Act 1980, 4.3.2; 6.4.1; 15.2.1; 17.1.2; 17.1.3
Local/national expenditure ratios, 4.4.2; 4.4.7; 4.4.13
London, 2.1; 2.2.3; 5.1.2; 6.4.4; 8.4.1; 11.1.3; 12.1.1; 12.4.2; 12.4.4; 12.4.5; 13.1.2
London Probation Service, 12.3.1
London Government Act, 17.1.1
London Transport Executive, 8.4.1

MAFF, 15.3.2; 15.3.3
Magistrates, 12.4.2; 12.4.3
Manchester, 2.2.1
Manpower, 1.4.2; 5.4.1; 10.1; 10.2; 10.3; 17.1.2
Manpower classifications, 10.1.5; 10.1.9; 10.2.3
Manpower, Administration of Justice, 12.4.7
Manpower, Cemeteries and crematoria, 14.5.2; 14.5.4
Manpower, Construction, 10.1.2; 10.2.1; 15.2.1; 15.2.2
Manpower, Courts, 12.4.13
Manpower, Direct labour organisations, 15.2; 15.2.4
Manpower, Environmental, 14.2.4; 14.2.5; 14.6.3; 14.6.6; 14.6.7; 14.6.8
Manpower, Fire service, 12.2.5
Manpower, Highways and Roads, 13.1.3; 13.2.2
Manpower, Libraries, 15.1.2; 15.1.3; 15.1.6; 15.1.8; 15.1.9
Manpower, Museums and galleries, 15.1.2; 15.1.3
Manpower, Police, 12.1.4; 12.1.7
Manpower, Probation service, 12.3.4
Manpower, Refuse collection/disposal, 14.3.3; 14.3.5
Manpower, Waste disposal, 14.4.3
Manpower Watch, Joint, 10.1.3; 10.1.4; 10.1.5; 10.1.7; 10.1.8; 10.1.9; 10.1.10; 10.1.11; 10.2.1; 12.3.4; 12.4.1; 13.1.3; 13.2.2; 14.2.4; 14.5.2; 14.6.3; 14.6.4; 14.6.12; 14.6.14; 14.6.18
Market price data, 4.4.13
Markets, 2.5.1; 3.3.4
Merseyside, 2.2.1
Metropolitan police, 6.4.4; 12.1.1
Metropolitan areas, 2.1; 2.2.1; 12.2.1
Metropolitan County Councils, 1.3.1; 2.1; 2.2.1
Metropolitan districts, 2.1; 11.1.2
Ministry of Development (NI), 3.4.2; 5.6.5; 6.5.2; 7.5.1
Miscellaneous Fire data, 12.2.5; 12.2.10; 12.2.11
Museums and galleries, 2.5.1; 15.1; 15.5.1
Museums and galleries finance, 15.1.2; 15.1.3
Museums and galleries manpower, 15.1.2; 15.1.3
Museums and galleries, Scotland, 15.1.7; 15.1.8; 15.1.9

National Accounts, 3.1.1; 3.4.3; 4.4.1; 4.5.6
National expenditure, 4.4.6; 4.4.7; 4.4.8; 4.4.10; 4.4.13
National Income, 4.4.6; 4.4.12; 17.2.2
National Park Boards, 16.1.1
National Park Supplementary Grant, 6.2.2; 6.2.4
Needs Element, 6.4.9; 6.4.11; 6.4.12; 6.4.13; 6.4.14; 6.4.15; 6.5.2
Net expenditure, 4.2.3
Net financial flows, 7.1.2; 7.4
Net financial transactions, 7.4
Net indebtedness, 7.5.2
New Earnings Survey, 10.2.1; 10.2.2; 10.2.3
New employment, 17.2.3
New loans, 7.2.1; 7.2.2; 7.5.1
New Town Development Corporations, 15.2.1
NLGSLC, 17.1.1
Non-trading capital consumption, 4.4.10
Northern Ireland, 1.3.1; 1.4.3; 2.1.1; 2.5; 3.1.4; 3.4; 4.4.1; 4.4.3; 4.4.5; 4.4.10; 4.4.11; 4.4.12; 4.5; 5.2.5; 5.6; 6.5; 7.5; 8.7; 9.3; 11.4; 12.1.5; 12.6; 13.5.1; 14.1.2; 14.7; 15.5; 16.2; 17.3.2
Northern Ireland Act 1974, 16.2.1
Northern Ireland district councils, 2.5.1; 4.4.3; 4.4.12; 4.5.2
Northern Ireland Housing Executive, 2.5.1
Notional charge on ratepayers, 4.5.5
NSC, 10.1.3

Occupational groupings, 10.2.1
Occupational Health and Safety, 14.2.6
Offences, 12.1.3
Ombudsmen, Local, 16.1.1
Operational costs, 14.4.4
Orkney, 2.3.1
Outstanding debt, 7.3; 7.5.2; 17.2.2
Outturn income, Actual, 5.2.3; 5.2.4; 5.3.8; 5.6.3; 6.2.5; 6.3.2; 6.4.2; 6.4.7; 6.4.12; 6.5.3; 6.5.4; 8.1.2; 8.4.1; 8.5.1; 8.7.3
Outturn data, Actual, 3.1.2; 3.2.3; 4.2.4; 4.2.5; 4.3; 4.4; 4.4.4; 4.5.2; 4.5.4; 17.1.4

Outturn summaries, 4.2.4
Outurn expenditure, Actual, 3.2.3; 4.2.2; 4.2.10; 4.3.3; 4.3.4; 4.3.5; 4.3.6; 4.4.2; 4.4.3; 4.4.4; 4.5.2; 4.5.5

Parking, Car, 4.3.2; 13.1.1; 13.1.2
Parks, 2.5.1; 3.3.2; 6.5.2
Passenger transport, 2.1; 2.2.1; 2.2.2; 2.2.3; 2.3.1; 2.5.1; 3.2.2; 3.3.4; 3.4.4; 4.4.11; 13.1.1
Passenger Transport Executive, 4.4.11; 8.4.1
Payments, 4.5.5
Penalty provisions, 6.4.13
Penny Rate Product, 9.2; 9.3.3
Performance indicators, 12.3.4; 13.1.5; 17.1.3
Performance of funds, 8.6.4
Performance, Waste disposal, 14.4.3; 14.4.4
Personal Social Services, 2.2.1; 2.2.2; 2.2.3; 2.3.1
Piers, Ports and, 3.3.4; 13.1.1; 13.3
Planned expenditure, 4.4.5; 4.5.4
Planned Supplementary Grants, 6.2.4
Planning, 2.1; 2.2.1; 2.2.3; 2.3.1; 3.2.2; 3.3.2
Planning, Local, 2.2.1; 2.2.2; 2.2.3; 2.3.1
Planning, Strategic, 2.2.2
Police, 2.1; 2.2.1; 2.2.2; 2.3.1; 2.4.1; 3.2.2; 3.3.5; 4.4.1; 6.2.1; 8.6.1; 8.6.3; 10.1.1; 10.1.8; 10.2.1; 12.1; 13.1.6
Police finance, 12.1.5; 12.1.7; 12.1.11
Police manpower, 12.1.4; 12.1.7
Police resources, 12.1.4; 12.1.10
Police workload, 12.1.3; 12.1.9
Police custody, Deaths in, 12.1.6; 12.1.12
Police, Metropolitan, 6.4.4; 12.1.1
Pollution control, 14.2.6
Port health, 14.2.4; 14.2.6
Ports and piers, 3.3.4; 13.1.1; 13.3
Poundage, Grant related, 5.3.9
Precepts, 5.1.2; 5.2.1; 5.3.3
Prevention of terrorism, 12.1.3; 12.1.9
Price changes, 4.2.7; 4.2.9; 4.4.5; 4.5.3
Privatisation, 14.3.8
Probation Service, London, 12.3.1
Probation Service, 2.1; 6.2.1; 12.3; 12.4.1
Probation service finance, 12.3.1; 12.3.4
Probation service manpower, 12.3.4
Probation service workload, 12.3.4; 12.3.5
Property classification, 9.1.2
Property net annual value, 9.3.1; 9.3.2; 9.3.3
Public expenditure, 4.1.3; 4.4.1; 4.4.2
Public conveniences, 14.1.2; 14.2.2; 14.2.4; 14.6.5; 14.7.1; 14.7.2
Public Health, 14.2.6
Public Health Hygiene Acts, 14.2.2
Public Works Loan Board, 7.1.1

Quality of data, 3.1.2; 17
Quarterly Borrowing and Lending Inquiry, 7.2.4; 7.2.6; 7.3.4; 7.4.1
Quarterly Employment Enquiry (NI), 10.3.2
Quarterly Employment Survey, 10.1.4; 10.1.11

Railways, 13.1.2
Rate bill, Average, 5.3.1; 5.3.8; 5.6.4; 6.4.11; 17.2.2
Rate collection, 3.2.3; 5.1.2; 5.4; 5.6.5; 9.2.1
Rate collection costs, 5.4.1; 5.4.3; 5.6.5
Rate Demand Notes, 5.3.6; 5.3.7; 5.3.8
Rate, District, 5.6.1
Rate Fund, 3.3.1; 3.3.2; 3.3.3; 3.3.4; 4.2.5; 4.2.6; 6.4.5; 8.7.2; 8.7.3; 13.1.1; 13.2.1
Rate Fund Capital Account, 7.2.1; 8.2.3; 8.4.2; 13.1.6
Rate Fund Revenue Account, 6.4.1; 8.2.1; 8.2.2
Rate Fund Services, 3.2.3; 4.2.4; 4.3.3; 4.3.5; 6.2.1; 8.2; 13.1.4
Rate Fund Services Account, 3.3.1; 3.3.2; 3.4.3; 13.1.6; 14.6.9
Rate, General, 5.1.3
Rate income, 5.2
Rate income estimates, 5.2.1; 5.6.3
Rate poundage, 5.3
Rate poundage, Average, 5.3.3; 5.3.5
Rate Product, Penny, 9.2; 9.3.3
Rate rebate amounts, 5.5.3
Rate rebate grant, 1.3.2; 5.2.3; 5.5; 6.1.2
Rate rebate recipients, 5.5.3
Rate rebates, 5.5
Rate, Regional, 5.6.1
Rate, Special, 5.1.3; 5.3.3
Rate Support Grant, 5.1.4; 5.3.8; 6.1.1; 6.4; 9.2.1; 17.1.2
Rateable values, 5.3.8; 9.1; 9.3.2; 9.3.3
Rates, 5
Rating Disabled Persons Act 1978, 5.5.2; 5.5.3
Rating and Valuation Association, 5.2.1; 5.2.2; 5.3.2; 5.3.6; 5.3.7
Refuse collection/disposal, 3.3.2; 14.3; 14.6.2; 14.7.1; 14.7.2
Refuse collection/disposal finance, 14.3.2; 14.3.5; 14.6.2
Refuse collection/disposal manpower, 14.3.3; 14.3.5
Refuse collection/disposal resources, 14.3.5; 14.3.8
Regional rate, 5.6.1
Regional statistics, 3.1.4; 3.2.3; 3.4.2
Rent rebates, 4.4.11
Rents, 8.1.2; 8.3.1; 8.7.1

Repairs and Renewals Fund, 3.3.5; 3.4.3; 3.4.4
Research and Intelligence Units, 17.1.1
Resources, Cemeteries and crematoria, 14.5.4
Resources element, 6.4.9; 6.4.12; 6.4.13; 6.4.14; 6.5.1; 6.5.2
Resources, Fire service, 12.2.4; 12.2.8; 12.2.11
Resources, Libraries, 15.1.2; 15.1.4
Resources, Police, 12.1.4; 12.1.10
Resources, Refuse collection/disposal, 14.3.5; 14.3.8
Restaurants, Civic, 3.3.4
Revaluations, General, 9.3.4
Revenue expenditure, 4.2; 4.5.5
Revenue/current account, 3.1.3; 3.4.1; 3.4.3; 6.2.1; 6.2.2; 6.5.3; 8.1.2; 8.3; 8.4.1; 8.5.3; 8.7.2; 8.7.3
Revised estimates, 4.2.8; 5.2.1
Road accidents, 13.1.3; 13.1.5; 13.1.6; 13.2.4; 13.2.5
Road construction, 13.1.1; 13.1.2; 13.2.1
Road lengths, 13.1.3; 13.1.5; 13.1.6; 13.2.4; 13.2.5
Road lighting and maintenance, 13.1.2; 13.2.1; 15.2.1
Road passenger transport, 10.1.2
Road safety, 13.1.1; 13.1.2
Roads, 3.3.2; 4.3.2; 6.5.2; 13
Roads finance, Highways and, 13.1.3; 13.1.4; 13.2.2; 13.2.3; 13.2.5
Roads manpower, Highways and, 13.1.3; 13.2.2
Roads workload, Highways and, 13.1.5; 13.2.2; 13.2.4
Roads, Highways and, 2.1; 2.5.1; 13.1; 15.2.2
Rodent extermination, 14.2.2

Sales, 8.1.1; 8.1.2; 8.2.1; 8.2.3; 8.3; 8.4.1; 8.4.2
Sales, Asset, 4.3.2
Sample Borrowing Survey, 7.2.5
School crossing patrols, 12.4.1; 12.4.9; 12.4.12; 12.4.13
Scope of review, 1.2
Scotland, 1.3.1; 1.4.3; 1.4.4; 2.1.1; 2.1.2; 2.3; 3.1.4; 3.1.5; 3.2.2; 3.2.3; 3.3.1; 3.3.5; 4.2.2; 4.2.4; 4.2.6; 4.2.9; 4.2.10; 4.3.2; 4.3.5; 4.3.8; 4.4.3; 4.4.4; 4.4.5; 4.4.10; 4.4.11; 4.4.12; 5.1.2; 5.1.3; 5.1.4; 5.2.1; 5.2.3; 5.2.5; 5.3.4; 5.3.5; 5.3.7; 5.3.8; 5.4.3; 5.5.3; 6.2.3; 6.2.5; 6.4.9; 6.4.10; 6.4.11; 6.4.12; 7.2.1; 7.2.6; 7.3.1; 7.3.2; 8.2.1; 8.2.2; 8.2.3; 8.3.1; 8.4.2; 8.5.2; 8.5.3; 8.6.1; 9.1.6; 9.2.4; 9.2.5; 10.1.8; 10.1.9; 10.1.11; 11.1.4; 11.2.1; 12.1.8; 12.1.9; 12.1.12; 12.2.7; 12.3.2; 12.3.3; 12.3.5; 12.4.9; 12.4.10; 12.4.11; 12.4.12; 12.5.4; 13.4.2; 16.1.1; 17.1.3; 17.1.4; 17.3.2
Scotland DLO's, 15.2.1; 15.2.4
Scotland highways, 13.2
Scotland libraries, 15.1.7; 15.1.9
Scotland museums and galleries, 15.1.7; 15.1.8; 15.1.9
Scotland Environmental Services, 14.1.2; 14.6
Scottish Office, 1.4.4; 3.2.1; 3.2.3; 4.2.3; 4.2.6; 4.3.6; 5.3.5; 5.3.7; 5.5.3
SCT, 1.4.1; 3.2.2; 4.2.4
Seasonal adjustments, 10.1.7; 10.1.8
Service indicators, 3.2.3
Service, London Probation, 12.3.1
Services, 1.4.2; 5.6.2
Sewerage, 2.3.1; 2.5.1; 14.1.2; 14.6.1; 14.6.8; 14.7.2; 15.2.1; 15.2.2
Shetland, 2.3.1
Shire county councils, 2.2.2
Shop Act, 14.2.6
Smallholdings, 3.2.2; 15.3
Smallholdings finance, 15.3.2
Social Security and Housing Benefits Act 1982, 5.5.2; 5.5.3
Social Services, 3.2.2; 8.2.1
Society of Metropolitan Treasurers, 3.2.2
Source of borrowing, 7.2.6; 7.3.1; 7.3.2; 7.3.3; 7.5.2
South Yorkshire, 2.2.1
Special rate, 5.1.3; 5.3.3
Special funds, 3.3.1; 3.3.5; 3.4.3; 4.2.10; 4.3.8; 6.2.1; 8.5
Specific grants, 6.1.1; 6.2.1; 6.2.5; 6.5.1; 6.5.2; 6.5.4
Spending controls, 4.3.2
SSLG, 17.2.2; 17.2.3
Staff car loans, 8.2.3
Standard Industrial Classification, 10.1.12
Statistical developments, 17.1
Statistics of Trade Act 1947, 10.2.1
Strategic planning, 2.2.2
Street cleaning, 14.1.2; 14.2.2; 14.6.2; 14.7.1; 14.7.2
Subsidies, 4.4.11
Subsidies and grants, 6; 15.3.2
Subsidies, Transport, 13.1.2; 13.2.1
Summary data, 3.1.1; 3.2.1; 3.2.3; 4.2.4; 4.2.5; 4.3.3; 4.4.7; 5.2.3; 5.2.4; 8.1.2
Superannuation Funds, 3.3.5; 4.2.10; 4.3.8; 4.4.6; 8.6; 12.2.5
Supplementary grants, 6.1.1; 6.2.2; 6.2.5
Supplementary rate, 5.1.3
Survey Control Unit, 15.4.1

Surveys, 15.4
Swimming baths, 2.5.1; 3.3.2; 4.3.2

Teachers, 3.3.5; 8.6.1; 10.1.1; 10.2.1; 10.3.1
Term of borrowing, 7.2.6; 7.3.1; 7.3.3; 7.5.2
Time lags, 17.2.4
Total net borrowing, 7.2.2; 7.2.3; 7.2.4; 7.2.5; 7.2.6; 7.5.1; 17.2.2
Tourism, 2.3.1
Town and Country Planning Act, 17.1.1
Trading, 4.4.11; 8.7.1
Trading Services Account, 3.3.1; 3.3.4; 3.4.4; 4.2.10; 4.3.3; 4.3.5; 6.2.1; 8.4; 13.1.1; 13.1.4; 13.1.6; 13.2.1; 14.6.9
Trading Services Capital Account, 7.2.1
Trading standards, 2.1; 3.2.2; 12.5
Transfer payments, 3.4.1; 4.4.6; 4.4.9; 4.4.11
Transport subsidies, 13.1.2; 13.2.1
Transport Executive, Passenger, 4.4.11; 8.4.1
Transport, Passenger, 2.1; 2.2.1; 2.2.2; 2.2.3; 2.3.1; 2.5.1; 3.2.2; 3.3.4; 3.4.4; 4.4.11; 13.1.1
Transport personnel, 10.2.1
Transport, Road passenger, 10.1.2
Transport Supplementary Grant, 6.2.2; 6.2.4
Transport Trading Services, 13.1.2
Treasury, 7.2.7
Tunnels, Ferries and, 13.1.1; 13.1.2
Tyne and Wear, 2.2.1
Types of expenditure, 4.1.2

Unemployment, 6.5.1

Valuation Assessors, Land, 5.3.8; 9.1.1; 9.1.7
Valuation officers, 9.1.1
Value, Property net annual, 9.3.1; 9.3.2; 9.3.3
Values, Rateable, 5.3.8; 9.1; 9.3.2; 9.3.3
Vehicle licensing, 13.1.3; 13.1.5; 13.1.6; 13.2.4; 13.2.5

Wages and Earnings, 4.4.10; 10.2.1; 10.2.2; 10.3.3; 12.1.6; 12.1.12; 12.2.5; 12.2.10
Wales, 1.3.1; 1.4.3; 2.2.2; 3.1.4; 3.1.5; 3.2.3; 4.2.2; 4.3.3; 4.4.5; 4.4.12; 5.2.3; 6.2.2; 6.2.5; 6.4.7; 7.2.2; 7.2.6; 9.1.4; 16.1.2
Waste collection, 2.2.1; 2.2.2; 2.2.3; 2.3.1; 2.5.1; 14.1.1; 14.1.2; 14.3
Waste collection charges, 14.3.5; 14.3.7
Waste disposal, 2.1; 2.2.1; 2.2.2; 2.2.5; 2.3.1; 2.5.1; 14.1.1; 14.1.2; 14.3; 14.4
Waste disposal finance, 14.4.2; 14.4.3
Waste disposal manpower, 14.4.3
Waste disposal performance, 14.4.3; 14.4.4
Waste disposal workload, 14.4.3
Water, 2.3.1; 2.4.1; 2.5.1; 8.4.1; 14.1.2; 14.6.1; 14.6.8; 14.6.9; 14.6.10; 14.7.2; 15.2.1; 15.2.2
Water rate, Domestic, 5.1.3; 5.3.5; 5.3.8; 8.4.2; 14.6.9; 14.6.13
Water resources, 14.6.15
Welsh Office, 3.2.1; 3.2.3; 4.2.5; 4.2.7; 4.3.3; 5.2.1; 5.3.2; 5.3.6; 17.2.1
West Midlands, 2.2.1
West Yorkshire, 2.2.1
Western Isles, 2.3.1
Workload, Fire service, 12.2.3; 12.2.8; 12.2.11
Workload, Highways and Roads, 13.1.5; 13.2.2; 13.2.4
Workload, Libraries, 15.1.2; 15.1.4
Workload, Police, 12.1.3; 12.1.9
Workload, Probation service, 12.3.4; 12.3.5
Workload, Waste disposal, 14.4.3